Electron–solvent and anion–solvent interactions

Electron–solvent and anion–solvent interactions

Edited by

Larry Kevan
Professor of Chemistry,
Wayne State University,
Detroit, Michigan, U.S.A.

and

Brian C. Webster
Lecturer in Theoretical Chemistry,
Glasgow University, Gt. Britain

ELSEVIER SCIENTIFIC PUBLISHING COMPANY
Amsterdam–Oxford–New York 1976

ELSEVIER SCIENTIFIC PUBLISHING COMPANY
335 JAN VAN GALENSTRAAT, P.O. BOX 211,
AMSTERDAM, THE NETHERLANDS

AMERICAN ELSEVIER PUBLISHING COMPANY, INC.
52 VANDERBILT AVENUE, NEW YORK, NEW YORK 10017

Library of Congress Cataloging in Publication Data

Main entry under title:

Electron–solvent and anion–solvent interactions.

    Includes bibliographical references and index.
    1. Solvation. I. Kevan, L. II. Webster, Brian C.
QD561.E482   541'.372   75–43929
ISBN 0–444–41412–6

WITH 123 ILLUSTRATIONS AND 45 TABLES

PRINTED IN THE NETHERLANDS

CONTRIBUTORS TO THIS VOLUME

R. Catterall — Department of Chemistry and Applied Chemistry,
University of Salford,
Salford, Gt. Britain

P. Delahay — Department of Chemistry,
New York University,
4 Washington Place, New York, U.S.A.

N. R. Kestner — Chemistry Department,
Louisiana State University,
Baton Rouge, Louisiana, U.S.A.

A. Mozumder — Radiation Laboratory,
University of Notre Dame,
Notre Dame, Indiana, U.S.A.

W. F. Schmidt — Hahn-Meitrer-Institut für Kernforschung Berlin GmbH,
Ber. Strahlenchemie, Berlin, Germany

P. Schuster — Institut für Theoretische Chemie,
Universität Wien,
Währingerstrasse, Wien, Austria

H. B. Steen — Department of Biophysics,
Norsk Hydro's Institute for Cancer Research,
Montebello, Oslo, Norway

M. C. R. Symons — Department of Chemistry,
The University,
Leicester, Gt. Britain

D. C. Walker — Chemistry Department,
University of British Columbia,
Vancouver, Canada

PREFACE

Nature presents us, in the phenomenon of electrolysis, with a single definite quantity of electricity which is independent of the particular bodies acted on.... This definite quantity I shall call E_1. Although presented in 1874 to the British Association, the paper by Johnstone Stoney, *On The Physical Units of Nature*, from which this passage is taken, remained unpublished until 1881. Within the following ten years Johnstone Stoney came to refer to this definite quantity of electricity as the electron. In 1863, Weyl reported the preparation of a blue solution on dissolution of an alkali metal in liquid ammonia. Since that time there has been intensive study pertaining to species formed in metal–ammonia solutions and it is now accepted that the blue colouration of the dilute solutions arises from solvated electrons.

Electrons are observed to become solvated in both polar and non-polar media. Experimental studies of considerable ingenuity have been pursued in order to elucidate the chemical behaviour and physical properties of this solvated species. Concurrent with such observations, considerable theoretical work has focused on explanations of the structure and properties of solvated electrons. In this book we draw together current knowledge of these species regarding their importance as reactive intermediates in radiolysis and as prototype examples of ionic solvation, electron tunneling processes and electronic properties of disordered systems.

The central theme of this book is electron–solvent interactions, contrasted where appropriate to anion solvation. It is the intention through the content, sequence and cross referencing of the selected contributions, to define more clearly the nature of electron–solvent interactions, and to show that this subject, which possesses much variety in theoretical and experimental technique, has an underlying unity. That there are certain omissions is recognized, but it is not our prime purpose to provide a set of review articles extending over the whole of the subject. Each contribution reveals the current state of the topic treated, and as such may be read by the research worker, postgraduate or graduate student, or final year Honours graduate or under-graduate alike. Only when the book is read in totality, however, will it be found whether our initial intention to emphasize the unity of this subject has been attained.

We should like to express our sincere thanks to our contributors for their interest and co-operation. We are also grateful to Sue London, Maureen Murphy and Agnes Wirtz for their help in maintaining some consistency of format in the illustrated and written material of the various chapters.

CONTENTS

xii

Chapter 6. Formation mechanisms and primary reactions of excess electrons in condensed polar media, by H. B. Steen   175

Chapter 1

THEORETICAL STUDIES OF ELECTRON–SOLVENT INTERACTIONS: SOLVED AND UNSOLVED PROBLEMS

N. R. KESTNER

I. Introduction

When extra electrons are introduced into a fluid, one can observe a variety of resulting "products", depending upon the type of fluid. Most of the "products" are not chemical species in our usual sense of the term since their properties depend strongly on various parameters such as temperature and pressure. It is the purpose of theoretical calculations to explain the nature of the medium including the extra electron (or electrons), i.e. the "products". However, in dealing with this subject one cannot expect theoretical calculations of extremely high accuracy since our knowledge of fluids is still primitive. Our theories of fluids, especially polar fluids, are far from sufficient to yield quantitative agreement of the properties of pure fluids, let alone fluids containing charged species. Thus, as used in this chapter, theoretical studies essentially mean "model" calculations. The models can be of various degrees of sophistication but they will all contain some major approximation. In this chapter we will examine some of the present models, from the most naïve to the most complicated versions which have recently been used. However, in all cases the results are approximate and therefore one can never hope to obtain precise agreement with experimental data. If all of the important features are properly included, then the theoretical results should be close to experimental observations. Much more important, however, the trends of the experimental results as a function of the many parameters (type of molecule, solvent density, temperature, pressure, etc.) should be reproduced. Most importantly, theory is only worthwhile if the calculations allow one to understand the experiments, and if the theory can predict new experimental results.

If the electron does not react with anything in the fluid to yield negative ions or radicals, it can exist in one of two basic forms: in extended, delocalized states or in localized states. The distinction is to be made on the basis of the wave function of the electron. In the former case, the electronic wave function of the extra electron is of the form of a plane wave (or, more properly, an orthogonalized plane wave) and extends over the entire fluid or at least over a macroscopic volume. In the localized case the electronic wave function decreases exponentially away from one point in the fluid and thus is more likely to be found near this point (the trapping

1

site) than away from it. These are the two general types of electronic states we will consider in this chapter. However, there are also time-dependent aspects to this problem. When one first introduces an electron into the medium it may exist for a short time (10^{-6} to 10^{-12} s) in a non-equilibrium state. The non-equilibrium refers to the fluid configuration. It takes the fluid some time to respond to the presence of the electron and thus the initial state of the electron may be modified with time. These points will be discussed in other chapters of this book. In this chapter we will only consider those "long time" situations after which the fluid has adjusted to the presence of the electron. We must also consider that this time is shorter than the time to react with the solvent, which in some cases places severe limits on the maximum time over which our theoretical calculations are valid.

These states can be characterized experimentally in many ways. As a general rule, the extended species will have a high mobility ($\sim 10^2$ cm^2 V^{-1} s^{-1}), as opposed to the low mobility of the localized states ($\sim 10^{-1}$ cm^2 V^{-1} s^{-1}). To be valid such a comparison should be made at an equivalent temperature and pressure, say the boiling point of the fluid. These arguments should be used rather cautiously since there are some cases which, for various reasons, do not obey these simple rules. The localized states will also have some type of optical spectrum and this has been extensively studied, especially in polar fluids.

There are two general classes of fluids: non-polar and polar. In order to understand the electron interaction with polar fluids we must first consider the simpler case of non-polar fluids.

II. Electrons in non-polar fluids

II.A Nature of electronic states

The mobility of electrons in non-polar liquids ranges from values such as 5×10^{-3} cm^2 V^{-1} s^{-1} in liquid hydrogen, through values of about 1 cm^2 V^{-1} s^{-1} for liquid hexane, to values as high as 2200 cm^2 V^{-1} s^{-1} in liquid xenon[1-3]. These values alone suggest that the state of the electron can be either localized or delocalized in non-polar fluids. Optical data in most non-polar fluids have not been obtained because the methods often used to introduce electrons do not permit sufficient electron concentrations. In a few hydrocarbon fluids the optical spectrum has been obtained indicating the presence of localized states or electron traps[4]. In order to survey this range of behavior let us first consider the factors which influence electron–rare gas interactions, as these are the simplest.

II.B Rare gas and small molecule systems

II.B.1 Electron–molecule interactions

At low energies the interaction of an electron and a simple molecule can be expressed in terms of a scattering length, a, which is defined in terms of a scattering potential $V(r)$ as[5]

$$a = \frac{2m}{\hbar^2} \int_0^\infty rV(r)U_0(r)\mathrm{d}r \tag{1}$$

where U_0 is the zero energy solution of the Schrödinger equation for the scattering problem. At low energies and using the Born approximation we find that, approximately

$$a \approx \frac{2m}{\hbar^2} \int_0^\infty V(r)r^2 \mathrm{d}r \tag{2}$$

The Born approximation is not quantitatively applicable here[5,6] but the importance of eqn. (2) is that the sign of a is the effective sign of V. In dealing with electrons, due to the requirements of antisymmetry, $V(r)$ is not a real classical potential, but rather an effective potential or pseudopotential, V_{ps}, which incorporates the effect of the Pauli exclusion principle[7]. Thus the scattering length, and especially its sign, indicate whether the essential part of the potential is repulsive ($a > 0$), i.e. dominated by the Pauli principle, or whether the interaction is attractive ($a < 0$), i.e. dominated by charge-induced multiple moments. More elaborate arguments which do not use the Born approximation confirm these conclusions[5,6].

In Table 1 we list some scattering lengths for typical electron–neutral molecule interactions. In the case of helium the scattering length is large and positive, indicating that the electron–helium atom interaction is dominated by the Pauli effect with the attractive part very minor due to the low polarizability of the helium atom[7]. On the other hand, in the case of xenon we have a large negative value, indicating dominance of the attractive interactions. We can approximate the pseudopotential as

$$V_{ps}(r) \cong Ae^{-\delta r} - \frac{e^2 \alpha}{2(r^2 + d^2)^2} \tag{3}$$

where α is the dipole polarizability and A, δ, and d are parameters which vary from system to system for r greater than some very small value. In addition there are higher order attractive interactions and other more subtle effects. The first term arises

TABLE 1

Some gas phase scattering lengths of electrons by neutral atoms and molecules

Atom or molecule	a (atomic units) (expt.)
He	$+1.15 \pm 0.05$[a]
Ne	$+0.39$[a]
Ar	-1.70[a]
Kr	-3.7[a]
Xe	-6.5[a]
H_2	$+1.25$[b], $+1.51$[b]

[a]T. F. O'Malley, Phys. Rev., **130** (1963) 1020.
[b]R. L. Wilkins and H. S. Taylor, J. Chem. Phys., **47** (1967) 3533.

primarily from the Pauli principle. In the case of helium the first term dominates the potential for most values of r, while in the case of xenon the first term is important only at small separations. The constant d, concerning the size of the atom, is introduced as a device to ensure that the potential does not become unreasonable at short separations. In the case of xenon and all other systems with negative scattering lengths, the second term dominates at most of the interaction distances. This simple form is adequate to explain the interaction of one atom (or molecule) and an electron.

II.B.2 Electron–medium interaction

(a) Quasi-free state   If the electron is introduced into a medium, two major modifications occur in the effective interaction. Firstly, the electron is no longer only scattered from one atom at a time; there are multiple scattering effects. Secondly, the long range interactions (r^{-4} effects) are screened by the medium. The proper function to use in screening the long range potential has been developed by Lekner[8]. In a medium of number density, ρ_n, in the region where $r > d$ and the long range potential can take on its simple unscreened form of $-e^2\alpha/2r^4$, Lekner showed that the proper screened function to use is

$$-\frac{e^2\alpha}{2r^4} f(r) \tag{4}$$

where

$$f(r) = 1 \qquad\qquad r < \sigma$$

$$f(r) = \frac{1}{1 + (8\pi/3)\alpha\rho_n} \quad r > \sigma \tag{5}$$

where σ is roughly an atomic hard-core diameter, using the approximation of Springett et al.[9].

To evaluate the effects of multiple scattering, we can use a variation of the Wigner–Seitz method[9-11]. To do this we replace each atom in the liquid by an equivalent atomic sphere of radius r_s where

$$r_s = (3\rho_n/4\pi)^{1/3} \tag{6}$$

Furthermore, we will assume that in the neighborhood of any given atom the average potential seen by the electron is firstly spherically symmetrical about that atom, and secondly characterized by the average translational symmetry. Using the standard Wigner–Seitz treatment, the wave functions in these Wigner–Seitz spheres are subject to the boundary conditions

$$(\partial\psi^0/\partial r)_{r=r_s} = 0 \tag{7}$$

while inside each sphere the wave function satisfies the basic equation

$$[-(\hbar^2/2m)\nabla^2 + V_m(r)]\psi^0(r) = V_0\psi^0(r) \tag{8}$$

In the above equation we have used the standard terminology V_0 to refer to the energy of an excess electron in the unperturbed fluid. The potential V_m consists of both the short range (or Hartree–Fock) contributions, V_{HF}, and the long range contributions which originate from eqn. (4), U_p; i.e.

$$V_m = V_{HF} + U_p \tag{9}$$

The long range polarization contributions can be evaluated in two parts, one outside the Wigner–Seitz sphere and the other inside where the potential is assumed to be constant. Springett et al.[9] showed that we can approximate U_p as

$$U_p = -\frac{3\alpha e^2}{2r_s^4}\left(\frac{8}{7} + \frac{1}{1 + (8\pi/3)\alpha\rho_n}\right) \tag{10}$$

The electronic wave function must be orthogonal to the wave functions of the atomic electrons. This can be accomplished as shown by Cohen and co-workers[9,10] if the above equation is replaced by an equivalent one

$$[-(\hbar^2/2m)\nabla^2 + V_m]\phi_0 = V_0\phi_0 \tag{11}$$

where ϕ_0 is smooth inside the Wigner–Seitz core, having the oscillations of ψ_0 removed and placed into the effective potential V_m, where it is assumed to cancel V_{HF} except within a hard-core radius $\tilde{a}$. The new function must obey the same boundary conditions as ψ_0, i.e. eqn. (8). Thus the solution of eqn. (11) is

$$\phi_0 = \sin k_0 (r - \tilde{a})/r \tag{12}$$

where

$$k_0 = [(2m/\hbar^2)(V_0 - U_p)]^{1/2} \tag{13}$$

The value of k_0 is obtained by using the boundary condition to obtain

$$\tan k_0(r_s - \tilde{a}) = k_0 r_s \tag{14}$$

From which we obtain

$$T = \hbar^2 k_0^2/2m \tag{15}$$

and

$$V_0 = T + U_p \tag{16}$$

The calculated values of V_0 agree with experimental results in most cases, as shown in Table 2. There is a large discrepancy only for the case of H_2 but this may be due to the neglect of electron–quadrupole interactions, a point not yet explored theoretically. If this idea is correct we would expect complications in the case of all diatomic and polyatomic molecules. Little experimental or theoretical work has been done on this point.

(b) Localized state The electronic energy of the localized state can be treated in the same framework as above[9] if we assume that the electron in this case exists in a

TABLE 2

The ground state energy of the quasi-free electron state in non-polar fluids

	$\tilde{a}_0^e$ (Å)	R_0(eV) Theory[a]	Expt.
^{4}He	0.751	+1.32	+1.05 ±0.05[b]
Ne	0.556	+0.46	0.5[c]
Ar	0.794	−0.63	−0.33 ±0.05[d]
Kr			−0.78 ±0.2[c]
H$_2$	1.0	2.2	−1[c]

[a]For original references, see ref. 9.
[b]M. A. Woolf and G. W. Rayfield, Phys. Rev. Lett., **15** (1965) 235.
[c]See ref. 2.
[d]B. Halpern, J. Lekner, S. A. Rice and R. Gomer, Phys. Rev., **156** (1967) 351.
[e]These parameters are obtained from calculations of the scattering lengths of electrons from these atoms and molecules when long range polarization effects are not included. They represent an effective hard core radius. The atomic values are from the work of Massey reported in ref. 6, while the value for H$_2$ is from K. Imui, Proc. Phys. Math. Soc., Jap., **20** (1938) 790. The use of experimental scattering lengths is inappropriate and would lead to very different results, i.e. for Ar the experimental scattering length is −0.9 Å, since they also include the effects of long range polarization.

cavity with void radius R and furthermore, that the electronic wave function has the form

$$\phi(r) = f(r) \qquad r < R$$
$$\phi(r) = f(r)\phi_0(r) \quad r > R \tag{17}$$

The wave function must satisfy

$$-(\hbar^2/2m)\nabla^2 f(r) = E_e f(r) \qquad\qquad r < R$$
$$\left[-(\hbar^2/2m)\nabla^2 + V_m\right]f(r)\phi_0(r) = E_e f(r)\phi_0(r) \quad r > R \tag{18}$$

if we ignore effective mass corrections. E_e is the ground state electronic energy of the localized state. Neglecting effective mass corrections is equivalent to assuming that $f(r)$ varies slowly over the Wigner–Seitz sphere. With this assumption one obtains the new equations[9].

$$(p^2/2m - E_e)f(r) = 0 \qquad r < R$$
$$(p^2/2m + V_0 - E_e)f(r) = 0 \qquad r > 0 \tag{19}$$

One can make the assumption that $f(r)$ varies rapidly within the Wigner–Seitz sphere and arrive at the same conclusion. It is therefore assumed that eqn. (19) holds under all conditions of $f(r)$ (see ref. 9).

The important point of eqn. (19) is that it corresponds to a particle in a spherical well of depth V_0. These equations can be readily solved to obtain the electronic energy of the electron in the localized state. However, the total energy, E_T of the localized electron includes another contribution, namely the medium rearrangement energy E_M[3,10]

$$E_T(R) = E_e(R) + E_M(R) \tag{20}$$

where

$$E_M = E_{PV} + E_{ST} + E_{pol}$$
$$E_{PV} = (4\pi/3)R^3 P$$
$$E_{ST} = 4\pi R^2 \gamma \tag{21}$$

assuming that macroscopic laws apply to pressure volume, E_{PV}, and surface tension, E_{ST}, energies (γ is the surface tension which should also depend on R). The polarization correction is the interaction of the trapped electron with the remainder of the fluid.

Since all terms in E_M are positive, we can only have any form of a localized state if $V_0 > 0$. Even in these cases the localized or bubble state may not be stable unless

$$E_T(R_0) < V_0 \tag{22}$$

where R_0 is the void radius with the lowest value of E_T, i.e.

$$\left(\frac{\partial E_T}{\partial R}\right)_{R_0} = 0 \tag{23}$$

The rules and general formulae for predicting the stability of localized states in various media have been worked out by Springett et al.[9]. They predict that localized electrons should be energetically stable in ^{4}He, ^{3}He, H_2, D_2, and probably Ne. The following experimental results confirm this in all cases. However, there exists one discrepancy, while the theoretical values for V_0 in H_2 and D_2 support localization; the experimental values do not. In this case theoretical values are subject to large uncertainties and possibly the experimental results also need more analysis. Studies on related systems such as N_2 are needed to resolve this.

When we compare the mobility of electrons in rare gas liquids with the V_0 values, we obtain further confirmation that electrons in helium are localized (low mobility), while electrons in argon, krypton and xenon are delocalized (high mobility)[3]. Further evidence of the correctness of this model is found from other experimental data in helium. The size of the cavity is in accord with Stokes' law and with evidence from the "lifetime edge" in captive cross-sections for negative ions trapped by quantized vortex lines[3]. More important, Northby and Sanders[12] observed photo-ionization of the trapped electrons in liquid helium. A photocurrent was observed only with photons of energy greater than 1.1 eV and, while the exact mechanism for this is not known, it does confirm the localized nature of the electron.

II.B.3 Effect of temperature, pressure and density

The energy of both the quasi-free and the localized states depends on the density of the solvent, pressure and temperature. It is thus possible to vary these and check the predictions of theory by experimental studies. Temperature effects have not been observed directly. However, if one assumes that the mobility of an electron in helium ($R_0 \sim 15$ Å) obeys Stokes' law, then one can explain the temperature-dependence of the mobility and the difference between the mobility of localized electrons in ^{3}He and ^{4}He[10].

The most dramatic predictions can be made for the effects of density or pressure. If we take a localized electron in a fluid and begin to decrease the density, the energy of the localized state decreases faster than the energy of the quasi-free state. Thus it should be possible to observe the transition from a localized state to a delocalized state by decreasing the density. Experimentally, it is easier to study the system as the density is increased. At low densities the medium rearrangement energy is simply the pressure volume contribution. Since both V_0 and E_T can be calculated as a function of density, one can predict the density at which this transition could occur. This was done for electrons in helium gas first by Jortner et al.[10] and Levine and Sanders[13]. A better calculation by Springett et al.[9] led to a value of about 10^{20} atoms cm^{-3} (the older value was $\sim 0.9 \times 10^{21}$ atoms cm^{-3} or 1.5×10^{-3} moles cm^{-3}). In Fig. 1 we show the experimental results from the mobility of electrons in helium gas obtained by Levine and Sanders[13,14], along with the transition point calculated by Jortner et al.[10]. From the dramatic change in the mobility with density, it is apparent that a transition from a quasi-free to a localized state has occurred.

The above models are somewhat naïve since there are large fluctuations in low density gases and these fluctuations can lead to regions favorable or unfavorable for electron localization even though the average density would be far from the transition region. These fluctuations are responsible for the width of the mobility curve in Fig. 1. In order to determine the transition range for mobility one needs to develop a theory for mobility directly[15]. To do this one divides the medium into regions and calculates the probability that each particular region can support a localized state, i.e. has a sufficiently high density. Then one assigns to the localized and delocalized species a mobility appropriate to that density. Finally, using percolation theory[16], one calculates the probability that there will be enough contiguous regions of high mobility to lead to the observation of a high mobility species. This was done by Eggarter and Cohen[15,17]. The resulting mobility curve agrees quite well with the experimental curves. Hernandez[18] has extended this work and was able to study the temperature-dependence of the mobility, low field Hall mobility, and also the field-dependence of the drift velocity. There is also evidence from the experimental work of Harrison and Springett[19] that a similar transition from a high mobility species to a low mobility species can be observed as a function of density in gaseous H_2.

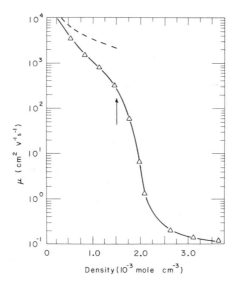

Fig. 1. Electron mobility data for electrons in helium gas at 4.2 K. These experimental results are taken from ref. 14. The dashed curve corresponds to the mobility of the quasi-free electron. The arrow indicates the calculated critical density for the transition from the delocalized·to a localized state.

There is a more complicated density-dependence in the case of atoms with a negative scattering length due to the domination of long range electron–atom attractions. These long range interactions are screened in a fluid leading to strongly density-dependent V_0 values or equivalently, to effective scattering length values. For liquid argon at ordinary densities Lekner[8] has calculated that the effective scattering length in the liquid is positive, being +0.76 Å at the triple point versus the zero density limit of −0.63 Å. In both regions, however, the electrons are in quasi-free states (unperturbed liquid structure) since V_0 is negative. For quasi-free electrons the zero field mobility is inversely proportional to the square of the scattering length. Since the scattering length goes from negative values at very low densities to positive values at higher densities, there must exist a density at which the scattering length goes through zero[20]. At that density one expects a very large mobility. As we have seen earlier, however, fluctuations in the liquid density must be considered and therefore, even at this critical density, most regions are either above or below this point. Lekner[20] has calculated the mean-square value of the effective scattering length and predicted the results to be expected. The first evidence of these mobility maxima were detected in liquid krypton by Schnyders et al.[21,22], but the most definitive work is that on fluid argon by Jahnke et al.[23]. They found the large mobility maximum but could not agree exactly with details predicted by Lekner on the higher density side. Thus more quantitative work is needed in this problem to supplement the original Lekner predictions[24].

Several interesting results originate by considering the pressure-dependence. First of all, in a localized state such as electrons in helium, the cavity size should decrease as pressure is increased. In the case of helium this has been observed[25-27]. Furthermore, it should be possible to make the quasi-free state more stable by sufficiently increasing the pressure. This pressure is so high that the liquid would solidify[9]. However, electron bubbles exist in solid helium and Jortner and Cohen[28] have predicted that at high pressures this same transition should occur. However, Dionne et al.[29] did not observe such a collapse at pressures up to 6660 atm. for solid helium. Theoretical results suggest it could occur at a still higher pressure.

III. Polar fluids

III.A Introduction
In polar liquids the stable state of the electron seems to be localized, judging from the vast amount of data which has been accumulated on the optical spectrum of trapped electrons in a vast variety of solvents. In the case of metal–ammonia solutions there is also the opportunity to study the effect of high concentrations of solvated electrons, a feature not usually possible since the solvated electron often reacts rapidly with the solvent. The very nature of the polar liquid makes it impossible to determine the bottom of the conduction band, i.e. V_0, by simple injection experiments and thus we do not know the precise energies of the localized and delocalized states. Polar liquids do offer a major challenge to theory since our understanding of pure polar liquids is so meagre and since the solvated electron can be observed in such a vast array of pure and even binary polar fluids under a variety of conditions of temperatures, pressures and densities.

III.B Electron–solvent interaction

III.B.1 Quasi-free electrons
Since it is impossible to determine V_0 experimentally, the only estimates have been provided by theoretical calculations. When the electron is in a plane wave (or similar delocalized state) it samples a large volume of the fluid. In such a case the electron should see a random array of dipole moments and thus V_0 should be unaffected by the polar character of the medium since

$$\left\langle \sum_i \frac{\mu(r - r_i)}{|r - r_i|^3} \right\rangle_{\text{average}} = 0 \tag{24}$$

i.e. "the effective electron–dipole interaction is zero". In this case, neglecting other multi-polar interactions, we should be able to calculate V_0 using the same formulae which applied in the case of inert gases as given in the last section. Until some experimental means is found to check this we shall assume these are the correct values, subject to rather large uncertainties.

III.B.2 Localized electrons in a continuum
The presence of a localized electron in a polar medium causes the molecules to start to exercise non-random fluctuations in their motions in such a way that they lead to a stronger interaction between the electron and the fluid. This problem was

first tackled by Landau[30] with regard to electrons trapped in ionic lattices. If r is the distance from the trapping site, the interaction is

$$\frac{-\beta e^2}{r} \tag{25}$$

where

$$\beta = \frac{1}{\epsilon_{op}} - \frac{1}{\epsilon_s}$$

and ϵ_{op} is the optical and ϵ_s is the static (low frequency) dielectric constant. This model is the so-called adiabatic version in which the binding originates from the inertial or low frequency polarization.

The wave function of an electron trapped in a localized state in these continuum models is arrived at by solving the Schrödinger equation

$$[-(\hbar^2/2m)\nabla^2 + V(r)]\Psi_{nl} = \epsilon(nl)\Psi_{nl} \tag{26}$$

where n and l refer to the quantum numbers of the electronic state.

In Jortner's[31] original work on electrons in ammonia, he modified Landau's potential to allow for a cavity of size R

$$\left.\begin{array}{ll} V(r) = -\beta e^2/r & r > R \\ V(r) = -\beta e^2/R & r \leq R \end{array}\right\} \tag{27}$$

Thus the electron was assumed to be completely localized in a well of constant depth. The total energy of the trapped electron was then calculated to be

$$E(nl) = \epsilon(nl) + S_{nl} + \pi_{nl}^a \tag{28}$$

where S_{nl} is the contribution of the electronic polarization to the binding and π_{nl}^a is the energy required to polarize the medium to the level required for binding the electron. Jortner was able to explain many of the properties of metal–ammonia solutions using this model which contained but one arbitrary parameter, R. However, it was not possible within this model to justify the value of R or even why the cavity existed. In addition, various temperature- and pressure-dependences had to be arbitrarily assigned to R.

Jortner[32] also developed an improved version of the adiabatic treatment in which the electron was not completely localized within the cavity. In that case the potential becomes

$$V(r) = +\frac{e\beta}{2}f(r) \tag{29}$$

where

$$\nabla^2 f = 4\pi e|\Psi|^2 \tag{30}$$

since f is the electrostatic potential due to the charge distribution $e|\Psi|^2$. The evaluation of the energy requires the choice of a wave function, calculation of f, and the substitution of these into eqns. (26) and (29), the former usually being solved by

12

variational methods. The specific formulae for the calculation of the energy to polarize the medium if the cavity size or start of the continuum is R have been derived by Land and O'Reilly[33]

$$\pi_{nl}^a = \frac{\beta}{2}\left[\int_R^\infty G_0(r)\Psi^2 r^2 dr + G_0(R)P(R)\right] \tag{31}$$

where

$$G_0(r) = \frac{1}{r}\int_0^r \Psi^2 t^2 dt + \int_r^\infty \Psi^2 t^2 dt \tag{32}$$

and

$$P(R) = e\int_0^R \Psi^2 t^2 dt \tag{33}$$

When the electron is completely localized in the cavity, then $P(R) = e$, $G_0(R) = e/R$, etc.

When the binding energy of the electron to the medium becomes very large it is not accurate to neglect the electronic polarization in binding the electron. In that case a new theory needs to be developed. Jortner[32] calculated the energy of an electron bound to a dielectric by a self-consistent or independent particle scheme. In that case one does not separate the static and high frequency polarizations. The proper potential to use is

$$V(r) = e(1 - 1/\epsilon_s)f \tag{34}$$

using the same definition of f as before (eqn. (30)). The energy to polarize the medium must also change in the same manner to

$$\pi_{nl}^{scf} = \frac{e}{2}(1 - 1/\epsilon_s)\langle\Psi|f|\Psi\rangle \tag{35}$$

where the matrix element is identical to that used to derive eqn. (31). Jortner et al.[34] have shown that this self-consistent field approach is better in describing electrons in water.

In treating excited states the procedure is similar, except that one must take into account the Franck–Condon principle and thus the medium polarization of the excited state is that of the ground state (except for high frequency polarizations which can adjust rapidly) and thus the formulae must be slightly modified.

In the self-consistent treatment the potential to use for the excited state i is

$$V(r) = e(1 - 1/\epsilon_{op})f_i + e(1/\epsilon_s - 1/\epsilon_{op})f \tag{36}$$

where f_i is the electrostatic potential due to the excited electron and f is that of the ground state. In the adiabatic approximation $V(r)$ is the same as eqn. (29), i.e. only the ground state potential enters.

While the continuum models have yielded initial successes in developing better theories, they have not been able to prove the existence of localized states: they contain arbitrary parameters, they require ad hoc adjustment of parameters to explain temperature and pressure effects and they cannot even begin to explain theoretical properties. More important, we know that the treatment of a liquid as a continuum is a very approximate procedure and cannot be a valid representation of electrons in liquids. In the case of ions in water we insist on solvation effects and in the case of an electron such phenomena should be even more important.

A more elaborate version of the continuum model has been developed by Tachiya et al.[35]. They can obtain optical line shapes and other properties but their model still neglects the very important short range electron–molecule interactions.

III.B.3 Localized electrons—short range effects

A number of studies have been made of localized electrons using semi-empirical molecular orbital theories to evaluate the short range electron–solvent interaction. These studies usually consider species such as $(NH_3)_4^-$ or $(H_2O)_4^-$ in various degrees of sophistication. Some studies have also concentrated on dimer models: $(H_2O)_2^-$ and $(NH_3)_2^-$. Some recent work in these areas includes Raff and Pohl[36]; Natori and Watanabe[37]; Ray[38]; Aldrich et al.[39]; Ishimaru et al.[40]; and Weissman and Cohen[41]. The most successful and elaborate semi-empirical calculations have been made by Howat and Webster[42] using the INDO–MO formulation. In their stable tetramer case the cavity size was smaller for water than for ammonia but not zero in either case. The excitation energies were 2.08 eV for water and 0.72 eV for ammonia, in good agreement with the experimental values of 1.72 and 0.8 eV, respectively. No studies were made on the line shape. They also did other calculations on mixed solvents which are equally encouraging.

These studies, although semi-empirical, are able to introduce all the factors which are important in electron–molecule and molecule–molecule interactions in the clusters. They are also capable of considering questions such as optimum orientations of the molecules. There are two difficulties with these calculations: firstly, they are sufficiently complicated so that it is difficult to do one system well and even more difficult to extrapolate the results to other systems since the major factors cannot be identified separately, and secondly, we know that there are long range polaron interactions which can affect the energy and even the wave function of the electron. Cohen and Weissman[41] have introduced such effects as a perturbation but it is still not clear if that is sufficient. The best approach, although limited in generality, seems to be the addition of long range polarization effects to a molecular orbital treatment of short range effects.

III.B.4 Localized electrons—long and short range components

The best models, and possibly the only valid ones, are those which include both the short and long range components of the electron–molecule interaction. Each of the interaction components are most important in regard to some property and thus one can only explain all properties by including all of the essential interactions.

The ideal model would be a quantum-statistical treatment of the molecular fluid containing an excess electron as an impurity including all interactions between the species present. This was done in only one case and that was with the rather simple case of electrons in ^{4}He at low temperatures[43]. Even in that case the treatment contained many approximations. Thus, it is unlikely that a similar calculation could be done on a polar liquid in the near future. Since our theory of polar liquids is the weakest link, we might as well approximate some of the fluid as a continuum. However, we must appreciate that the molecules near the trapping site are greatly affected by the electron and thus should be singled out for special treatment. This suggests the first aspect of a reasonable model: a few molecules will be treated as a discrete solvation layer around the trapping site and the medium beyond will be treated as a continuum.

In Fig. 2 we introduce the basic notation. The distance to the center of the molecules in the first coordination layer is r_d and the continuum fluid begins beyond r_c. The simplest long range electron–medium interaction in the continuum consists of the polaron contributions plus V_0, the energy of the quasi-free electron. As we showed in the case of inert gas liquids, the effect of multiple scattering and the screened electronic polarizability interactions of the electrons outside the cavity can be reproduced by simply adding V_0 to the potential outside the trapping site or, in our case, in the continuum region (see eqn. (19)). Since the electron in the ground state is almost completely localized within the cavity or at least within r_c, the long range potential of the trapped electron will be very closely approximated by

$$V(r) \cong -\frac{\beta e^2}{r} + V_0 \qquad r > r_c \tag{37}$$

However, it is not very difficult to use the full form of the adiabatic or, even better, the self-consistent field version.

The short range electron–molecule interactions require more careful considera-

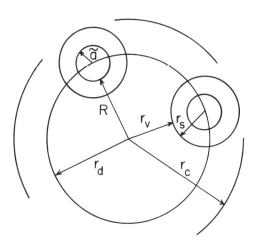

Fig. 2. Definitions of the distances involved in the CKJ model. r_v is the void radius of the cavity, r_s is the effective solvent radius and $\tilde{a}$ is the effective hard-core of the molecules located at a distance r_d from the center of the cavity. The continuum begins at r_c.

tion. In the case of inert gases the net interaction was subsumed into a pseudo-potential which was used in the form of V_0. In our case we have two major complications, firstly, the molecules are polar and thus electron–dipole moment interactions are very important, and secondly, because the molecules are polar they are non-spherical and have strongly angular-dependent intermolecular forces between them. All these features could be taken into account with a proper molecular orbital treatment of the systems $[S_n^- + \text{continuum}]$ where S is the solvent molecule and n would generally be in the range of 4, 6, 8, 12 or so. The continuum treatment would be as discussed above. Calculations can be made to find the most stable "species" of this type.

At this point it is worthwhile emphasizing two basic assumptions made here for which corrections may be necessary later. Firstly, in this work we usually use energy as a criterion for stability and not the more proper free energy. This was not serious when dealing with rare gases since the temperatures were so low, but it becomes increasingly important when dealing with room temperature liquids, especially strongly hydrogen-bonded liquids in which solvation entropies of molecules around the trapping site can be large. Under various circumstances these entropies can be positive or negative. We will later see cases in which this distinction is crucial. Under normal conditions, however, this point need not concern us. The second point is more subtle. In the above calculations we assume a sharp break between the first coordination layer and the remainder of the liquid. Surely this cannot be correct and some ESR evidence[44] shows that even in glasses the second coordination layer is quite strongly affected by the trapped electron. The effect of this on our calculations is hard to assess since we do include some aspects of this in our long range inter-actions but undoubtedly not all. The obvious solution to this problem is to work with larger and larger clusters of molecules (plus the continuum) until no further changes in properties are observed. Such detailed calculations are probably impossible so we can do little more than to be aware of their effect.

Only very recently have detailed molecular orbital calculations of the $[S_n^- + \text{continuum}]$ system been attempted. Newton[45], in work still in progress, has cal-culated the ground states of the water and ammonia systems. These calculations are exceedingly complex and time-consuming due to the number of variables involved and the small contribution one is after. The electron is only bound by a few electron volts relative to the approximately ten thousand electron volt energies of the water and ammonia molecules. Thus calculations must be done very carefully and very accurately to be sure that the results are meaningful. Newton's work is at the double zeta level of accuracy. We will refer to Newton's results again later but it is sufficient to say here that he found it essential to include the long range effects. Without the long range contributions $(H_2O)_4^-$ is not stable to thermal dissociation. The heat of solution from his model has been calculated to be ~ 1 eV for water, based on an extra electronic contribution of -1.25 eV, with an optimum value for r_d of 2.45 Å for a fixed molecular orientation[45]. His recent work, as yet unpublished, has included studies of molecular reorientation and initial studies of excited states. Furthermore, his initial studies on the ammonia system suggest that the cavity size there is only slightly larger than that in water.

These calculations are extremely important but, because of their complexity, suffer from some of the same faults discussed in Sect. III.B.3. It is very hard to generalize the results to other molecules and each individual calculation is so involved that it is hard to give rapid guidance to experimentalists. Furthermore, the critical factors are hard to identify without doing large numbers of systems. In addition, these calculations cannot be rapidly modified to study temperature, pressure and other effects. So, while we need a few of these very precise calculations, there is an increasing need to have a less accurate model but one which is easy to use and adapt to a wide variety of circumstances.

Such a model was developed by Copeland et al. (CKJ)[46], improving on some of the basic ideas in the earlier work of Land and O'Reilly[33]. The CKJ model treated the continuum by the adiabatic model. The related work by Fueki et al. (semi-continuum model)[47] used the better self-consistent version. In most work where the results can be compared, the two methods lead to similar results when other factors are similar. The reason for the agreement is that the electron is quite well localized and in such cases the differences between the methods disappear. Since the two models yield such similar results we will emphasize the CKJ model and its improved versions. The semi-continuum model has been discussed in detail by Kevan[4].

In both models one assumes that the essential feature of the molecule in the first coordination layer is its dipole moment. We assume that the coordination shell consists of N molecules (usually 4, 6, 8, or 12) with dipole moment μ_0 oriented at an angle θ to the radius vector. Although theoretically there are reasons to leave θ or really $\cos\theta$ in the basic potential, we prefer to use the thermal average of $\cos\theta$, the Langevin result, and introduce the temperature-dependence into the potential. Performing the thermal average later leads to the same result in most cases. Thus the potential which traps the electron is

$$\left.\begin{aligned}
V(r) &= -N\mu\, e/r_d^2 - \beta e^2/r_c & 0 < r < R \\
&= -N\mu\, e/r_d^2 - \beta e^2/r_c + V_0 & R < r < r_d \\
&= -\beta e^2/r + V_0 & r_d < r
\end{aligned}\right\} \tag{38}$$

with

$$\mu = \mu_0 \langle \cos\theta \rangle \tag{39}$$

and

$$\langle \cos\theta \rangle = \coth X - X^{-1} \tag{40}$$

with

$$X = \mu_0 e C_{1s}/k_B T r_d^2 \tag{41}$$

where C_{1s} is the charge enclosed in the cavity of radius R. The $1s$ denotes that we want the charge determined by the ground state wave function.

Putting this into the Schrödinger equation yields an energy W_i to which is added the contribution from the electronic polarization to yield the electronic energy of the state in question $\varepsilon(nl)$.

The total energy, $E(nl)$, must include the energy to modify the medium, the medium reorganization energy, $E_M(nl)$

$$E(nl) = \epsilon(nl) + E_M(nl) \tag{42}$$

where

$$E_M(nl) = E_{ST} + E_{PV} + E_{d-d}(nl) + \pi(nl) + E_{HH} \tag{43}$$

The above equation contains several terms not included when dealing with non-polar media. The term E_{d-d} represents a classical dipole–dipole repulsion which originates when one attempts to align all dipole moments; the adjacent dipoles then repel one another. Specifically the expression to use is

$$E_{d-d}(nl) = \frac{D_N \mu_T^2}{r_d^3} \tag{44}$$

where

$$\mu_T = \mu_0 \langle \cos \theta \rangle + e\alpha C_{nl}/r_d^2 \tag{45}$$

and C_{nl} is the charge enclosed in a radius R due to the wave function of state nl. The constants, D_N, are listed in Table 3. The term $\pi(nl)$ is the same as that discussed previously in Sect. III.B.2 (eqn. (31)).

The E_{PV} and E_{ST} terms may be neglected under normal conditions. It is not even clear what should be used for a surface tension in E_{ST}. At higher pressures the term E_{PV} is important but the volume to be used is not simply the void volume due to the disorganization of the medium. Most important of all terms is E_{HH}. This represents the quantum mechanical intermolecular repulsion of the molecules in the first coordination layer. It is called E_{HH} because it is dominated by the repulsion of hydrogen atoms in most cases. For example, when ammonia forms the coordination shell it is oriented so that all of its hydrogens point toward the center of the cavity. The same is true of adjacent molecules and thus strong steric repulsions of the hydrogens originate. This term is very crucial in describing trends in the behavior of

TABLE 3

Constants for medium rearrangement energy

In general		For ammonia		
N	$D_N{}^a$	C_{HH}^N (eV)	A_N	$B_N(\text{Å})$
4	2.2964	2602.4	1.633	0.471
6	7.1140	5204.7	1.414	0.600
8	12.820	6940.0	1.155	0.752
12	41.074	10416.0	1.000	0.843

[a]Values from A. D. Buckingham, Discuss, Faraday Soc., **24** (1967) 151.

electrons in homologous series but it is also very hard to calculate well empirically. The molecular orbital calculations automatically include such effects in their total energy. For ammonia and water we have used the form

$$E_{HH} = C_{HH}^N \exp(-4.60(A_N R - B_N)) \cdot \langle \cos \theta \rangle \tag{46}$$

with the ammonia constants defined in Table 3.

For water, it was our original feeling that the water molecule could rotate when forming the cavity to reduce the intermolecular repulsion to such an extent that we could assume

$$E_{HH}^{CKJ}(H_2O) = 0 \tag{47}$$

Gaathon[51] used another form for the repulsion due to Kamp

$$E_{HH}^{GJ}(H_2O) = I_N A(\sigma/C_N r_d)^9 \tag{48}$$

where $\sigma = 2.8$ Å, I_N is the number of interacting pairs, C_N is a distance scaling parameter, and $A = 4$ kcal mol^{-1} is an empirical energy parameter. This leads to significant water–water repulsions in the hydrated electron cavity. The recent a priori work of Newton[45] supports this latter form. Logan and Kestner[48] have also made calculations of the water interactions using a form similar to that of eqn. (46) which leads to cavity sizes similar to those of Newton. The work of Fueki et al. ignores this term[4,47]. Regardless of the success of the theories more work is needed to define this term more precisely. The last factor involving cos θ was added to account for the fact that with random orientations, i.e. $\langle \cos \theta \rangle = 0$, we do not want to count the repulsions of this term. This is also an unsatisfactory aspect of the present approximation which is currently being improved upon.

In other molecular systems the term E_{HH} is often very important. Fueki et al.[4,48,49] have played down its emphasis since it is so hard to approximate but Jou and Dorfman[50] used a variation of eqn. (46) to obtain, via the CKJ model, reasonable results for tetrahydrofuran. However, Fueki et al. also obtained reasonable agreement with spectral data alone by neglecting E_{HH}[49]. Thus, independent means are required to verify the need for this term.

Gaathon and Jortner (GJ)[51] have improved upon the CKJ model and made it more realistic in regard to the electron–molecule interaction in the first coordination layer. Their potential is (using the same notation as before)

$$
\begin{aligned}
V_{GJ}(r) &= -N\mu e/r_d^2 - \beta e^2/r_c & 0 < r < R \\
&= -N\mu e/r_d^2 - \beta e^2/r_c + V_{os} & R < r < r_d \\
&= -\beta e^2/r_c + V_{os} & r_d < r < r_c \\
&= -\beta e^2/r + V_0 & r > r_c
\end{aligned}
\right\} \tag{49}
$$

In this work all electron–molecule interactions (except the electron–dipole part) have been replaced by an effective V_0 for the shell. This V_{os} is calculated based on the density of the molecules in the first coordination layer. This density is not the same as that in the bulk medium and thus V_{os} is not necessarily similar to V_0.

III.C Results of model calculations—non-optical properties

We will now review some of the current results of model calculations of localized electrons in fluids. For the most part we will emphasize the more semi-empirical models such as those of Copeland et al., Fueki et al., and Gaathon and Jortner, since they have been able to explain the widest range of experimental data. Primarily, this mass of data which has been generated is due to the ease of doing the calculations. The detailed molecular orbital calculations have only been done for a few systems so far. We shall also emphasize in this chapter the CKJ and GJ models due to greater personal familiarity with them but the semi-continuum results are complementary to this work and will be referred to when other calculations are not available.

In solving for one particular cavity model, i.e. one particular solvent, a choice of a number of molecules, and the other physical parameters, one needs to solve the appropriate Schrödinger equation for the trapped electron, eqn. (26) where $V(r)$ is of the form of one of the three models, e.g. eqns. (38) or (49). In most of the work published ψ_{nl} has been assumed to be hydrogenic, i.e. the ground state is like a hydrogen $1s$ orbital with exponent ζ and the first excited state a hydrogenic $2p$ orbital with some different orbital exponent. The reason for this choice is primarily historical, partly computational. Recall that the simple form of the polaron potential is indeed Coulombic. Also, hydrogenic-type orbitals are rather easy to work with and quantum chemists are experienced with their integrals. In defense of this choice it should be said that they are not really too bad in regard to leading to reasonable predictions of most properties. However, Carmichael and Webster[52] have shown that the correct numerical wave function is rather different from the hydrogenic one, even for the Jortner version of the polaron model eqn. (27). Feng et al.[53] have shown some improvements in using simple Gaussian functions. Kestner et al.[54] have pointed out that the sum of two Slater-type (exponential) functions (of the $1s$ and $2s$ form) leads to a reasonably good solution of the equations. Logan and Kestner[48] have been continuing this work. An alternative solution was proposed by Gaathon and Jortner[51] who used a combination of spherical Bessel functions and exponentials. This latter form is like a particle in a box solution and indeed the energy level scheme of the solvated electron in the case of ammonia[55] and water is closer to that of a particle in a box than that of a hydrogen atom since the next highest spherically symmetric state (call it $2s$) is located *above* the $2p$ level, contrary to what is found with atomic energy levels. For the same historical reasons the first p orbital is labelled $2p$, a notation we will also adopt here. Nevertheless most calculations to be reported here will be of the elementary type (single Slater exponential) unless otherwise noted. In the future more attention will have to be paid to this aspect of the problem.

In Table 4 we present some of the results of Gaathon and Jortner[51] for water and ammonia using their more elaborate wave functions, the Kamp repulsion (eqn. (48)) between water molecules, and the exponential repulsion between ammonia molecules (eqn. (46)). Also included in the table are calculations of the oscillator strength of the first allowed transition but these values were calculated from the GJ model using wave functions made up of the sum of two Slater exponents[54]. The close agreement of the momentum and dipole length results suggests that the trial wave functions is fairly good. The ground state energies so calculated were comparable,

TABLE 4

Solvated electron results

	NH$_3$		H$_2$O	
	Experiment[a]	Theory	Experiment[a]	Theory
Transition energy ($1s \rightarrow 2p$)(eV)	0.8	0.89[b]	1.72	2.0[b]
$\triangle_{1/2}$(eV)	0.46	0.18[b] + 0.07[c]	0.92	0.42 + 0.09[c]
Temperature coefficient of transition energy (cm^{-1} deg^{-1})	-0.2×10^{-3}	-0.3×10^{-3d}	-1.0×10^{-3}	-0.3×10^{-3d}
Equilibrium radius r_d^0 (Å)		2.87[b]		2.29[b]
Heat of solution $\triangle H = -E_t(1s)$(eV)	1.7 ±0.7	1.02[b]	1.7	1.5[b]
Photoelectric threshold (eV)	1.6[e]	1.94[b]		3.0[b]
V_0(eV)[a]		−0.22		+0.19
Oscillator strength ($1s \rightarrow 2p$)	0.77		0.65	
Momentum formula		0.89[d]		0.80[d]
Dipole length		0.64[d]		0.73[d]
Transition energy ($1s \rightarrow 2s$)(eV)		1.34[f]		

[a]For original experimental references, see ref. 46.
[b]GJ model, see ref. 51.
[c]Polaron contribution via Levich formula, see ref. 69.
[d]See ref. 54.
[e]This is Häsing's value, Ann. Phys. (Leipzig), **37** (1940) 509. Recent work by H. Aulich, B. Baron, P. Delahay and R. Lugo confirm a threshold of about this size, J. Chem. Phys., **58** (1973) 4439.
[f]See ref. 55.

and in most cases slightly better than the original Gaathon calculation. When the same calculations were done with single term Slater functions it was not uncommon for the two numbers to differ by a factor of two or three. Finally in Table 4 we see the predictions for the location of the $1s \rightarrow 2s$ dipole forbidden transition[55]. This separation was calculated using three term wave functions carefully optimized. It might be possible for this state to be observed via two photon absorption in ND$_3$.

Of course, the results in Table 4 are only for the equilibrium or lowest energy configuration. The actual calculations are done as a function of some radius parameter. Thus we determine the energy of the system as a function of the cavity radius for each of the states. In Fig. 3 we present one of the typical curves of the CKJ model for the ammoniated electron[56].

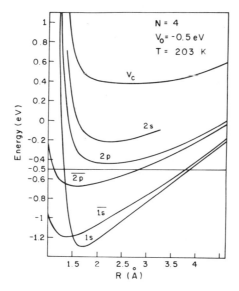

Fig. 3. Electrons in ammonia. Configurational diagrams for the total energy as a function of the radius R for various electronic states when $N = 4$, $V_0 = -0.5$ eV and $T = 203$ K. The ground state is denoted by $1s$ and the two bound excited states and the continuum level which exist when the polarization field is determined by the ground state are denoted by $2s$, $2p$, and V_c, respectively. The $2p$ state is the lowest energy p state when the polarization is determined by the $2p$ excited state, and the $\overline{1s}$ state is the lowest energy level under the same polarization.

The models are capable of predicting a great many properties of the solvated electron. Unfortunately, except for the ammoniated and hydrated electron, we know very little experimentally except for the optical spectrum and in some cases how it depends on parameters such as temperature and pressure. We shall therefore first examine the variety of predictions of the models for the water and ammonia systems, later returning to comparison of models and experimental data on the optical properties in other solvents. Earlier theories are reviewed elsewhere[34,51,58,59].

The following are some of the results for water and ammonia:

(1) A *stable cavity* is predicted which has a finite non-zero radius. Thus the model is consistent with the existence of this "species". The model also predicts stable localized species in a wide variety of solvents. This is a point which the earliest models such as Ogg[60] and Jortner[31] were not capable of answering.

(2) The most common *coordination number* seems to be four or six. The results in Table 4 are for the most stable configuration, namely four. Nuclear magnetic resonance data on ammonia solutions have been interpreted by Catterall[61] to be consistent with $3 \leq N \leq 13$. The work of Pinkowitz and Swift[62] which suggests much higher values must include contributions from the second layer. This is consistent with the recent double resonance work of Kevan and co-workers[63] in alkaline ice where they find a number of about four for the first coordination layer, but substantial participation of molecules further into the medium. Furthermore, the high

frequency permittivity of metal–ammonia solutions has been interpreted by Breitschwerdt and Radsheit[64] as due to about fifteen ammonia molecules irrotationally bound to some charge center per dissolved sodium atom. All of this is consistent with our model. Most molecular orbital studies also assume $N = 4$ in their calculations.

(3) *Volume expansion* data in metal–ammonia solutions is consistent with the predicted cavity size. In interpreting experimental volume expansion data several factors need to be considered. There is some electrostriction about the sodium ion and the remainder of the volume change includes any perturbation of the solvent structure by the electron (structure-breaking). Jortner[31] has estimated that the effective radius of the solvated electron in ammonia is ~ 3.2 Å. However, since the density of molecules in the first coordination layer is less than that in the normal bulk liquid, the actual void radius, r_v, in Fig. 2 is a lower limit to the cavity size. Copeland et al.[46] showed that their result ($r_d^0 \cong 2.7$ Å) was equivalent to an effective cavity radius of ~ 3.0–3.1 Å, in good agreement with experimental data. This "structure-breaking" effect has also been found by Lepoutre and Demortier[65] in their study of the entropies of transfer and entropies of reaction of electrons in ammonia.

The situation in the case of water is confusing. Historically, it was assumed that the cavity was very small because the Jortner theories[31,34] could only explain the optical properties if the cavity size approached zero. The early CKJ and semi-continuum calculations also had small cavities of $r_d^0 \sim 1.5$ Å (about the size of a water molecule). However, the recent work of Gaathon and Jortner[51] gives $r_d^0 = 2.29$ Å, while Newton[45] obtains $r_d^0 = 2.45$ Å, values significantly larger. In studying ammonia–water mixtures, Logan and Kestner[48] found it necessary to assume cavity sizes near those of Newton[45] (i.e. had to increase water–water repulsions) to explain composition variation. The experimental data is not clear-cut. The recent double resonance work of Kevan and co-workers[63] in alkaline ice, however, lead to values of 2.4 Å, in good agreement with the Newton work but these values are subject to some interpretation. The critical test of the cavity size in both water and ammonia comes from a study of pressure-dependences of band maxima but we shall delay that discussion until later.

(4) *Mobility and viscosity* data are consistent with a very loose first coordination layer but nevertheless localized electron. The experimental mobility in ammonia[66] is 1.08×10^{-2} cm^2 V^{-1} s^{-1} and 2.5×10^{-3} cm^2 V^{-1} s^{-1} in water[67]. The viscosity of metal–ammonia solutions is very interesting in that it is considerably lower than that of the pure solvent[68].

(5) *Heat of solution data* is also in fair agreement with experimental data if we assume that

$$\Delta H = -E_t^0(1s) \tag{50}$$

where $E_t^0(1s)$ is the lowest value of the total energy. The ammonia value is uncertain due to our lack of knowledge of the absolute heat of solution of a proton in ammonia.

One assumption is buried in eqn. (50). Nowhere in our formalism do we consider the fact that in forming the cavity some hydrogen bonds are broken by the molecules forming the first coordination layer, with possible contributions from the second layer as well. This should be a small positive contribution to $E_t^0(1s)$ and would help bring about agreement with experiment. Since the model already contains other limitations we have not pursued this point but the models are becoming sufficiently accurate to consider these points. We will return to this subject when we consider mixed solvents.

(6) The *photoconductivity threshold* can be calculated in terms of the photoconductivity onset, I, which is simply the energy needed to go from the ground state to the vertical continuum level, V_C or

$$I = V_C(r_d^0) - E_t^0(1s) \tag{51}$$

Using the CKJ model[56] this number is ~ 1.9 eV for ammonia. The actual experimental value is likely to be less since we have calculated it at the equilibrium cavity radius and any thermal motion will smear such effects. Unfortunately this threshold, and even photoconductivity itself, has never been observed in these liquid systems.

(7) The *photoelectric threshold*, P, for electron emission into the gas phase is $-\epsilon(1s)$, provided that $V_0 > 0$. If $V_0 < 0$, additional energy must be provided to raise the electron to the vacuum level. The actual mechanism for this process is not known but the various possibilities are outlined in Chap. 4 by Delahay.

Knowing P and ΔH one can obtain an estimate of $E_m(1s) = \Delta H - P$. Thus if these experimental values can be accurately determined they will provide another test for the theory. In the case of electrons in ammonia the value is quite uncertain but since we suspect that $E_m(1s)$ is positive, the lower experimental value of ΔH (1.0 eV) is probably the more accurate.

(8) The *photoionization energy-dependence* can also be calculated. The shape of the bound-free transitions has been calculated by Kestner and Jortner[69] and by Fueki et al.[70] The former calculation was for ammonia while the latter group treated several alcohols, amines, and ice. The calculations in the latter case also agree with photobleaching and photoconductivity experiments. They also confirm the common hypothesis that the bound-free transitions lie at higher energies than the primary optical absorption, a point which had been questioned by a few researchers[71]. The results also suggest that the bound-free transition does not proceed via a higher bound–bound state followed by autoionization. Depending on the mechanism of photoemission, these curves may also explain the photoemission energy profiles of Delahay[72]. The shape of the curves are quite similar except at high energies.

(9) Associated with the curves in Fig. 3 are *totally symmetric vibrations* of the cavity. Using the R-dependence of the energy one can fit the curves to a harmonic oscillator function about the minimum

$$E_t(1s) = E_t^0(1s) + \tfrac{1}{2}K(r_d - r_d^0)^2 + \cdots \tag{52}$$

The value of K is equal to the $(\partial^2 E/\partial r_d^2)$. The frequency of the totally symmetric vibration, ν_s, is related to force constant K by

$$\nu_s = (1/2\pi)(K/\mu')^{1/2} \tag{53}$$

where μ' is the appropriate reduced mass, in this case the mass of one solvent molecule. Using the CKJ model we obtain a value of about 174 cm^{-1} for the ammonia ground state, 61 cm^{-1} for the ammonia $2p$ state, and 202 cm^{-1} for the water ground state. Earlier published numbers[16] contained several numerical errors.

Smith and Koehler[73a] have made an extensive Raman study of sodium–ammonia solutions. They were unable to find any evidence for this vibration or for any vibration from about 20–400 cm^{-1}. The most significant feature of the Raman spectrum was the decrease in intensity in the region of 120–300 cm^{-1} when sodium was added to pure ammonia. In fact, the entire spectrum became less interesting the more sodium that was present. The trends observed have been assigned to rotational structure being damped by the sodium ions and, or, the electrons. From comparisons with F-centers in solids, one would have expected to see the predicted vibration. Unfortunately, no simple clear-cut explanation has yet been provided for this null experiment. It could be due to cavity lifetimes, large distributions of cavity sizes, local heating, or many other causes. None the less, it remains a major mystery in this field of research.

(10) *Magnetic properties* are very hard to calculate by any model. In early work on metal–ammonia solutions[57] one of the major tests for the accuracy of the model was how well it could reproduce such quantities as spin densities on various nuclei. We now know from studies in various fields that these are extremely subtle properties and only by a complete detailed molecular orbital treatment can one hope to obtain such data. It is possible that the work of Newton[45] will be sufficiently accurate but all other calculations are probably not precise enough. This data is important and should be studied but we will need better models first.

III.D Results of model calculations—optical properties

III.D.1 Position of band maximum

The most obvious optical property is the one we have already discussed and on which most of the research has been done, namely the position of the band maximum. In Table 5 we present a survey of the results of the CKJ and semi-continuum models. When both are calculated with similar assumptions concerning hydrogen repulsions, etc., the two models agree quite well (as well as we can expect for these simple models). The most important feature of this table is that the position of the optical absorption of a wide variety of solvated electrons can be explained by the model calculations.

Furthermore, trends within homologous series generally fall into line with the steric considerations. For example, the bulky t-butanol has a much lower excitation energy than its straight chain counterpart (the experimental numbers are 1.46 eV vs. 1.83 eV). Such bulky solvent molecules lead to larger cavity sizes which, all other things being equal, lead to lower transition energies. Similar comments can be made concerning 2-propanol vs. 1-propanol (experimental values 1.51 eV vs. 1.67 eV). The bulky ethers have very low transition energies for steric and other reasons. Similar comments can be made concerning dimethyl sulfoxide.

TABLE 5

Summary of band maxima calculations

Ammonia	
Experimental (203 K)	0.80 eV
Theoretical	
Copeland, Kestner, Jortner	0.94[a]
Kestner, Logan	1.09[b]
Logan, Kestner	0.92[c]
Kevan, Fueki, Feng	1.18[d]
Gaathon, Jortner (300 K)	0.89[e]
Water	
Experimental (298 K)	1.72 eV
Theoretical	
Copeland, Kestner, Jortner	2.70[a]
Gaathon, Jortner	2.0[e]
Logan, Kestner	2.02[c]
Fueki, Feng, Kevan	2.15[d]
Ice[d]	
Experimental (77 K)	1.9
Theoretical	
Fueki, Feng, Kevan	1.84
Methanol[d]	
Experimental	2.3(77 K) 1.87(298 K)
Theoretical	
Fueki, Feng, Kevan	2.09 1.85(298 K)
Ethanol[d]	
Experimental	2.3(77 K) 1.80(298 K)
Theoretical	
Fueki, Feng, Kevan	2.15 1.79
Tetrahydrofuran[f]	
Experimental	0.585 eV (room temperature)
Theoretical	
Jou, Dorfmann	0.583 eV ($V_0 = -0.5$ eV)
2-methyl tetrahydrofuran[d]	
Experimental	1.0 (77 K)
Theoretical	
Fueki, Feng, Kevan	1.04
Triethylamine[d]	
Experimental	0.75 (77 K)
Theoretical	
Fueki, Feng, Kevan	0.83

[a] See ref. 46—CKJ model.
[b] See ref. 55—small cavity, two-term wave function—CKJ model.
[c] See ref. 48—large cavity for water—modified GJ model.
[d] See ref. 49—small water cavity—semi-continuum model.
[e] See ref. 51—large cavity for water—GJ model.
[f] See ref. 50—CKJ model.

Fueki et al. made a study of the effect of matrix polarity and found that the major trends can be explained by the magnitude of the dipole moment. Thus one can develop simple empirical rules which could be quantified by model calculations: first, consider the magnitude of the dipole moment and this should yield the general region of the spectrum in which the absorption should occur. To be more specific one should then estimate the steric effects which will result when the dipole moment of the molecule is properly aligned. The larger this effect the more the transition energy will be lowered.

III.D.2 Temperature-dependence of band maximum

In addition to the position of the band maximum we can also study its dependence on other experimental variables such as temperature. In the early work of Jortner[31] this dependence was explained by a combination of the temperature-dependence of the dielectric constants and by an ad hoc variation in the cavity radius. The present models, especially the semi-continuum models, should be capable of yielding this data without any ad hoc assumptions, since all factors relating to the property of the "species" are included. In Table 6 we list the predictions for the temperature-dependences for ammonia and water. It is important to realize one fact: theoretical values of temperature-dependences are often based on a constant density of solvent, whereas experimental values usually involve constant pressure. Therefore, the experimental results listed in Table 6 were converted to constant density values via

$$\left(\frac{\partial h\nu_{max}}{\partial T}\right)_\rho = \left(\frac{\partial h\nu_{max}}{\partial T}\right)_P - \left(\frac{\partial h\nu_{max}}{\partial \rho}\right)_T \left(\frac{\partial \rho}{\partial T}\right)_P \tag{54}$$

In Table 6 we list the experimental data which are needed in eqn. (54).

The agreement between experimental and theoretical values in Table 4 for ammonia and water is reasonably good. It is difficult to completely separate out the major causes for the observed dependences. However, it does appear that the most important single factor[46] is the change in dipole orientation which in the simplest

TABLE 6

Experimental data on temperature and pressure dependences

	Ammonia (240 K)	Water (300 K)
$[\partial h\nu_{max}/\partial T]_P$	-1.5×10^{-3} eV deg$^{-1a,b}$	-2.8×10^{-3} eV deg^{-1c}
$[\partial h\nu_{max}/\partial \rho]_T$	1.0 eV cm^3 g^{-1d}	2.5 eV cm^3 g^{-1a}
$[\partial \rho/\partial T]_P$	1.3×10^{-3} g cm^{-3} deg^{-1d}	5.5×10^{-4} g cm^{-3} deg^{-1d}

[a]See ref. 81.
[b]I. Hurley, T. R. Tuttle, Jr. and S. Golden, in J. J. Lagowski and M. J. Sienko (Eds.), Metal–Ammonia Solutions, Butterworths, London, 1970, pp. 503–511.
[c]See ref. 94; R. R. Hentz, Farhataziz and E. M. Hansen, J. Phys. Chem., **75** (1971) 4974.
[d]Int. Critical Tables, Vol. 3, McGraw-Hill, New York, 1933, p. 23.

models is defined by the Langevin factor. While the cavity size may change as a natural result of this and the temperature-dependences of the other physical parameters, it has a very minor contribution. The change in the dipole orientation has been referred to by Hentz and co-workers as the *local dielectric effect* as opposed to the bulk effects[73b].

Fueki et al. have calculated the temperature-dependences at constant pressure[74] for various alcohols (methanol, ethanol, and 1-propanol), using the semi-continuum theory. They were able to account for about 70% of the observed dependences over rather large temperature ranges. The results support the idea that the major factors are the change in dipole orientation and the changes in V_0 values (because the density changes), but they have also suggested that instead of the gas phase dipole moment it is more appropriate to use the effective dipole moment which originates from the Kirkwood treatment since the molecules in the first coordination layer are not completely free. This could explain why the values for water in Table 6 are not as good as those for ammonia. However, such a treatment has its own uncertainties and more work will be needed to justify its importance. Certainly the fact that the molecules are not completely free should enter into our models at some stage. This was first emphasized in an empirical manner by the correlations of Freeman[75].

As an extra feature of the above calculations, Fueki et al.[74] were able to calculate temperature-dependent relaxation times for dipole orientation. These values are now becoming experimentally available and the calculated values, at least for 1-propanol, appear to agree well with data of Gilles et al.[76] at high temperatures and Baxendale and Wardman[77] at low temperatures. This subject is beyond the scope of this chapter and will not be pursued here.

III.D.3 Pressure effects on the band maximum

The first calculation of the effects of pressure on band maxima were made by Feng et al.[78] on the methanol and ethanol system. They pointed out the importance of changes in V_0 values. They did not pursue the older questions of the role of cavity compression since the cavities in the solvents they treated were small and changed little with pressure.

Recently, with the new experimental data on ammonia and water at quite high pressures, Logan and Kestner[48] have pursued the pressure effects on these systems in more detail. In Table 7 we list the results for these systems calculated via the CKJ model for two choices of intermolecular repulsions in the water case. In ammonia there is a puzzling good agreement at the lowest pressure and the theoretical results at higher pressures parallel the experimental ones. In water the experimental and theoretical results also parallel one another, but the absolute values do not agree.

The calculations in Table 7 were done without including the E_{PV} term of the medium rearrangement energy for the column marked "no cavity size decrease" and with that term for the other column. In calculating the E_{PV} term for ammonia, we used the effective cavity radius of 3.2 Å calculated by Jortner from experimental data (and about the same as we would calculate (see Sect. III.B.4, eqn. (43)) i.e.,

$$E_{PV} = (4\pi/3)R_{eff}^3 P \tag{55}$$

We can see that in the range of 2–6 kbar, where this contribution is important, about half of the spectral shift originates from the decrease in cavity size with pressure. It should be emphasized that that is not an ad hoc assumption but originates as a natural result of the calculation.

In the case of water it appears that most, if not all, of the change in band maximum with pressure can be explained by the model without any cavity size decrease. Nevertheless, an effective cavity radius of 2.5 Å would improve the agreement with experiment leading to an extra 0.02 eV or so shift between 2 and 6 kbar.

Unfortunately, in the case of water this effective radius cannot be determined directly. From the volume of activation of reactions not controlled by diffusion, Hentz et al.[79] have estimated that the effective electron cavity volume is about 10 ml mol^{-1} if electrostriction contributes 3 ml mol^{-1}. This would be equivalent to an effective radius of 1.6 Å. With such an effective radius the models predict very little compression under pressure and very little shift due to the change in cavity radius. However, such a small effective radius is not predicted by some of the more recent calculations. It thus appears that the more recent calculations of Newton[45],

TABLE 7

Effects of pressure on the band maxima in ammonia and water

P(kbar)	V_0(eV)	$h\nu_{max}$(eV)		
		Theory		Experiment[b]
		No cavity size decrease	With cavity size decrease[a]	
Ammonia (296 K)	−0.5	0.662	$0.662(r_d^0 = 2.75$ Å$)$	0.67
0.009	−0.43	0.680	$0.703(r_d^0 = 2.67$ Å$)$	0.80
2.2	−0.29	0.745	$0.801(r_d^0 = 2.60$ Å$)$	0.91
Water (302 K)—medium strength repulsions between molecules				
0.001	0.29	2.56	$2.56(r_d^0 = 1.85$ Å$)$	1.71
2.1	0.47	2.65		1.84
6.3	0.72	2.76		2.00
Water (302 K)—strong repulsions between molecules[c]				
0.001	0.29	1.72	$1.72(r_d^0 = 2.50$ Å$)$	1.71
2.1	0.47	1.78	$1.83(r_d^0 = 2.45$ Å$)$	1.84
			$1.94(r_d^0 = 2.42$ Å$)$	2.00

[a]Based on an effective cavity radius of 3.2 Å for ammonia and 2.5 Å for water.
[b]Experimental results: NH$_3$: ref. 83; H$_2$O: ref. 80.
[c]Repulsions adjusted to yield the correct transition energy for water ($E^0(1s) = -1.26$ eV at 0.001 kbar).

Gaathon and Jortner[51], and those of Logan and Kestner[48], in which the hydrated electron is associated with a large void volume cannot be reconciled with the experimental studies at high pressures[80]. On the other hand, the value of the activation volume in ammonia[81,82] is consistent with the 3.2 Å effective radius we have been using in that case. That value is also predicted by the CKJ model[46].

Unless a more careful analysis of the relationship between the theoretical cavity radius and the effective radius can explain this discrepancy in water, these studies may suggest that some element of the model is incorrect[84]. We point out again that these two quantities are by no means identical, nor is the effective radius always larger[85].

III.D.4 Line shape

Once we have determined potential energy curves as a function of a nuclear coordinate such as R or r_d, these can be used to determine the line shape as well as the band maximum. However, at present this has only been done in detail for ammonia. These calculations can be done easily if two assumptions are met: (a) we can use the high temperature approximation, and (b) we can use the Condon approximation. Since the medium modes are estimated to be of the order of 1 cm^{-1}[86] and the totally symmetric vibration is similar to 175 cm^{-1} in ammonia, at temperatures of 200–300 K, we can safely use the high temperature limit. The semi-classical Condon approximation which assumes that the electronic transition moment is not a function of the nuclear coordinates can be tested in our model and it has been found to vary no more than 30% from one extreme of the band to the other[69]. This should then represent the maximum uncertainty of the line shape.

In the high temperature limit, Kubo and Toyozawa[87] have shown that the appropriate line shape function to use in the present problem is

$$F(E) = \frac{|M|^2}{Z} \int dX \exp[-E(1s, X)/k_B T] \delta(E + E(1s, X) - E(2p, X)) \tag{56}$$

where X is a set of coordinates appropriate to the present problem, Z is the ground state partition function, and M is the electronic transition moment.

If we neglect the role of the medium modes and assume that the dipole orientation, $\mu_0 \cos \theta$, can be replaced by its Langevin thermal average, the only coordinate involved is R and the potential curves we must use are those plotted in Fig. 3. In that case, the line shape function takes on a rather simple form

$$A(E) = \frac{|M|^2}{Z} \exp[-E_t(1s, R)] \left| \frac{dR}{dE_t} \right| \tag{57}$$

This line shape has been calculated for ammonia and in Fig. 4 we show one example. We refer to this as a temperature-dependent result since we have used the thermal averaged value of $\cos \theta$ in our potential. For comparison we present the line shape with dipole perfectly aligned, $\langle \cos \theta \rangle = 1.0$, independent of temperature.

The use of the Langevin average of $\cos \theta$ is obviously incorrect and thus one really needs to calculate the potential curves as a function of R and $\cos \theta$ and then use these to determine line shapes and other temperature-dependent results. We have done

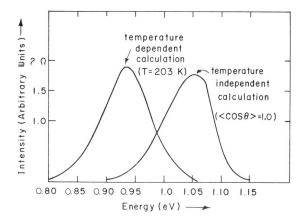

Fig. 4. Electrons in ammonia. Optical line shape for $1s \to 2p$ transitions calculated from the configurational diagrams with a single (R) mode. $N = 4$, $V_0 = -0.5$, and $T = 203$ K.

this[69] and for almost all properties the two calculations yield the same results, i.e. pre-averaging $\cos\theta$ over temperature is allowed. This is because the energy is a linear function of $\cos\theta$ appropriate to the ground state (~ 0.85) and averaging yields the Langevin result. Figure 5 shows examples of such curves for the two extreme values of $\cos\theta$ in ammonia. Using this set of curves in the general expression (eqn. (56)) where $X = (R, \cos\theta)$ we have calculated the line shape shown in Fig. 6.

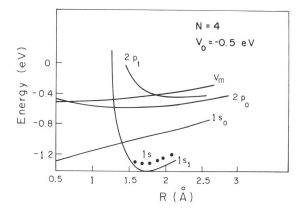

Fig. 5. Electrons in ammonia. Temperature-independent potential model configuration diagrams for the total energy as a function of the radius R for $N = 4$ and $V_0 = -0.5$ eV. The last subscript now refers to the value of $\cos\theta$. Only curves for two values of $\cos\theta$ are plotted. Each state has the bulk medium polarized according to its electron density (see text). The dots labeled $1s$ denote the temperature averaged result where $(\cos\theta)_{1s} = 0.8$–0.9. V_m is the medium or continuum level.

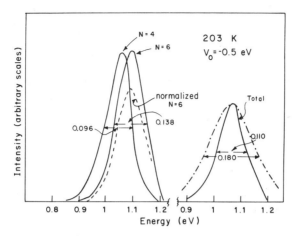

Fig. 6. Calculated line shapes for electrons in ammonia: for specific coordinations (left) and for the total sum (right). The dot–dash line shows the effect of including long range polaron fluctuations. The half-widths of each band are indicated in −0.5 eV.

By comparing Figs. 5 and 6 we can see that the $\cos\theta$ mode results in a small broadening at low energies and contributes very little to the band half-width.

In our calculations with the CKJ model[56] the $N = 4$ and $N = 6$ cavities in ammonia had similar energies and similar oscillator strengths, thus we can derive a composite total line shape from both "species"

$$F_T(E) = F_4(E) + \exp(-\triangle E^0/k_B T)F_6(E) \tag{58}$$

where $\triangle E^0$ is the difference between the ground state energies of the $N = 4$ and $N = 6$ results. This is plotted on the right hand side of Fig. 6. When the same calculation is done at 273 K, the line half-width increases from 0.110 to 0.165 eV, or roughly as the square root of the temperature[69].

There is another set of coordinates which contribute to the line width and shape, namely the low frequency medium modes. If these obey the harmonic approximation and if they can be approximated by a single frequency in both electronic states we can adopt the approach of Levich[88] and evaluate contribution to the half-width as

$$\Gamma_\nu = 4(k_B T E_s \ln 2)^{1/2} \tag{59}$$

where

$$E_s = \frac{1}{8\pi\beta} \int (\mathbf{D}_s - \mathbf{D}_p)^2 dV \tag{60}$$

where $\mathbf{D}_s$ and $\mathbf{D}_p$ are the electric displacement vectors for the $1s$ and $2p$ states, respectively. E_s has been calculated using the wave functions from the CKJ model as ~ 0.024 eV. Thus we find that the medium modes contribute about 0.07 eV to the half-width at 203 K and ~ 0.08 eV at 273 K. If the line widths are approximately

Gaussian, the complete half-width will be the sum of the individual half-widths. We have indicated in Fig. 6 the effect of these polaron modes on the line width.

Tachiya et al.[35] have developed an improved continuum model of the solvated electron. As one highlight of their treatment they are able to study the effects of fluctuations in the long range polarizations and within a linear theory have calculated the effect on line width.

The comparison of the present theory and experiment is demonstrated in Fig. 7. The position of the band maximum does not agree but that is less serious than the observations one can make concerning line width and especially line shape. The line width is underestimated by a factor of two or three (see also Table 4). Furthermore, the experimental line shape is definitely skewed toward the high energy side, contrary to the theoretical curve which is almost symmetrical. In addition, the experimental half-width, in the case of ammonia at least, is almost independent of temperature, in sharp contrast to the theoretical predictions. Furthermore, the line width does not appear to be dependent on solvent density ruling out any strong competition among cavities of various coordination numbers.

There are several possible explanations for this qualitative discrepancy but almost all can be eliminated, at least initially. These include a larger variety of coordination numbers than we have calculated here, with an almost continuous distribution of energies, and contributions of higher excited states, even the continuum. The latter can be estimated and should have almost no oscillator strength (see Fig. 7). Within our model their energies are simply

$$E(nl) = -\frac{\beta^2 e^2}{2n^2} + V_0 \quad \text{for } n \geq 3 \tag{61}$$

These energies can be calculated relative to the $2p$ state and are indicated in Fig. 7. These excited levels should also be greatly affected by the density of the

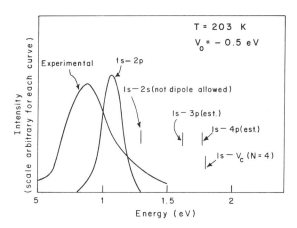

Fig. 7. Comparison of calculated and experimental line shapes as well as the position of higher excited states for electrons in ammonia. Similar comparisons exist for electrons in other solvents.

fluid since they depend directly on β and yet lines in very low density water vapor are as broad as those in the liquid (see Sect. IV).

Thus the question of the extremely broad lines remains. Two other explanations have been proposed which have not been fully evaluated as yet: asymmetric fluctuations (estimates indicate that asymmetric modes have a very small force constant; one calculates that at most they could contribute $kT \ln 2$ or 0.012 eV to the half-width) and permanent distortion of the cavity from its spherical shape (this could arise from the rapid exchange of solvent molecules between the first coordination layer and the bulk medium and at any one instant one would have a variety of cavity shapes). Although we cannot treat this latter problem easily without major changes in our model, one can argue that this could lead to asymmetric lines with little temperature-dependence if non-spherical cavities had larger transition energies as well as different transition energies for the $1s \rightarrow 2p_0$, $2p_{\pm 1}$ states (a fact which we have not been able to verify so far: similar results can be found in ref. 89). The problem of line shape in mixed solvents is even less well explained by our theoretical models (see Sect. V).

There has been one proposal for the observed line shape due to Kajiwara et al.[71] which yields a reasonable line shape but involves two drastic and extremely radical assumptions. They assume that the bound electron should be treated as a particle in a finite box with no long range interactions, and that there is only one bound state in the cavity. Both assumptions cannot be justified in any complete model. A more thorough discussion of the general line shape problem and the limitations of eqn. (61) has been given by Funabashi in ref. 71b.

Thus the problem of line shape remains the major puzzle of the theory. It is hoped that the problem is not so serious as to force us to abandon the present models but in any case it appears to indicate the need for major improvements. In this regard the reader is invited to consider the alternative proposals regarding higher excited states as discussed by Funabashi[71b].

III.D.5 Higher excited states

We have already indicated in Fig. 7 the position of some of the higher excited states of p character. We expect them to be even broader than the $2p$ state as well as being quite weak, and thus it is probably not too surprising that extensive studies of the spectra at large energies give no indication of these states. The most elaborate study of this type was done in ammonia by Hentz and co-workers[90].

Another state of some interest is the $2s$ state. Logan and Kestner[55], using three term Slater-type wave functions, calculated this for ammonia. The results of these calculations are presented in Table 8. The relative position of the $2s$ state is indicated in Fig. 7. The $2s$ state is not dipole-allowed and would not be seen in a one-photon absorption process. It could be seen in two-photon absorption and studies in ND_3 seem especially promising. The fact that the $2p$ state is close could also increase the intensity of such absorption. The model calculations which indicate that the $2p$ state lies below the $2s$ state clearly show that the lower excited states are dominated by a potential which is very non-Coulombic. In fact, the level spacing and orderings are much more characteristic of a particle in a box than of a Coulombic potential.

TABLE 8

The energy of the $1s$ and $2s$ states of the electron in ammonia solutions using a three-term wavefunction ($T = 203$ K)

V_0 (eV)	R_0 (Å)	Total energies (eV)		Transition energy (eV)	
		$1s$ state	$2s$ state	$1s \to 2p$	$1s \to 2s$
$N = 4$					
0.5	1.75	$-0.677(-0.537)^a$	0.869	1.328	1.546
0.0	1.70	$-1.053(-0.909)^a$	0.394	1.200	1.447^b
-0.5	1.70	$-1.454(-1.30)^a$	-0.114	1.093	1.340^c
$N = 6$					
0.0	2.15	$-1.156(-0.972)^a$	$+0.533$	1.369	1.689
-0.5	2.10	$-1.480(-1.294)^a$	$+0.0675$	1.211	1.548

[a]Calculated using a one-term wave function (CKJ model).
[b]Estimated half-line width is 0.124 eV with temperature dependence of the band maximum of -8.1 cm^{-1} K^{-1}.
[c]Estimated half-line width is 0.112 eV.

The highly excited states feel the effect of the Coulombic potential more strongly (eqn. (61)). Thus they provide a sensitive test of the theory.

III.D.6 Emission

After the electron is excited to the $2p$ state, it must lose its energy in some way. One possible relaxation mechanism is emission, although the lack of observation of any fluorescence suggests this is not the major pathway. To calculate the expected emission one needs to recalculate the $2p$ state, not in the Franck–Condon limit where the average value of $\cos \theta$ is determined by the $1s$ wave function, but where $\cos \theta$ is determined by the $2p$ state itself. This state is the relaxed $2p$ state and is labelled $\overline{2p}$ in Fig. 3. The emission is from this state down to the $1s$ level whose $\cos \theta$ value is the same as that of $\overline{2p}$. In Fig. 3 we label this relaxed Franck–Condon type $1s$ state $\overline{1s}$. Using the same techniques as in the previous sections we can predict the line shape and position of the emission. This is shown in Fig. 8. It is distinctly red-shifted from the absorption since the $2p$ charge density is rather diffuse.

IV. Dense polar gases

In Sect. I we presented evidence (see Fig. 1) for a type of "phase transition" of the electron mobility of electrons in gaseous helium. This has caused some of us to inquire if this might not be possible in other systems. Cole[91] has suggested that a sort of "phase diagram" might even by constructed (Fig. 9). Of most interest in the polar fluids are H_2O and NH_3, since neither has a stable negative ion. In Fig. 9 we have qualitatively indicated where these would be located on such a diagram.

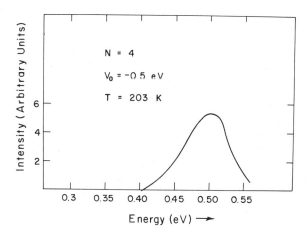

Fig. 8. $\overline{2p} \to \overline{1s}$ emission line shape calculated from the temperature-dependent one-mode model for the ammoniated electron.

Unfortunately, as yet no one has made any studies of electron mobility in the relevant density ranges but several major studies have been done on the solvated electron spectrum as a function of density in D_2O, NH_3, and ND_3. The studies on yields in ND_3 are probably the most important since they were done in the super-critical region. In that case Olinger et al.[92] found a dramatic decrease in electron yield for ND_3 densities of less than 10 M (0.2 g cm^{-3}). On the other hand, Gaathon et al.[93] have observed the solvated electron spectrum down to densities as low as

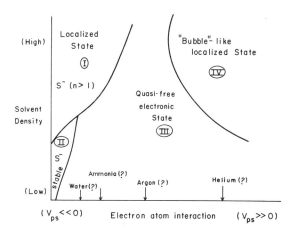

Fig. 9. A hypothetical "phase diagram" originally due to Milton Cole indicating the possibilities of transitions from quasi-free to localized states in a variety of solvents. This has been observed in helium (Fig. 1) and evidence exists for the transition in supercritical ammonia (see text).

0.02 g cm^{-3} for D_2O along the vapor–liquid curve. Other studies[94] have observed the solvated electron in supercritical D_2O to as low a density as 0.2 g cm^{-3}. In neither case was the transition to the quasi-free state at very low densities observed in D_2O. The experimental studies of Gaathon have also presented other problems, namely the half-width of the spectrum is almost constant as a function of solvent density and about the same as in the liquid. Whether this observation is free of clustering effects (i.e. the electron being trapped in pre-existing clusters of D_2O) is not clear and requires more supercritical studies. Gaathon and Jortner[51] have made theoretical studies of the spectra predicted (Fig. 10) for these cases. The results in ammonia agree reasonably well with experimental data. The water data cannot be directly compared.

In order to calculate the stability of the solvated electron, however, one needs to look at the total energy (i.e. heat of solution) and add to that the entropies of solvation. The stability criterion thus becomes

$$\triangle G < 0 \tag{62}$$

Kestner et al.[95] have tried to estimate these entropy terms using, firstly, conventional statistical mechanics of clusters, and secondly, entropy of localizing solvent molecules within a region of space. The results are very sensitive to the parameters used. In the first case it was predicted that hydrated electrons could be stable down to 0.02 g cm^{-3} with the ammoniated electron transition at about 0.05 g cm^{-3}. In the second case the transitions are predicted at much higher levels, between 0.1 and 0.2 g cm^{-3}. However, these calculations are far from complete and work is continuing to improve the estimations of entropies. Gaathan has found that the solvated electron is found when clusters of a certain size are likely to be found in the vapor.

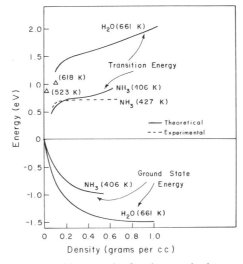

Fig. 10. The total energy and transition energies for electrons in dense water and ammonia vapors. Solid lines represent calculations of Gaathon and Jortner (see text)[51]. The dashed line shows the experimental results of Olinger and Schindewolf in ammonia gas[92]. The triangles are experimental results in water vapor by Gaathon et al.[93].

Another aspect of interest here is the pressure effect. In unpublished work Schinde-wolf[84] has observed curious volumes of activation due to the fact that in these dense supercritical gases the coordination layer is more dense than that of the bulk fluid (the gas). This area represents one of the most exciting for both theoretical and experimental work[96]. It is hoped that more studies will soon be forthcoming in supercritical regions including binary mixtures and mobility studies.

V. Electrons in mixed polar liquids

Another exciting new area of research is the study of electrons in mixed solvents. While studies have been done in polar and non-polar mixtures, we will only mention here a few studies in binary mixtures of polar liquids.

The experimental work has been summarized in a number of review articles by Dorfman[97-99]. Depending on the two liquids selected one can find behavior ranging from an almost linear-dependence of the excitation spectrum on composition (e.g. ammonia–water, and EDA–ether) to one in which one component dominates (e.g. water–ether where water dominates, and ethanol-hydrocarbon where ethanol dominates).

Two theoretical studies have been made of the ammonia–water system. Howat and Webster[42] calculated the energy of the $[(H_2O)_n(NH_3)_{4-n}]^-$ species by their semi-empirical molecular orbital theory. They find an almost linear dependence in the excitation energy vs. n. This work makes an implicit assumption which must be justified, if indeed it is true, namely that in a 50–50 mixture (or any other case) the cavity species contains 50% of each component in the first solvation layer.

Logan and Kestner[48] have calculated the ground state energy and the transition energy for all such species in all solvent compositions using a model similar to Gaathon and Jortner[51], including long and short range interactions. They found that it was necessary to assume some intermolecular repulsions between water molecules (and a large cavity size) in order to have all coordination species in a given bulk solvent have similar free energies. The free energy was obtained from the total energy by adding entropy effects from demixing (a simple combinational factor) and from entropies of solvation (using the Lepoutre and Demortier[65] values) and including estimates of the relative energies to break hydrogen bonds in ammonia and water (estimated as 0.99 kcal mol^{-1} for each H_2O which replaces an NH_3). In Fig. 11 we show the results for all cavity species and have connected those points with the lowest free energy according to the criteria mentioned above. The energy of all species, however, is very similar and thus all contribute to the observed line shape. Thus we would expect these models to predict rather broad line shapes at intermediate concentrations, a feature in sharp contrast to the experimental results of Dye et al.[100]. If the cavity size is allowed to shrink by reducing the repulsions between water molecules, cavities with high proportions of water molecules begin to dominate. This is true regardless of whether one or two term wave functions are used.

While all theoretical calculations in this area are preliminary they do emphasize the subtle features in such studies. To explain the experimental results properly the theory must be sure to calculate the relative energies and free energies of the

38

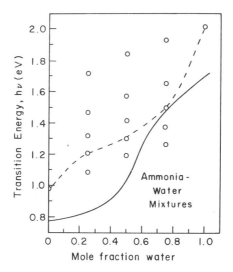

Fig. 11. Ammonia–water mixed solvents: band maximum versus water mole fraction. The circled points refer to calculations on various four-coordinated cavities: the highest energy point refers to an $(H_2O)_4$ cavity, the next highest to $(H_2O)_3NH_3$, and so on to the $(NH_3)_4$ cavity in these solvents. The dashed line connects points based on a preliminary minimum free energy criterion (see text). The experimental work of Dye et al.[100] is the solid line.

"species" involved, an aspect on which the theory has not been adequately tested until now.

VI. Two-electron cavity species

In metal–ammonia solutions, where the concentration of electrons can be increased to high values, it has been observed that the electron spins begin to pair up at higher concentrations[101]. One way for this to occur is to have a cavity species which contains two electrons. In various other systems such species have also been proposed but never distinctly proven. Feng et al.[102] have calculated the energy of such species in a variety of solvents using the semi-continuum theory. In all cases this species appears to be stable with respect to two one-electron cavities and therefore should be observed at higher electron concentrations. In ammonia they found that the one- and two-electron species had very similar electronic excitation energies. Copeland and Kestner[103] did a similar calculation using the CKJ model. There is a serious error in that result which has been corrected by Logan[104]. In correcting the error Logan also calculated the ground state and transition energies using more accurate wave functions. The results are presented in Table 9. The one-term results agree reasonably well with the work of Feng et al. except for larger transition energies for the two-electron species, than for the one-electron case. The Coulomb and exchange terms were treated as in the Feng et al. calculations, i.e. the exchange term was screened by the optical dielectric constant. Rather dramatic changes occurred in the two-electron case when two-term wave functions were used. It is well known

TABLE 9

Calculations of the two-electron species in ammonia at 203 K

		One-term wave function	Two-term wave function
One-electron cavity			
	ΔH_1(eV)	-0.941eV	-1.006eV
	$h\nu$(eV)	0.926	1.10
		($N = 4$)	($N = 4$)
Two-electron cavity			
	ΔH_2(eV)	-2.601	-2.828
	$h\nu$(eV)	1.29	1.61
		($N = 4$)	($N = 12$)
	ΔH_{21}(eV) $= \frac{1}{2}\Delta H_2 - \Delta H_1$	-0.359	-0.408

that the single Slater function places too much charge outside the cavity. This is not very serious in the one-electron case but in the two-electron case it leads to a small value for the electron–electron repulsions in the cavity. When the wave function is improved by using the two-term representation, the electrons repel each other more strongly. Thus the energy of the system increases. This effect can be reduced by allowing the cavity to expand and this results in the most stable two-electron cavity species having twelve molecules in the first coordination layer. The transition energy also increases because of the larger number of electron–dipole interactions. Again the two-electron transition energy is distinctly larger than the one-electron cavity result. The energy gained by the formation of a two-electron cavity species has increased in this better calculation but that should be regarded as a lower limit since more solvent hydrogen bonds must be broken to form the $N = 12$ cavity. In addition, this calculation is very sensitive to the long range electron–solvent interaction.

VII. Concluding remarks: future directions, basic problems

In this chapter we have reviewed some of the results as well as some of the disappointments of the theoretical model calculations of electrons in fluids. We have not dealt with electrons in glasses as these were recently discussed thoroughly in a review by Kevan[4].

The present models are quite good at explaining trends in the optical and other properties of electrons in fluids under most experimental conditions. One major problem in treating electrons in fluids appears to be the accurate introduction of the intermolecular potential between molecules in the first coordination layer. Only for ammonia are the results completely satisfactory. This problem appears to be at its worst in water. However, new intermolecular potential functions have been derived[105] and these should now be applied in these calculations. Only with the proper treatment of this contribution can we straighten out the uncertainties as to the cavity size in water and reconcile the experimental and theoretical pressure-dependences of the optical spectra.

All of these problems could be handled simply if a better molecular orbital treatment could be developed which is sufficiently easy to use. At present we have the very accurate but very time-consuming work of Newton[45] and the incomplete theory of Howat and Webster[42] (who neglect long range effects). This development should have a high priority in further theoretical studies.

The studies of dense gases and mixed solvents have convinced us that more attention should be paid to the solvent, its hydrogen bonds, entropies of solvation, and the like. This aspect has not been as important up until now because we have not needed such precise values for free energies of solutions of the solvated electron species.

The most serious problems reside in the line shape. More precise theoretical calculations are needed here, as well as a re-analysis of the assumptions used in calculating shapes. More attention should be paid to the suggestion of Delahay[72], that possibly the tails could involve auto-ionization. In addition, more experimental work on the optical properties is needed. The previous attempts at hole burning with the hydrated electron spectrum[106] should be repeated with more potent lasers and on other systems. In this regard some very recent work by Hager and Willard[107] on organic glasses is providing evidence that it can be done in some systems. Techniques such as these may be able to sort out the various contributions to the shape. In addition, the search for the $2s$ state by two-photon absorption would greatly help to pin down the details of the effective potential seen by the electron.

The theory of solvated electrons has made significant advances within the last few years, in large measure because of the close co-operation of experimentalists and theorists with a wide variety of interests. This co-operation is essential for continued progress.

References

1 H. T. Davis, L. D. Schmidt and R. G. Brown, in J. Jortner and N. R. Kestner (Eds.), Electrons in Fluids, Springer-Verlag, Heidelberg, 1973, pp. 393–410.
2 B. Raz and J. Jortner, in J. Jortner and N. R. Kestner (Eds.), Electrons in Fluids, Springer-Verlag, Heidelberg, 1973, pp. 413–422.
3 J. Jortner and N. R. Kestner, in J. J. Lagowski and M. J. Sienko (Eds.), Metal–Ammonia Solutions, Butterworths, London, 1970, pp. 49–103.
4 For a general review, especially of glasses, see L. Kevan, in M. Burton and J. L. Magee (Eds.), Advances in Radiation Chemistry, Vol. 4, Wiley, New York, 1974, pp. 181–305.
5 T. Y. Wu and T. Ohmura, Quantum Theory of Scattering, Prentice-Hall, Englewood, New Jersey, 1962.
6 N. F. Mott and H. S. W. Massey, Theory of Atomic Collisions, 3rd edn., Oxford University Press, London, 1965.
7 N. R. Kestner, J. Jortner, S. A. Rice and M. H. Cohen, Phys. Rev. A, **140** (1965) 56.
8 J. Lekner, Phys. Rev., **158** (1961) 130.
9 B. E. Springett, M. H. Cohen and J. Jortner, J. Chem. Phys., **48** (1968) 2720.
10 J. Jortner, N. R. Kestner, S. A. Rice and M. H. Cohen, J. Chem. Phys., **43** (1965) 2614.
11 B. E. Springett, M. H. Cohen and J. Jortner, Phys. Rev., **159** (1967) 183.
12 J. A. Northby and T. M. Sanders, Phys. Rev. Lett., **18** (1967) 1184.
13 J. L. Levine and T. M. Sanders, Phys. Rev., **154** (1967) 138.
14 J. L. Levine and T. M. Sanders, Phys. Rev. Lett., **8** (1962) 159.
15 T. P. Eggarter and M. H. Cohen, Phys. Rev. Lett., **25** (1970) 807; **27** (1971) 129.

16 V. K. S. Shante and S. Kirkpatrick, Advan. Phys., **20** (1971) 325.

17 T. P. Eggarter, Phys. Rev. A, **5** (1972) 2496.

18 J. P. Hernandez, Phys. Rev. A, **5** (1972) 635, 2696; **7** (1973) 1755.

19 H. R. Harrison and B. E. Springett, Chem. Phys. Lett., **10** (1971) 418.

20 J. Lekner, Phys. Lett. A, **27** (1968) 341; Phil. Mag., **18** (1968) 128.

21 H. Schnyders, S. A. Rice and L. Meyer, Phys. Rev. Lett., **15** (1965) 187.

22 H. Schnyders, S. A. Rice and L. Meyer, Phys. Rev., **150** (1967) 127.

23 J. A. Jahnke, L. Meyer and S. A. Rice, Phys. Rev. A, **3** (1971) 734.

24 J. A. Jahnke, N. A. W. Holzworth and S. A. Rice, Phys. Rev. A, **5** (1972) 463.

25 B. E. Springett, Phys. Rev., **155** (1968) 138.

26 R. M. Ostermeier and K. W. Schwartz, Phys. Rev. A, **5** (1972) 2510.

27 J. Poitrenaud and F. I. B. Williams, Phys. Rev. Lett., **29** (1972) 1230.

28 M. H. Cohen and J. Jortner, Phys. Rev., **180** (1969) 238.

29 V. E. Dionne, R. A. Yound and C. T. Tomizuka, Phys. Rev. A, **5** (1972) 1403.

30 L. Landau, Phys. Z. Sowjetunion, **3** (1933) 664. It was applied to these problems by Davydov (J. Exp. Theor. Phys. USSR **18** (1948) 913, Deigen, Zhur. Exp. Theor. Phys. USSR, **26** (1954) 300 in Platzman and Franck, Z. Phys., **138** (1954) 411).

31 J. Jortner, J. Chem. Phys., **30** (1959) 839.

32 J. Jortner, Mol. Phys., **5** (1962) 257.

33 D. E. O'Reilly, J. Chem. Phys., **41** (1964) 3736; R. H. Land and D. E. O'Reilly, J. Chem. Phys., **46** (1967) 4496.

34 J. Jortner, S. A. Rice and E. G. Wilson, in G. Lepoutre and M. J. Sienko (Eds.), Solutions Metal–Ammonia, Benjamin, New York, 1964, pp. 222–276.

35 M. Tachiya, Y. Tabata and K. Oshima, J. Phys. Chem., **77** (1973) 263. This is based on an improved model due to K. Iguchi, J. Chem. Phys., **48** (1968) 1735.

36 L. Raff and H. A. Pohl, Advan. Chem. Ser., **50** (1965) 173.

37 M. Natori and T. Watanabe, J. Phys. Soc. Jap., **21** (1966) 1573.

38 S. Ray, Chem. Phys. Lett., **11** (1971) 573.

39 M. S. Aldrich, L. P. Gray and L. C. Cusachs (unpublished work done in 1970).

40 S. Ishimaru, H. Kato, T. Yamobe and K. Fukui, J. Phys. Chem., **77** (1973) 1450; Chem. Phys. Lett., **17** (1972) 264; S. Ishimaru, H. Tomita, T. Yamabe, K. Fukui and H. Kato, Chem. Phys. Lett., **23** (1973) 106.

41 M. Weissman and N. V. Cohan, Chem. Phys. Lett., **7** (1970) 445; J. Chem. Phys., **59** (1973) 1385.

42 G. Howat and B. C. Webster, J. Phys. Chem., **76** (1972) 3714.

43 K. Hiroike, N. R. Kestner, S. A. Rice and J. Jortner, J. Chem. Phys., **43** (1965) 2625.

44 L. Kevan, in H. F. Adler, O. F. Nygaard and W. K. Sinclair (Eds.), Proc. Fifth Int. Congr. on Radiat. Res., Academic Press, New York, 1975.

45 M. Newton, J. Chem. Phys., **58** (1973) 5833 and unpublished work.

46 D. A. Copeland, N. R. Kestner and J. Jortner, J. Chem. Phys., **53** (1970) 1189.

47 K. Fueki, D-F. Feng, L. Kevan and R. Christofferson, J. Phys. Chem., **75** (1971) 2291.

48 J. Logan and N. R. Kestner (unpublished).

49 K. Fueki, D-F. Feng and L. Kevan, J. Amer. Chem. Soc., **95** (1973) 1398.

50 F. Y. Jou and L. M. Dorfman, J. Chem. Phys., **58** (1973) 4715.

51 A. Gaathon and J. Jortner, in J. Jortner and N. R. Kestner (Eds.), Electrons in Fluids, Springer-Verlag, Heidelberg, 1973, pp. 429–446.

52 I. Carmichael and B. Webster, J. Chem. Soc. Faraday Trans. II, **70** (1974) 1570.

53 D-F. Feng, D. Ebbing and L. Kevan, J. Chem. Phys., (1974) 249.

54 N. R. Kestner, J. Jortner and A. Gaathon, Chem. Phys. Lett., **19** (1973) 328.

55 J. Logan and N. R. Kestner, J. Phys. Chem., **76** (1972) 2738.

56 N. R. Kestner, in J. Jortner and N. R. Kestner (Eds.), Electrons in Fluids, Springer-Verlag, Heidelberg, 1973, pp. 1–25.

57 T. P. Das, Advan. Chem. Phys., **4** (1962) 303.

58 G. A. Kenny-Wallace and D. C. Walker, in A. J. Bard (Ed.), Electroanalytical Chemistry, Vol. V, Marcel Dekker, New York, 1971, pp. 1–66.

59 An excellent review of earlier work and some more recent molecular orbital work is found in B. C. Webster and G. Howat, Radiat. Res. Rev., **4** (1972) 259.

60 R. A. Ogg, J. Amer. Chem. Soc., **68** (1940) 155.

61a R. Catterall, Nature (London), **10** (1971) 229; b R. Catterall, in J. J. Lagowski and M. J. Sienko (Eds.), Metal–Ammonia Solutions, Butterworths, London, 1970, pp. 105–130.

62 R. A. Pinkowitz and T. J. Swift, J. Chem. Phys., **54** (1971) 2858.

63 M. K. Bowman, L. Kevan and I. M. Brown, Chem. Phys. Lett., **22** (1973) 16; H. Yoshida, D. F. Feng and L. Kevan, J. Chem. Phys., **58** (1973) 3411. The recent work is summarized by L. Kevan in ref. 4.

64 K. G. Breitschwerdt and H. Radscheit, Ber. Bunsenges. Phys. Chem., **75** (1971) 644.

65 G. Lepoutre and A. Demortier, Ber. Bunsenges. Phys. Chem., **75** (1971) 647.

66 C. A. Kraus, J. Amer. Chem. Soc., **43** (1921) 749.

67 K. H. Schmidt and W. L. Buck, Science, **151** (1966) 70.

68 C. A. Hutchison, Jr. and D. E. O'Reilly, J. Chem. Phys., **52** (1970) 4400.

69 N. R. Kestner and J. Jortner, J. Phys. Chem., **77** (1973) 1040.

70 K. Fueki, D-F. Feng and L. Kevan, J. Chem. Phys., **59** (1973) 6201.

71 For example, a T. Kajiwara, K. Funabashi and C. Naleway, Phys. Rev., **6** (1972) 808; b K. Funabashi, Advan. Radiat. Chem., **4** (1974) 103.

72 P. Delahay, this volume, Chap. 4; L. Nemec, B. Baron and P. Delahay, Chem. Phys. Lett., **16** (1972) 278; B. Baron, P. Delahay and R. Lugo, J. Chem. Phys., **55** (1971) 4180; H. Aulich, B. Baron, P. Delahay and R. Lugo, J. Chem. Phys., **58** (1973) 4439; H. Aulich, P. Delahay and L. Nemec, J. Chem. Phys., **59** (1973) 2354.

73 a B. L. Smith and W. H. Koehler, in J. Jortner and N. R. Kestner (Eds.), Electrons in Fluids, Springer-Verlag, Heidelberg, 1973, pp. 145–160; B. L. Smith and W. H. Koehler, J. Phys. Chem., **76** (1971) 2481; b R. R. Hentz and G. A. Kenney-Wallace, J. Phys. Chem., **78** (1974) 541; see also ref. 83.

74 K. Fueki, D-F. Feng and L. Kevan, J. Phys. Chem., **78** (1974) 393.

75 G. R. Freeman, J. Phys. Chem., **77** (1973) 7.

76 L. Gilles, J. E. Aldrich and J. W. Hunt, Nature (London) Phys. Sci., **243** (1973) 70.

77 J. H. Baxendale and P. Wardman, Nature (London), **230** (1971) 449; J. Chem. Soc. Faraday Trans. I, **69** (1973) 584.

78 D-F. Feng, K. Fueki and L. Kevan, J. Chem. Phys., **57** (1972) 1253.

79 R. R. Hentz, Farhataziz and E. M. Hansen, J. Chem. Phys., **57** (1972) 2959.

80 R. R. Hentz, Farhataziz and E. M. Hansen, J. Chem. Phys., **55** (1971) 4974.

81 U. Schindewolf and R. Olinger, in J. J. Lagowski and M. J. Sienko (Eds.), Metal–Ammonia Solutions, Butterworths, London, 1970, pp. 199–215.

83 Farhataziz, L. M. Perkey and R. R. Hentz, J. Chem. Phys., **60** (1974) 4383.

84 U. Schindewolf (private communication) points out that in dense polar gases the volume of activation can be of opposite sign to that observed in the liquid, since in this case the first coordination layer is more dense than the bulk gaseous medium.

85 N. V. Cohan, G. Finkelstein and M. Weissmann, Chem. Phys. Lett., **26** (1974) 93, suggest that the small cavity in water arises because the electron cannot break the hydrogen bonds.

86 R. R. Dogonadze, Ber. Bunsenges. Phys. Chem., **75** (1971) 628.

87 R. Kubo and Y. Toyozawa, Progr. Theor. Phys., **13** (1955) 160.

88 V. G. Levich, Advan. Electrochem. Electrochem. Eng., **4** (1966) 249.

89 M. Tachiya, Y. Tabata and K. Oshima, J. Phys. Chem, **77** (1973) 2286.

90 Farhataziz, R. R. Hentz and L. M. Perkey (unpublished) have evaluated the spectrum down to ~0.3 microns at 23°C.

91 M. Cole (unpublished).

92 R. Olinger, S. Hahne and U. Schindewolf, Ber. Bunsenges. Phys. Chem, **76** (1972) 349.

93 A. Gaathon, G. Czapski and J. Jortner, J. Chem. Phys., **58** (1973) 2648.

94 B. D. Michael, E. J. Hart and K. H. Schmidt, J. Phys. Chem., **75** (1971) 2798. They did not try to find the minimum density which would support a solvated electron.

95 N. R. Kestner, A. Gaathon and J. Jortner (unpublished). A. Gaathon and J. Jortner have recently developed a simple fluctuation theory which works well in the D_2O case (private communication).

96 This area of research is closely related to ionic solvation, a topic treated in Chaps. 8 and 9 of this volume. For a review of gas phase ionic solvation, see P. Kebarle, in B. E. Conway and J. O. M. Bockris (Eds.), Modern Aspects of Electrochemistry, Vol. 9, Plenum, New York, 1974 pp. 1–44. This topic is also important in heterogeneous nucleation in the vapor phase (see, for example, E. F. O'Brien and G. W. Robinson, J. Chem. Phys., 61 (1974) 1050).

97 L. M. Dorfman, F. Y. Jou and R. Wageman, Ber. Bunsenges. Phys. Chem., 75 (1971) 681.

98 L. M. Dorfman and F. Y. Jou, in J. Jortner and N. R. Kestner (Eds.), Electrons in Fluids, Springer-Verlag, Heidelberg, 1973, pp. 447–457.

99 L. M. Dorfman, in H. F. Adler, O. F. Nygaard and W. K. Sinclair (Eds.), Proc. Fifth Int. Cong. Radiat. Res., Academic Press, New York, 1975.

100 J. L. Dye, M. G. DeBacker and L. M. Dorfman, J. Chem. Phys., 52 (1970) 6251.

101 For a review, see J. J. Lagowski and M. J. Sienko (Eds.). Metal–Ammonia Solutions, Butterworths, London, 1970, especially the articles of J. L. Dye pp. 1–17 and R. Catterall pp. 105–130.

102 D.-F. Feng. K. Fueki and L. Kevan, J. Chem. Phys., 58 (1973) 3281.

103 D. A. Copeland and N. R. Kestner, J. Chem. Phys., 58 (1973) 3500.

104 J. Logan and N. R. Kestner (unpublished). The error in ref. 103 is that the individual electron medium interaction in the Hamiltonian was not made double the value in the one-electron cavity case. This must be done since the medium is now polarized by two electrons.

105 See, for example, A. Ben-Naim and F. H. Stillinger, Jr., in R. A. Horne (Ed.), Water and Aqueous Solutions: Structure, Thermodynamics, and Transport Processes, Wiley, New York, 1972.

106 G. Kenney-Wallace and D. Walker, Ber. Bunsenges. Phys. Chem., 75 (1971) 634.

107 S. L. Hager and J. E. Willard, J. Chem. Phys., 61 (1974) 3244.

Chapter 2

GROUND STATES OF EXCESS ELECTRONS IN LIQUIDS AND GLASSES: MAGNETIC RESONANCE STUDIES

R. CATTERALL

I. Introduction

I.A Magnetic resonance spectroscopy

Magnetic resonance studies by their very nature are concerned with the properties of a single electronic state. In the vast majority of studies this state is necessarily the electronic ground state although excited electronic states can be studied if artificially populated (e.g. by optical pumping). Since magnetic resonance is a branch of spectroscopy and therefore concerned primarily with transitions between energy levels, two necessary conditions must be fulfilled before the technique can be applied. Firstly, the (ground) electronic state must be degenerate and secondly, we must be able to lift the degeneracy slightly. The systems most frequently studied which satisfy these conditions are either singlet ground states comprising degenerate nuclear spin states, or doublet ground states, in which case we are primarily concerned with the degenerate electron spin states. In both cases we lift the degeneracy of spin angular momentum states by applying an external magnetic field to the system. The first example quoted above is the well known case of nuclear magnetic resonance of diamagnetic materials and will not concern us further. The second example comprises both electron and nuclear spin resonance of paramagnetic materials as typified in this chapter by excess electron states. Fortunately, the magnetic fields required to lift the degeneracy of electron or nuclear spin states differ by more than 10^3, so that we can generally treat our systems by a simple first- or second-order perturbation approach. This is not always the case, however, and in Sect. III.D we will consider excess electron states in which the electron and nuclear spin angular momenta are so strongly coupled that the assumption of separate quantum numbers to describe electron and nuclear spin functions is no longer valid, and we must treat the total spin angular momentum of the coupled electron–nuclear system. The experimental techniques of simple electron and nuclear spin resonance are extensively described in a number of standard texts[1].

I.B Excess electrons

It remains to say a few words of introduction about the systems we will be studying, i.e., excess electrons in liquids and glasses. As is often the case, it is easy to quote obvious examples; solutions of sodium in liquid ammonia contain excess electrons, but if benzene is added to the solutions, producing benzene anions, then there are no longer any excess electron states in solution. We propose the following definition in the spirit of usefulness rather than comprehensiveness. Excess electron states in liquids and glasses are the bound states of additional electrons introduced into fluid or solid non-crystalline media, whose dominant electronic feature is a set of completely filled σ-bonding and possibly non-bonding molecular orbitals together with a set of empty antibonding molecular orbitals. An additional feature will generally be the absence of any π orbitals. Thus, solvents such as water, ammonia, alcohols, amines, ethers, and even hydrocarbons, satisfy our definition completely whilst we "stretch" the last qualification to include a system containing $P = O$ but to exclude many (but not all) systems containing $C = C$. There has been considerable dispute[2,3] in the literature about trapped electrons in methyl cyanide, but this now appears to be resolved in terms of dimeric negative ions, at any rate we exclude this system from our discussion.

I.C Structural information from magnetic resonance studies

The magnetic parameters associated with the resonant transfer of energy between a radiation field and a set of non-degenerate spin levels give detailed information about the molecular structure and the spatial part of the wave function of the state being observed. The total one-electron wave functions representing each of the spin states is factored in the usual way into a common spatial function and a particular spin function. The magnetic parameters we derive from resonance experiments yield information about the spin functions. In the simple one-electron approximations, with which we will be primarily concerned, the spin density in the spin function is necessarily equal to the electron density in the spatial part and we have a direct measure of electron delocalization through space. As we shall see, the extent of delocalization is one of the most crucial questions in any study of excess electrons. This one-electron approximation, of course, is not always valid, and we shall repeatedly meet the question of "negative spin density" at solvent hydrogen nuclei where we cannot simply equate spin and electron densities (to do so would imply $\Psi^2 < 0$ in the spatial part of the wave function).

In summary, the parameters which we endeavour to extract from the magnetic resonance results will be measures of (1) stoichiometry (how many nuclei of a particular sort?), (2) geometry (how far, on average, is the excess electron from a particular nucleus or set of equivalent nuclei?), (3) lifetimes (over what time scale do the magnetic interactions persist?), and (4) delocalization (electron spin) density maps (what proportion of its time does the excess electron spend in a particular spatial position?). In general, we will be satisfied with effective or average values of these parameters, but in one instance we will be able to derive the distribution function of a spin density parameter which gives a detailed picture not only of an "average" excess electron, but also of the extent to which particular excess electrons differ from each other in an amorphous medium.

Where molecular species deviate from spherical symmetry magnetic resonance experiments can also yield additional geometrical parameters which describe angular variations in the spin functions. In all cases of excess electrons in fluids these aniso-tropic interactions are averaged to near-zero by rotational motion (tumbling) and contribute only to spin relaxation, whilst for all reports of excess electrons in the solid state it appears that the assumption of spherical symmetry is more than adequate to describe the experimental results.

One further magnetic parameter, which we will in general ignore, is a measure of the degree of quenching of admixtures of orbital angular momentum with the spin angular momentum of the excess electron. Only for the states described in Sect. III does the electronic g-factor deviate significantly from the value associated with spin-only angular momentum.

1.D Electron spin pairing

From the point of view of structural analysis of excess electron states, it is unfortunate that solvated electrons show a marked tendency to pair their spin angular momenta. This property is discussed in its own right in Sect. IV but enters as a complicating factor into all earlier sections on fluid metal solutions, where all magnetic interactions are averaged over all nuclei in the sample. Thus all the equations involve the term R/α where R is the mole ratio of solvent to metal, and α is the ratio of unpaired electrons to metal cations. In the limit of low metal concentrations, $\alpha \rightarrow 1$, i.e. each metal atom donates a single electron. At all finite concentrations, $\alpha < 1$ and must be determined experimentally (often by non-resonant magnetic techniques). In Sects. II and III we assume, where necessary, the values of α given in Sect. IV.

1.E Crystalline analogues

One of the greatest theoretical problems involved in an understanding of excess electron states in disordered media is the lack of translational symmetry. In formulat-ing qualitative models of excess electron states two examples of excess electron states in crystalline media have served as paradigm examples. The first of these, the F-center, is an electron trapped at an anion vacancy (Fraenkel defect) in, typically, an alkali halide crystal. Electron resonance and electron–nuclear double resonance (ENDOR) experiments have yielded a particularly detailed electron spin density map of electron delocalization, in some cases out to the twelfth shell of ions surround-ing the center. These centers will not be discussed in detail, but the reader is referred to the definitive review by Gourary and Adrian[4] and to more recent books by Markham[5] and by Schulman and Compton[6]. More recent work has added only marginally to that of Gourary and Adrian.

The second crystalline example is the "classical" semiconductor; a silicon crystal doped with phosphorus atoms. In this case the excess electron is weakly bound to a P^+ center at a normal lattice site and localized states give way to extended states even at very low temperatures. The bound state of the excess electron is extensively delocalized and includes something like 200 silicon atoms in its orbit. The reader is referred to the classic ENDOR studies of Feher[7] and to a recent book[8] for further details.

I.F Delocalization

As mentioned above, one of the most important questions we ask about excess electron states is the extent of delocalization. The theoretical approach to solvated electrons has been reviewed in depth in Chap. 1, but we introduce at this point two models which have dominated the theoretical approach to "impurity states" which we loosely describe as states centered on, and derived from, a foreign impurity species embedded in the host medium. Impurity states differ from trapped or solvated electrons in much the same way as the centers in phosphorus-doped silicon differ from F-centers in alkali halides. The first approach is the extreme of a "tight-binding center" in which the excess electron state is adequately described by minor perturbations of the electronic wave function of the isolated impurity species. Sodium atoms trapped in an argon matrix are adequately described by this approach[9].

In direct contrast to the "tight-binding" approximation we have "weak-binding" states in which the excess electron is delocalized so completely from the impurity donor species that its state is described as a property of the host matrix, a screened hydrogenic or Mott–Wannier function[10] with no parentage in the impurity species. Delocalization to this extent is generally only encountered in excited states (excitons) but in Sect. III.E. we shall describe ground states of excess electrons which approach very closely to this limiting model. In general, however, most of the states described in that section fall squarely between the two limits and we discuss some attempts to generate models for these intermediate states.

I.G Cavity formation

Almost all theoretical descriptions of solvated electrons have based the physical picture of the state upon a cavity in the solvent, a small region of empty space in an otherwise particulate or continuous medium. The importance of this concept is adequately brought out by the discussions in Chap. 1. Whilst the concept of cavity formation (or, at least, dilution of the medium) in the vicinity of excess electrons is adequately established for the case of an excess electron embedded in a pure medium, the same cannot be said for states derived from impurity donors. One of the most remarkable features about solutions of alkali metals in ammonia is that the cation plays such a minor role (see Sect. III.A,B.). In other solvents the role of the cation is of considerably greater importance and we must consider the excess electron states resulting from a competition between cavity formation and binding to the donor species. A wide variety of such states, apparently all of similar energy, are discussed in Sect. III.D.

II. Structure of the ground state of isolated solvated electrons

II.A Fluid solutions

II.A.1 Solutions of alkali metals in liquid ammonia

(a) Electron spin resonance studies The early definitive work of Hutchison and Pastor[11] has been extended by several groups[12–31], notably the ^{15}N isotopic experiment by Pollak[12] and the extensive spin-echo work by Cutler and Powles[13–15]. Although qualitatively simple, good quantitative measurements are technically difficult but could be achieved with modern equipment (almost all serious studies date from the 1950's). Unfortunately, the dominant problem remaining is one of chemical analysis.

The most obvious feature is that to a very good approximation the g-factors are very close to the free spin value (2.002319 for spin-only angular momentum) and independent of the nature of the metal, its concentration, the temperature, or the frequency of measurement. Closer examination[32], however, shows a slight trend to lower g-factors with increasing atomic number of the metal and increasing temperature. These slight shifts were observed for solutions whose concentrations indicate extensive ion-pairing between solvated electrons and cations, and cannot be attributed to isolated solvated electrons. This proximity of the g-factor to the free spin value indicates that all orbital angular momentum is effectively quenched and that an s-function might be expected to approximate closely to the wave function of the ground state of the solvated electron.

The electron spin resonance line width of moderately concentrated sodium and potassium solutions is one of the narrowest observed. Unlike the g-factors, the line widths show a clear-cut dependence on the nature of the metal, metal concentration, temperature, and isotopic composition. In most cases however, extrapolation to infinite dilution is not too difficult whilst the relatively low concentration dependences for solutions <0.1 M strongly suggest that the magnetic interactions between the unpaired electron and the solvent molecules are very similar in the isolated solvated electron and its ion-pair with a solvated cation. Hutchison and O'Reilly[30] have shown that the change in line width on deuterium substitution correlates satisfactorily with changes in solvent viscosity, whilst the marked change[12] on substituting ^{15}N identifies the dominant magnetic interaction as between the unpaired electron and nitrogen nuclei. A more extensive investigation of solutions in ^{15}NH$_3$ to confirm and extend Pollak's important "one-off" result[12] would be well worthwhile. Further discussion of line widths is deferred to the section on electron relaxation.

The line shapes observed for potassium solutions[17] approximate very closely to the Lorentzian function predicted for rapidly modulated magnetic interactions over the concentration range 0.01–0.2 M and from 240 to 300 K. Similar observations have been made by the author for all the alkali metals over the temperature interval 198–300 K. Under certain conditions electron resonance line widths provide a direct measure of the rates at which perturbed electron spin distributions return to their equilibrium values. It turns out that these conditions are generally satisfied extremely well for fluid metal–ammonia solutions, but this fortunate conclusion can only be arrived at after direct measurement of relaxation rates using pulsed micro-wave techniques. The continuous wave saturation experiments of Hutchison and Pastor[11] and particularly of Hutchison and O'Reilly[30] also provide very strong evidence in a less direct manner that these conditions are satisfied. Unfortunately, these measurements on potassium show a still unexplained anomalous behavior in the electron spin relaxation mechanism (see Sect. II.A.1.e). Early direct measurements by Blume[28] and Pollak[12] have been comprehensively extended by Cutler and Powles[13–15].

For the purpose of this section the most important conclusion is that the electron relaxation rate is effectively constant for concentrations <0.1 M for any given metal, whilst differences between metals are very minor. This further substantiates our thesis that the magnetic interactions with solvent molecules in solvated electrons are hardly affected by ion-pairing to solvated cations. Despite the problem of estimating unpaired spin concentrations, the temperature-dependence of the limiting relaxation times for dilute solutions correlates very well with changes in solvent viscosity, whilst Pollak[12] showed that the major change observed in ^{15}N ammonia correlates extremely well with the difference in nuclear magnetic moment. In general, $T_1 = T_2$ for all metals investigated except for potassium where $T_1 > T_2$ is observed independent of frequency. This point is discussed further below (Sect. II.A.1.e).

The observed shifts of the electron resonance spectra from the free electron value contain, in addition to the admixture of orbital angular momentum already mentioned, a contribution resulting from the unequal population of the nuclear spin states of any magnetic nuclei in solution. The difference in population results in an effective local magnetic field at the electron which modifies the applied field and results in an apparent g-shift (Overhauser shift). Since the magnetic fields associated with spinning nuclei are $\sim 10^3$ less than the fields associated with electrons, the magnitude of these shifts is small. They have, however, been measured by Lambert[33] and the results justify our previous neglect of this contribution to the resonance frequency.

(b) Nuclear magnetic resonance studies Measurements of shifts of nuclear magnetic resonance spectra arising from the proximity of an unpaired electron to the nucleus in question can only be made if the lifetime of electron–nuclear interactions is much shorter than the time taken to observe a transition between nuclear spin states. If these times are comparable, the lifetime of any nuclear spin state is shortened by as much as six to eight orders of magnitude with a corresponding decrease in the precision with which the resonance frequency can be measured. In all solutions of alkali metals in ammonia so far examined, the observation of a single narrow nuclear resonance line is a clear indication that the above condition is fulfilled. Whilst this condition makes possible the observation of nuclear resonance signals, it also implies an averaging out of the electron–nuclear interactions over all nuclei in the sample. The interaction is manifested in the nuclear resonance spectra in two ways, firstly as a shift in the nuclear resonance frequency (the Knight shift), and secondly as an additional mode of nuclear spin relaxation. Since both these modifications are

reduced by the ratio, number of unpaired electrons to number of nuclei, it is obvious that we have an inherent lack of sensitivity in making measurements of nuclear resonance parameters. Thus, whilst electron spin resonance studies have been made at concentrations down to 10^{-5} M, with a correspondingly simple extrapolation to infinite dilution, there is an effective lower concentration limit of 10^{-2} M for nuclear resonance studies. Ion-pairing and even spin-pairing processes are already well established at this concentration and the application of nuclear resonance data to the analysis of the structure of isolated solvated electrons can be a rather subjective matter. Basically there are only two possibilities: that electron–solvent nuclei interactions are different in the isolated solvated electron and the ion-pair, or that they are (at least approximately) equal. We shall argue that the latter is the most reasonable assumption and the later analysis will proceed accordingly. In this section we will be concerned exclusively with nuclear resonance properties of solvent nuclei, i.e. hydrogen and nitrogen. No significant isotopic studies have been reported and thus our attention is focused exclusively on ^1H and ^{14}N.

Knight shifts are effectively the inverse of the Overhauser effects just described. The inequality of the population of the electron spin states results in an effective local magnetic field at the nucleus which modifies the applied field. Knight shifts for nitrogen nuclei have been measured by McConnell and Holm[34], Acrivos and Pitzer[35], and O'Reilly[36]. Of these reports, those of O'Reilly are the most extensive and extend to lower concentrations. Where the investigations overlap, there is no substantial disagreement in the measurements, although interpretations have differed. Although the results demonstrate a good linear relationship on a log–log basis[36] between the observed shift (in p.p.m.) and the mole ratio $NH_3:M$ for all the alkali metals (M = metal), this linearity has no theoretical significance. The magnitude of the Knight shift, $K(N)$ is given by

$$K(N) \propto \chi_m^{obs}/R$$

where χ_m^{obs} is the molar magnetic susceptibility of a solution of mole ratio $NH_3:M = R$. This implies that a plot of $RK(N)$ against χ_m^{obs} should be linear if the interaction between the unpaired electron and nitrogen nuclei is independent of concentration. Furthermore, this line should pass through the origin. Any deviations from these predictions could be taken to indicate a concentration-dependence of the coupling which might be interpreted as evidence for a modification of the electron–nitrogen interaction when solvated electrons aggregate with solvated cations. The data are plotted in this way in Fig. 1, where the χ_m^{obs} are those corresponding to the smoothed out data (see Fig. 10). A more direct plot of experimental values cannot be made since molar susceptibilities and Knight shifts were measured on different samples. Figure 1 appears to show good linearity with no metal-dependence and extrapolates well to the origin. We take this as further justification of our assumption that magnetic interactions are little modified by ion-pairing. As shown below, the magnitude of the electron–nitrogen interaction can be obtained from this plot.

In direct contrast to the large, easily measurable shifts obtained in nitrogen nuclear resonance spectroscopy, the shifts observed for proton resonances[37,38] are small, like normal chemical shifts, and they also occur in the "wrong" direction.

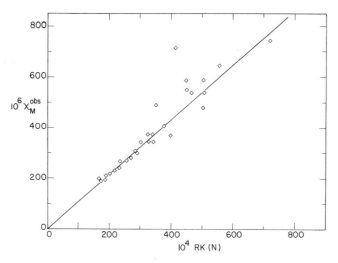

Fig. 1. Nitrogen Knight shifts at room temperature (ref. 39). The slope of the line corresponds to a total hyperfine coupling between the unpaired electron and nitrogen nuclei of $na_N = 110.0$ G.

An attempt to treat the proton resonance shifts in the manner applied to the nitrogen data[39] resulted in a very obvious breakdown. Hughes[37] interpreted this data by assuming the interaction between solvated electrons and protons to be zero for the isolated solvated electron and small, but finite, for the ion-pair. An alternative analysis[39] interpreted the observed shift as the sum of a concentration-independent proton Knight shift common to both the solvated electron and the ion-pair, together with a concentration-independent chemical shift of protons involved in the spin-paired species. Whilst this latter interpretation is the most reasonable, the number of adjustable parameters guarantees a fit to the observation. Nevertheless, the magnitude of the interaction agrees well with that found independently by lithium–ethylamine solutions (see Sect. III.A). No experimental measurements have been reported for nitrogen nuclear relaxation in solutions of alkali metals in liquid ammonia. Proton relaxation measurements[40,41] and the calculation of relaxation rates for nitrogen is discussed elsewhere[39,44].

(c) Double resonance experiments　There are only a few results on nuclear Overhauser enhancement[26,33,41] for both hydrogen and nitrogen nuclei. The data for nitrogen nuclei are of poor precision and contribute little to the subsequent analysis since the enhancement is dominated by a single interaction. In contrast, the results for protons are of somewhat higher precision and provide the means of separating[39] various mechanisms of proton relaxation. The results of ENDOR study of fluid solutions of sodium in ammonia have been given in a thesis[42], but have not been interpreted to date.

(d) Structure of the solvated electron in ammonia　In this section we attempt to interpret the extensive magnetic resonance results obtained for solutions of alkali metals in liquid ammonia with a view to elucidating the structure of the ground state

of the solvated electron in ammonia. The approach follows closely that given in my plenary lecture[39] at the Colloque Weyl 2 and cross-reference is made to the published report of that lecture for some of the detail. Since that lecture one significant new experiment has been reported by Pinkowitz and Swift[43] which apparently comes to a somewhat different conclusion. This work is discussed and it is concluded that discrepancies between the two analyses are not significant. A more recent review by Lambert[44] at Colloque Weyl 3 retraces a part of the ground covered in my Colloque Weyl 2 lecture[39] without making the assumption of similar magnetic interactions in the electron and its ion-pair. That author concludes that his approach introduces too many parameters to allow extraction of any significant new information.

Any analysis of the magnetic data is dependent upon a knowledge of the spin-pairing fraction α although, since R/α enters into the various magnetic interactions between electron and solvent, it is often possible to cancel it out when results from two techniques are combined. This latter approach is used whenever possible[45]. When estimates of α are unavoidable we use the smoothed-out values presented in Fig. 10. The applicability of these data for potassium solutions to solutions of other metals is, however, questionable since the literature contains conflicting reports of the metal-dependence of α.

We summarize the important expressions and the symbolism introduced to describe the magnetic interactions between the solvated electron and the solvent[39]. It should be stressed that all interactions are described in terms of average or effective values which are taken to be representative of normal variations in dis-ordered media (e.g. different solvation numbers and, or, geometries for particular species) and of non-equivalent nuclei (e.g. summations over weakly interacting second and third solvation shells and strongly interacting primary shells). The analysis is carried out in terms of two parameters, an effective solvation number, n, and an effective radial distance, b_{eH}. No attempt is made to probe distribution functions for these parameters.

There is a wide variety of possible interactions which could be responsible for each of the magnetic properties, e.g. the observed transverse electron relaxation rate, T_{2e}^{-1}, can be expressed as a sum of several contributions from rapid modulation of magnetic interactions such as scalar (Fermi contact) hyperfine coupling between electrons and nitrogen, hydrogen and metal nuclei, or of corresponding magnetic dipole interactions between the separated magnetic particles, or of spin–orbit interactions. These possibilities were considered in detail in the Colloque Weyl 2 lecture[39] and final conclusions about dominant interactions will be quoted without further discussion. Lambert[44] came to the same conclusions in his Colloque Weyl 3 review.

Electron spin relaxation is dominated by the rapid modulation (fluctuation) of isotropic hyperfine contact interactions between electrons and nitrogen nuclei. If the modulation is extremely rapid, the two electron rates T_{1e}^{-1} and T_{2e}^{-1} are equal and given by

$$T_{1e}^{-1} = T_{2e}^{-1} = a_N n a_n^2 \tau_c \tag{1}$$

where $a_N = 2x_N I_N(I_N + 1)/3\hbar^2$, n is the effective number of solvent molecules solvating the electron, x_N is the number of nitrogen nuclei in each molecule (in this case $x_N = 1$), a_N is the isotropic hyperfine coupling constant to a single nucleus, τ_c is the correlation time for modulation of the interaction, and I_N is the nuclear spin of nitrogen. From Fermi's equation we have

$$a_N = (8\pi/3)g_e g_n \mu_B \mu_n |\Psi(0)|_N^2 \tag{2}$$

where g_e, g_n, μ_B and μ_n are g-factors and magnetons for electrons and nuclei, and $|\Psi(0)|_N^2$ is the square of the unpaired electron wave function at the center of the nucleus in question.

Proton spin relaxation is more complicated and we identify three significant contributions to the longitudinal relaxation rate T_{1H}^{-1}, an intermolecular dipolar coupling associated with solvent–solvent interactions, and rapidly modulated electron–proton dipolar and contact interactions. The first of these may be approximated by the relaxation rate in pure solvent[36] since the majority of such interactions will occur between molecules distant from any unpaired electrons. The last two terms are of approximately equal importance and given by

$$T_{1H}^{-1} = T_{1H}^{-1}(\text{pure NH}_3) + T_{1H}^{-1}(\text{contact}) + T_{1H}^{-1}(\text{dipolar}) \tag{3}$$

where

$$T_{1H}^{-1}(\text{contact}) = (C\alpha\tau_c/nR)(na_H)^2$$
$$T_{1H}^{-1}(\text{dipolar}) = (C\alpha\tau_c/nR)B_H^2(nb_{eH}^{-3})^2$$
$$B_H = g_e g_H \mu_B \mu_n$$
$$C = 2S(S + 1)/3\hbar^2$$

a_H is the isotropic hyperfine coupling constant to a single proton, and g_e and g_H are the electronic and proton g-factors, respectively.

In principle, the two contributions to proton relaxation should be separable by high frequency measurements, where $T_{1H} > T_{2H}$ and the correlation time in eqn. (3) must be replaced by, for example, a frequency-dependent Fourier transform, $J(\omega_0, \tau_c)$, of an autocorrelation function, $G(\tau)$, for the modulation of contact interactions

$$G(\tau) = \exp(-\text{abs}(\tau)/\tau_c)$$

so that

$$J(\omega_0, \tau_c) = 2/(1 + \omega_0^2 \tau_c^2)$$

where $\omega_0 = \omega_e - \omega_n$ is the difference in resonance frequencies of the electron and the nucleus at the magnetic field of the experiment. In practice, the frequency at which the interactions are modulated in these solutions requires frequencies and magnetic fields which are above the normal working ranges (60–100 MHz) and thus the frequency-dependence cannot be used to separate the contributions to proton

T_1 relaxation. This separation can, however, be achieved by using the observed Overhauser enhancement L_H of the proton resonance signal when the system is simultaneously irradiated at the electron resonance frequency, since contact and dipolar coupling mechanisms lead to enhancements of opposite sign.

$$L_H = 1 + [E_H/T_{1H}^{-1}][T_{1H}^{-1}(\text{contact}) - T_{1H}^{-1}(\text{dipolar})/2]$$
$$E_H = g_e\mu_B/g_H\mu_n \tag{4}$$

The Knight shift of the solvent nuclear resonance is a simple measure of the electron–nuclear contact interaction averaged over all nuclei in solution, e.g., for nitrogen we have

$$K(N) = DE_N na_N \alpha/RT \tag{5}$$
$$D = -S(S + 1)/3k_B$$
$$E_N = g_e\mu_B/g_N\mu_n$$

where k_B is Boltzmann's constant.

Thus from the slope in Fig. 1 we get

$$na_n/g_e\mu_B = 110\text{G}$$
$$n|\Psi(0)|_N^2 = 6.44 \times 10^{24} \text{ cm}^{-3}$$

for the total unpaired electron density at the nitrogen nuclei. This can be used in the relaxation equations in the following way. The product $na_N^2 \tau_c$ is generally invoked in the relaxation process, whilst from the Knight shift we obtain na_N. Squaring this we can substitute $(na_N)^2$ into the relaxation equations to yield the term $(na_N)^2 (\tau_c/n)$.

We are now in a position to derive some structurally significant information about the solvated electron from the magnetic properties, nitrogen Knight shift, proton spin-lattice relaxation, proton Overhauser enhancement, and electron relaxation. Thus from eqns. (1) and (3)–(5) we get the density of solvent molecules in the electron's solvation shell

$$(nb_{eH}^{-3})^2 = [3a_N C/(B_H DE_N)^2][(R/\alpha)^3(K_N T)^2/T_{1e}^{-1}]$$
$$\times[T_{1H}^{-1} - T_{1H}^{-1} (\text{pure NH}_3) - (L_H^{-1})T_{1H}^{-1}/E_H] \tag{6}$$

For $R > 300$ we find the solvent shell density is approximately constant, $nb_{eH}^{-3} \sim 0.18 \text{ Å}^{-3}$ (Table 1).

A relation similar to the above can also be obtained using the proton Knight shift instead of the Overhauser enhancement.

$$(nb_{eH}^{-3})^2 = [(RT/\alpha)^2/2(B_H D)^2]\{[(T_{1H}^{-1} - T_{1H}^{-1}(\text{pure NH}_3)) (R/\alpha)(K_N^2/T_{1e}^{-1})]\}$$
$$\times (A_N/CE_N^2) - K_H^2/E_H^2\} \tag{7}$$

However, as we saw, the interpretation of proton shifts is problematic and we prefer to rearrange eqn. (7) to provide an independent estimate of na_H

TABLE 1

Magnetic resonance parameters and calculations for solvated electrons in metal–ammonia solutions

M	R	α	$10^6 K_N$	$10^{-5}T_{1e}^{-1}$	$10^{14}\tau/n^a$	$\Delta(T_{1H}^{-1})$	L_H	nb_{eH}^{-3b}	$-x_H na_H/g_e \beta_e^c$
(M)				(s^{-1})	(s)	(s^{-1})		$(\overset{\circ}{A}{}^{-3})$	(G)
0.01	3546.0	0.862	29.1	3.02	5.38	0.170	11.0	0.209	16.2
0.02	1772.0	0.707	42.4	3.06	6.92	0.312	20.0	0.184	15.2
0.04	884.6	0.535	61.5	3.15	7.77	0.522	36.0	0.176	14.7
0.06	588.9	0.429	76.5	3.23	7.49	0.652	48.5	0.179	15.2
0.08	441.1	0.340	89.5	3.35	6.35	0.727	59.0	0.197	16.9
0.10	352.4	0.325	100.9	3.47	7.40	0.770	67.5	0.170	16.3
0.20	174.9	0.231	147.0	4.27	8.81	0.750	98.0	0.125	12.8
0.30	115.8	0.183	183.5	6.15	11.70	0.670	124.0	0.092	9.9
0.50	68.5	0.133	243.3	31.60	51.40	0.500	215.0	0.032	4.0

Mean values for $3546 > R > 352$: [a]6.89; [b]0.186; [c]15.8.

$$(na_H)^2 = \left[a_N/3C(DE_N)^2\right]\left[(R/\alpha)^3(K_N T)^2/T_{1e}^{-1}\right]$$
$$\times\left[T_{1H}^{-1} - T_{1H}^{-1}\ (\text{pure NH}_3) + 2(L_H - 1)T_{1H}^{-1}/E_H\right] \tag{8}$$

Again, for $R > 300$ we find na_H is approximately constant (Table 1) and the total coupling to protons for electron $x_H na_H \cong 15.8\ \text{G}$. Again we note that our constant coupling to protons is consistent with both our interpretation of a constant coupling to nitrogen from analysis of nitrogen Knight shifts and also with our tentative interpretation of proton Knight shifts (Sect. II.A.1.b). It should be noted that the constancy of the proton coupling derived in this way is not forced in anyway by our initial assumptions.

A further piece of valuable structural information can be obtained from a comparison of nitrogen Knight shifts and electron relaxation rates.

From eqns. (1) and (5) we get

$$\tau_c/n = (DE_N/Ta_N^{1/2})^2(T_{1e}^{-1}/K_N^2)(\alpha/R)^2 \tag{9}$$

Again we find τ_c/n approximately constant for $R > 300$ (see Table 1).

We have now determined several parameters of structural importance, the solvent density in the vicinity of the solvated electron, n/b_{eH}^{-3}, the correlation time per solvent molecule in the unit, τ_c/n, and the magnitude of the total isotropic hyperfine coupling constants to both proton and nitrogen nuclei. We have shown that all these parameters are effectively constant over relatively wide concentration ranges and can, with reasonable justification, extrapolate these values to infinite dilution and use them to describe the ground state of the solvated electron in liquid ammonia. But what reliance can we place on the numerical values?

Firstly, we have used the spin-pairing fraction, α, for potassium solutions, which are fairly well defined experimentally, to analyse Knight shifts for all the alkali metals and magnetic relaxation data obtained from sodium solutions. Since α often appears as a square, or even a cubic term, any inaccuracies in α are greatly enhanced.

In mitigation, however, it would be reasonable to expect any errors to cause deviations from the constancy of the calculated parameters but unreasonable to attribute the constancy to errors in the data. Secondly, all the measurements of magnetic parameters are themselves subject to error as indeed is the metal concentration term R. The smoothing involved in getting values of all observations at the same concentrations would be expected firstly, to iron out these errors to some extent and secondly, if the smoothing is badly carried out, to introduce systematic errors which again would be most reasonably expected to cause deviations from constancy of the calculated parameters.

We conclude, therefore, that the calculations have been carried out in a self-consistent manner and that we can attach some weight to the results.

One further useful task is to derive the squares of the unpaired electron wave function at the various nuclei, this is done using Fermi's equation (eqn. (2)), and our final estimate of the significant parameters are

$$\tau_c/n \cong (6.89 \pm 1.0) \times 10^{-14} \, \text{s}$$

$$nb_{eH}^{-3} \cong (0.186 \pm 0.02) \times 10^{24} \, \text{cm}^{-3}$$

$$3na_H \cong -(15.8 \pm 1.0) \, \text{G}$$

$$3n|\Psi(0)|_H^2 \cong -7.20 \times 10^{22} \, \text{cm}^{-3}$$

$$na_N \cong (110.0 \pm 1.0) \, \text{G}$$

$$n|\Psi(0)|_N^2 \cong 6.44 \times 10^{24} \, \text{cm}^{-3}$$

As it stands, this analysis is incomplete, we cannot obtain estimates of the fundamental parameters n, b_{eH}, $|\Psi(0)|_N^2$ and $|\Psi(0)|_H^2$ without an independent determination of the correlation time τ_c. In principle one should write separate correlation times for each of the separate modulation processes discussed. Arguments for equating these were given at Colloque Weyl 2[39] and will not be repeated here.

In the absence of any direct determination of the correlation time it is possible to put upper and lower bounds on the probable time scale for the modulation of magnetic interactions. From a consideration of a variety of properties such as ammonia inversion frequency, dielectric relaxation, electrical conductivity, and nitrogen quadrupolar relaxation, a correlation time $2 < 10^{13}\tau_c < 8$ would appear reasonable. O'Reilly[47] has suggested that spin–orbit coupling to solvent might contribute up to 20% of the electron relaxation even in dilute solutions which would have the effect of raising the limits to around $3 < 10^{13}\tau_c < 10$. From $\tau_c/n = 6.9 \times 10^{-14}$ s we get $4 < n < 14$ and $2.7 < b_{eH} < 4.3$ Å. Whilst these values are obviously reasonable, they do not have the precision required to answer what is perhaps the most fundamental questions about the ground state of the solvated electron: how localized is the electron? Thus an effective coordination number of 4 would strongly suggest localization on a primary solvation shell, whilst a number of 14 would only be compatible with relatively extensive delocalization onto second or even third solvation layers.

To resolve this question Pinkowitz and Swift[43] attempted to measure the correlation time for magnetic interactions directly by exploiting the frequency-dependence

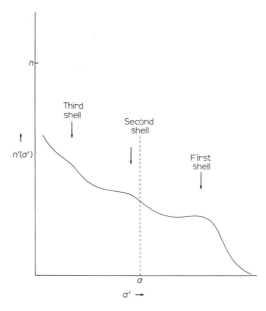

Fig. 2. Schematic distribution functions for a realistic physical situation and for the effective approximation adopted in the analysis (see text). $n'(a')$ is the number of nuclei having a given coupling constant a'.

of the transverse proton spin–spin relaxation, T_{2H}, in potassium–ammonia solutions. They did find a frequency-dependent relaxation at 60 and 100 MHz and derived effective solvation numbers varying from 15 to 47 in the concentration range $1.5–4 \times 10^{-3}$ M for the temperature interval -37 to $+28$ °C. No significance can be attributed to the range of effective coordination numbers which fluctuated in a manner independent of both metal concentration and temperature, and although these results seem to imply a greater delocalization than the previous analysis ($n \sim 40$ gives $b_{eH} \sim 5.9$ Å, suggesting strong second and even third shell interactions), a value of n between 15 and 20 would not strain either analysis significantly. Although the latter analysis uses total, coupling constants to nitrogen ranging from 99.8 to 160.7 G dependent upon both concentration and temperature (after suitable interpolation from the Knight shift result) without any smoothing, the fluctuations in n show *no* correlation with the total coupling constant and must stem from the curve fitting procedure.

The models described in Chap. 1 do not allow any serious comparisons to be made between calculated and observed distributions of unpaired electron (spin) density; the wave functions employed are too oversimplified. Nevertheless, the presence of high spin density on nitrogen and near-zero density on hydrogen is mirrored adequately in some treatments.

The most obvious interaction between theory and experiment lies in the stoichiometry of the model[56]. Analysis of magnetic resonance results strongly suggests a solvation number of around 10–20 so that any serious calculations will have to take second solvation shell molecules into account. The origin of the negative spin density at hydrogen is beyond the capabilities of all calculations so far envisaged*. Obviously, spin polarization is involved, but the often-quoted analogy with aromatic anions is not really relevant. The effect is minor but so surprisingly constant from one solvent to another (negative densities have been demonstrated for ammonia[37], mcthylaminc[49], cthylamine[50], HMPA[51], and diglyme[52]), that it may turn out to be more significant than is currently recognized.

(e) Future developments Almost all work has shown that for potassium solutions alone T_{1e} is substantially greater than T_{2e}. O'Reilly[36] suggested an explanation for his earlier observation of T_1/T_2 at low frequency, but the later work of Cutler and Powles[15] at higher frequency gave the same T_1/T_2 which contradicts O'Reilly's explanation[48] (or any other reasonable one). The effect is obviously genuine, obviously specific to potassium solutions, and obviously in need of explanation.

One of the key problems in the analysis of magnetic data is the determination of the spin-pairing fraction, α. In the preceding discussion and in previous papers, the dominant relaxation mechanisms for electron and nitrogen relaxation have been identified, and a relatively simple approach to the determination of α would be a direct comparison of electron and nitrogen relaxation rates

$$\alpha = I_N(I_N + 1)/S(S + 1) \ x_N R(T_{1N}^{-1})/(T_{1e}^{-1})$$

Alternatively of course, α/R could be determined in this way for use in eqns. (2)–(8) which would yield a considerable improvement in the structural parameters.

II.A.2 Solutions of alkali metals in hexamethylphosphoramide (HMPA)

In many respects solutions of the alkali metals in HMPA show many similarities[53] to ammonia solutions, solubilities are high, optical spectra are dominated by an IR band, and electron spin resonance spectra comprise an intense, very narrow, singlet close to the free spin. Problems of solvent purification, solution preparation, and alkali metal analysis are considerably exaggerated in this solvent, but Stodulski[51,54] has attempted to provide the range of magnetic resonance studies required for structural analysis of solvated electrons in HMPA [$OP(N(CH_3)_2)_3$]. In this solvent we have the additional possibility of significant magnetic interactions between unpaired electrons and phosphorus nuclei. Measurements of Knight shifts and relaxation rates and the resulting structural parameters are given in Table 2.

The most surprising feature of the structure of the solvated electron in HMPA is the lack of any significant involvement by the phosphorus nuclei and the P = O group. The phosphorus nuclear resonance results suggest very little spin density at the phosphorus, and it is difficult to envisage any appreciable density on oxygen without an appreciable leak onto phosphorus. The total coupling to hydrogen nuclei is strikingly similar to that in ammonia in both size and sign despite the alkyl nature of the protons. In contrast, the total spin density at nitrogen nuclei is lower than in ammonia by a factor of ~ 5.

In the absence of any frequency-dependent relaxation data, or Overhauser enhancements, we are unable to proceed further with an analysis.

---

*Note added in proof: Ab initio calculations reported by M. Newton at Colloque Weyl V (Michigan State University, June 1975) have reproduced the sign of the spin density at hydrogen in aquated electrons.

TABLE 2

Correlation times, hyperfine coupling constants, and unpaired electron spin densities at nuclei in sodium–HMPA

Item		HMPA
$\dfrac{\tau_c}{n}$ (s)	(i) from T_{2p}^{-1} and $K(P)$	$\sim 0.98 \times 10^{-11}$
	(ii) from T_{2H}^{-1} and $K(H)$	$< 0.83 \times 10^{-11}$
	(iii) from T_{2e}^{-1} and $K(N)$	$> 0.23 \times 10^{-11}$
τ_D(s)	(Debye correlation time)	0.75×10^{-9}
$|n|\Psi(0)|_N^2|(\text{cm}^{-3})$	(i) from $K(N)$	$< 7.98 \times 10^{23}$
	(ii) from T_{2N}^{-1}	$< 3.9 \times 10^{23}$
$|x_N n|\Psi(0)|_N|(\text{cm}^{-3})$		$< 1.7 \times 10^{24}$
$|n|\Psi(0)|_H^2|(\text{cm}^{-3})$ from $K(H)$		-2.99×10^{21}
$x_H n|\Psi(0)|_H^2 (\text{cm}^{-3})$		-5.68×10^{22}
$|n|\Psi(0)|_p|(\text{cm}^{-3})$	(i) from $K(P)$	$< 1.53 \times 10^{22}$
	(ii) from T_{2p}^{-1}	1.74×10^{22}
T_{2e}^{-1} (contact e–N)(s^{-1})		$< 4.54 \times 10^5$
T_{2e}^{-1} (contact e–H)(s^{-1})		1.15×10^4
T_{2e}^{-1} (contact e–P)(s^{-1})		3.56×10^3
$\Sigma\, T_{2e}^{-1}$ (s^{-1})		$< 4.69 \times 10^5$
T_{2e}^{-1} (obs.)(s^{-1})		5.35×10^5
$x_N n(A_N/g_e\beta_e)$(G)		± 21.36
$x_H n(A_H/g_e\beta_e)$(G)		-12.75
$x_p n(A_P/g_e\beta_e)$(G)		± 1.66

II.A.3 Solutions of alkali metals in other solvents

Electron spin resonance spectra have been reported for solutions of alkali metals in a variety of fluid amines and ethers. In all cases the spectra contain a single narrow line which has been attributed to isolated solvated electrons. In none of these instances is it possible to carry out much serious structural analysis although one attempt has been made to estimate the total unpaired spin density at nitrogen nuclei in ethylamine solutions[50].

II.B Solid solutions

As we saw above in Sect. II.A one of the main obstacles in the structural analysis of solvated electrons was the problem of the correlation time, the characteristic time associated with the fluctuation of the magnetic interactions in fluid solutions. An alternative approach to the structural problem is to freeze-out the fluctuations and thus eliminate the problem entirely. Ideally, if we could freeze a solution of sodium in ammonia we would get a rigid (independent of time) snapshot of a particular instant in the fluid system, no averaging of magnetic interactions would take place and we might hope to derive a detailed picture of a solvated electron together with distribution functions for all significant parameters which would describe the variations which occur from one solvated electron to another.

II.B.1 Frozen solutions of alkali metals

The "obvious" experiment of freezing a solution of sodium in ammonia in order to observe electron spin resonance spectra in the solid state was first carried out by Clark et al.[55]. Unfortunately, ammonia is a relatively strongly hydrogen-bonded medium and exhibits a strong tendency to crystallize out when solutions are frozen. Indeed, Levy[20] had used the freezing technique to study small particles of colloidal alkali metals. In an attempt to achieve a vitreous sample, Clark et al.[55] added a structure-breaking electrolyte (sodium iodide) before freezing. The resulting blue, apparently homogeneous, frozen solution had a single Gaussian electron resonance line of width 4.6 G, clearly distinguished from the Lorentzian line of width 2 G characteristic of colloidal metal particles. The Gaussian shape strongly suggested unresolved hyperfine coupling to magnetic nuclei, but when pursued further[32], the freezing process proved to be extremely difficult to reproduce. Out of a total of some thirty solutions of sodium in ammonia of various concentrations of metal and sodium halides, repeated re-freezing yielded only three spectra[57] similar to the one reported in the earlier work[55].

Attempts to avoid crystallization by freezing metal solutions in less associated solvents such as aliphatic amines[58] and ethers[59] ran into another problem. The solutions could be frozen satisfactorily, but the strongly temperature-dependent spin-pairing process resulted in extremely weak or undetectable signals at the low temperatures required. An early paper by Berry and co-workers[60] described a successful freezing and electron resonance study of a solution of lithium in a complex ether–amine–hydrocarbon solvent mixture, and more recent work[61,62] has shown that lithium solutions in general, are easier to freeze without crystallization and yield considerably higher unpaired spin concentrations. Very recently it has been established[63] that frozen solutions of the heavier alkali metals in HMPA can also be studied by electron spin resonance, and these systems are discussed in some detail in Sect. III.D.

Unpaired electron spin concentrations in ammonia solutions are generally high, but the problem of solvent crystallization is great. A wide variety of techniques for shock cooling ammonia solutions with and without structure-breaking additives was tested[64,65] and a technique developed for freezing metal solutions containing more than 0.1 M metal. A typical spectrum of a frozen solution at 77 K is given in Fig. 3. The dominant feature is a very sharp singlet close to the free spin position whose Lorentzian line shape and microwave saturation characteristics[64,65] are typical of electrons experiencing rapidly modulated magnetic interactions. The conclusion cannot be escaped; freezing solutions of these concentrations has no effect on the fluctuation of magnetic interactions. We can only conclude that the modulation mechanism is not dependent upon molecular motion in the solution, but is basically

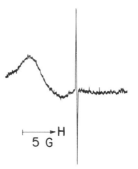

Fig. 3. Electron spin resonance spectra (77 K) of a rapidly frozen solution of potassium in liquid ammonia containing 0.4 M metal and 1.0 M KBr (ref. 64).

electronic in nature. A very similar effect is observed[66] in HMPA solutions (without any structure-breaking additives), but here the accessible concentration range is much wider. It is observed that the motional narrowing of the signal from unpaired electrons persists down to concentrations of $\sim 10^{-2}$ M in metals. Below this concentration the modulation is frozen out, the line broadens rapidly to about 6 G while the line shape changes to the Gaussian function characteristic of static interactions. The isotopic dependence of this electron resonance signal was investigated for ^{6}Li and ^{7}Li solutions; a narrowing from 6 to 4 G was observed when ^{6}Li was used, suggesting that an appreciable interaction with metal nuclei was involved, so that this species cannot be identified with isolated solvated electrons. Of these signals from frozen metal solutions, therefore, we must conclude that the narrow signal is dependent upon electron–electron interactions, while the broader line arises from a species in which there is interaction between the unpaired electron and metal nuclei.

There is some evidence for a considerably broader line (~ 35 G for $\lesssim 100$ K) in frozen solutions of alkali metals in ammonia[65] which becomes narrower as the temperature is raised towards the melting point (~ 200 K), whilst the unexpected maximum line width observed by Levy[20] and by Catterall and Edwards[67] at ~ 100 K in frozen potassium solutions appears to stem from the same cause. It is not clear, however, whether this signal is to be identified with isolated solvated electrons or not, although changes in line width with temperature, far below the freezing point, argue strongly for some electron–electron interactions.

There has been one report[68] of an 11 G single line electron resonance spectrum from a frozen solution of sodium in ammonia which does not correlate with any of the previous work. We have found[69], however, that partial decay of solutions of alkali metals in ammonia leads to signals (possibly from superoxides) of this order of line width.

II.B.2 Matrix isolation techniques

An alternative approach to the production of frozen solvated electrons was pioneered by Bosch[70], and consists of the condensation from the gas phase onto a cold finger (4 K) of ammonia molecules and sodium atoms in suitable proportions. The deposition is generally done by condensing alternate layers of solvent and solute. The work by Bosch was a study by optical spectroscopy of the blue solids and need not concern us further, except to note that both solvated electrons and metal-dependent species were formed and identified. More recently, a closely parallel electron resonance study was carried out by Beuermann[71], but the results show no trace of solvent hyperfine structure; all spectra were single lines with no differences between NH_3 and ND_3 samples or different alkali metals.

Similar experiments have been carried out by Bennett et al.[72,73] on sodium or potassium in water systems. In this case the deposits are also blue and the results demonstrate conclusively that we may attribute the electron resonance signal to the isolated solvated electron. There is no significant change in the electron resonance line when potassium metal is used in place of sodium, whilst the use of D_2O instead of H_2O achieves an almost three-fold reduction in line width. On warming from 77 K, no change is observed for D_2O samples, whilst above 140 K H_2O samples begin to show resolved hyperfine structures (Fig. 4) which can be unambiguously attributed to isotropic hyperfine coupling to six (or possibly eight) protons. The sign of the coupling constant cannot be determined in the simple electron resonance experiment, but the magnitude of the total hyperfine coupling (i.e. the total s-electron spin density) to hydrogen is closely similar to that observed in solutions of alkali metals in ammonia. As we saw in ammonia the spin density at hydrogen represents only a few percent of the unpaired electron, the majority ($\gtrsim 95\%$) is accommodated in an s-state on nitrogen. It seems likely, therefore, that the coupling to hydrogen observed by Bennett et al.[72,73] for sodium in water represents an equally minor fraction of the unpaired electron. Unfortunately, the normal ^{16}O isotope is non-magnetic and any spin density at oxygen will have no direct effect on the electron resonance signal. Experiments on ^{17}O enriched water would be informative.

An alternative approach by Marx et al.[74] was to irradiate water vapour with high energy ionizing radiation immediately prior to deposition on a cold finger at 77 K. An electron resonance signal closely similar to that observed by Bennett was found at low temperatures, with a similar narrowing in D_2O, but no resolution of hyperfine structure was achieved when the samples were warmed.

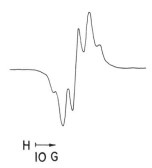

H ⊢——→
10 G

Fig. 4. Electron spin resonance spectra for a deposit of potassium metal in water at 173 K (ref. 159).

II.B.3 Radiation-induced electrons

So far we have been almost exclusively concerned with the generation of excess electrons by ionization of alkali metals, but the technique of Marx et al.[74] indicates that the ionization of the solvent itself by high energy radiation is equally capable of producing electrons that can be trapped under suitable conditions. The technique of gas phase irradiation, followed by deposition on a cold finger, is technically complex, and simpler forms of stabilizing radiation-induced electrons have been studied in some detail. We divide these studies into solvent vacancy trapping and anion vacancy trapping, somewhat arbitrarily but representative of a genuine, if ill-defined difference. This technique introduced a new problem reminiscent of the crystallization problem we met in freezing solutions of metals in ammonia. Again, it appears that the production of electrons in crystalline media is energetically unfavourable, and significant yields of solvated electrons are only attained in vitreous or amorphous media. It is this point which leads to the concept of electron trapping at vacancies—voids of approximately molecular sizes which are assumed to be present in amorphous solids in relatively high concentrations, but are almost absent in molecular crystals. In contrast, Shottky defects (cation and anion vacancies) in ionic crystals are well known.

(a) Solvent vacancy trapping When a wide variety of alcohols, ethers, amines, or even hydrocarbons, are frozen (rapidly if necessary) they form "glasses", amorphous media with a high degree of short range order (no broken or dangling bonds, for example), but with little or none of the long range order (e.g. translational symmetry) which is characteristic of the crystalline state. If these vitreous molecular solids are irradiated with high energy radiation, we can picture the primary process as being ionization of a single solvent molecule to yield a mobile hole in the valence band of the matrix and a "hot" electron in the conduction band. At any stage in the subsequent description, electron–hole recombination can, and probably does, occur to reverse the ionization process, but if the kinetic energy of the electron is appreciable (i.e. many times kT) the importance of this recombination will not be significant.

The trapping of the "hot" electron can be envisaged, firstly as a succession of collision processes in which the kinetic energy of the electron is continuously degraded into the thermal modes of the lattice until we are left with "thermal" electrons. At this stage we picture the important process as being one of competition by a variety of electron traps for the thermal electrons. Among these traps we can include impurities, container walls, solvent molecules themselves, and voids or vacancies of molecular dimensions. The first two are of no interest to us since they represent mechanisms whereby our objective of solvating electrons is frustrated. We might reasonably expect that chemical purity and normal sample size would render these processes of little significance. Of the third process, the formation of molecular anions, we observe that the solvents mentioned are basically filled-shell molecules with large energy gaps to the first vacant molecular orbital. Any molecular anions formed we would therefore expect to be energetically unstable. Nevertheless, we must admit that when an excess electron does interact with a solvent molecule it must enter these high energy molecular orbitals. The fourth possibility listed above, vacancy trapping, can be pictured as an attempt to avoid putting excess electrons into high energy, possibly antibonding, molecular orbitals. The vacancy may be thought of crudely as a region of vacuum bounded by potential walls representing the Pauli exclusion forces trying to keep the excess electrons away from the filled molecular orbitals of the solvent molecules. In the simplest approximation, therefore, we may approximate the state of the electron in a cavity by a hydrogenic s-like orbital, in the limit completely isolated from solvent by surrounding potential walls, but more realistically penetrating the solvent to some extent and entering the high energy vacant molecular orbitals. This picture represents a brief summary of almost all theoretical descriptions (see Chap. 1) of the orbital which the electron occupies. The additional questions of stabilization and localization are generally answered by invoking the concept of a physical polarization of the medium around the trapped electron. Thus, although there is a trapping potential based on a cavity, the electron is only relatively weakly bound until it has dug its own trap deeper by physically polarizing the medium around the cavity to represent a configuration adapted to a central negative charge. On an electronic scale this last adaption must be a slow process, inherently temperature sensitive. The importance of this medium polarization process is obviously dependent upon the polarity of the medium, thus for alcohols with a highly polar OH group, the matrix reorganization should contribute significantly to the nature of the final trapping site, for less polar ethers, medium polarization should be much less important, whilst for the nominally non-polar hydrocarbons one only expects slight physical polarization due to bond dipoles.

The polarization process at a molecular level can be thought of as a contraction of the medium onto the electron resulting from a charge–dipole attraction opposed primarily by a Pauli repulsion between the electron and the solvent molecules. In order to optimize the charge–dipole interactions it is necessary to rotate solvent molecules so that randomly oriented molecular dipoles are lined up pointing towards the electron. In terms of the electron resonance spectra we might expect both the contraction and the preferential orientation to result in stronger magnetic interactions between the unpaired electron and specific magnetic nuclei. In the case of alcohols the polarization can be envisaged as a combination of a general molecular reorientation accompanied by a rotation around the CO bond to a final configuration in which the OH bond points towards the electron at the center of the vacancy.

Qualitatively at least, this picture of the polarization process is amply confirmed by electron resonance results for ethanol[75,76]. If ethanol and its hydroxyl or alkyl deuterated variants are irradiated at 4 K, a single line electron resonance signal of width 4–6 G is obtained. The effect of deuteration indicates that the interaction of all hydrogens is approximately equal. If the irradiated ethanol is warmed to 77 K, the line width increases significantly to about 14 G in the fully protonated form, whilst results for the deuterated form indicate clearly that the increased line width stems primarily from the hydroxyl proton interacting more strongly. The polarization process is evidenced by the changes in the electron resonance spectra and is accompanied by equally drastic changes in the electronic spectra[77] in the visible and near-IR regions. At 4 K, a broad band is observed peaking at ~ 7000 cm^{-1}: on warming to 77 K the band narrows and shifts strongly to higher energies.

In contrast to these dramatic changes in the spectra of solvated electrons in the polar ethanol medium, the changes in the corresponding spectra of the much less polar medium 2-methyltetrahydrofuran (MTHF) are relatively insignificant. For all temperatures at which the solvated electron is observed, the band widths and positions of electron spin resonance and electronic absorption bands are approximately constant[75,77], indicating that polarization of the medium around the electron is not particularly important in determining the strength of the electron–solvent interactions. Similar changes in optical spectra for fluid[78] (200 K) and solid[79] (77 K) ethanol have been observed on a microsecond time scale following pulsed-radiolysis. Short time relaxation in non-polar media like 3-methylpentane[79,80] is accompanied by only slight changes in the optical band.

Other experimental studies of irradiated alcohols,[81,87,100] amines,[86,97–100] ethers,[84–91] and hydrocarbons[86,87,91–99] bear-out this general picture of electron solvation in these media, but, with two exceptions, contribute little additional information. Firstly, there has been an unconfirmed report of resolved hyperfine structure in the electron resonance spectra of solvated electrons in γ-irradiated methanol[101], the spectrum is consistent with a unit comprising five equivalent methanol molecules with the structure stemming from a dominant isotropic hyperfine interaction with the hydroxyl protons. The overall coupling constant is not too dissimilar to that observed in ammonia and water.

In contrast, the electron resonance spectra of γ-irradiated N-deuteropyrrolidine[102] clearly show the existence of isotropic hyperfine coupling to four equivalent nitrogen nuclei, whilst other deuteration results suggest that coupling to hydrogen is equally small and similar to that observed in ammonia, water, and alcohol. In this case, as in fluid HMPA solutions, the total coupling to nitrogen is $\lesssim 20\%$ of that observed in ammonia and the question remains unanswered as to the whereabouts of the remaining 70–80% of the unpaired spin density. The results for γ-irradiated triethylamine[103] (line width 4.0 G) suggests that in this case the coupling to nitrogen may be an order of magnitude less, whilst coupling to protons is equally small, with an even greater difficulty in locating the unpaired spin.

One possibility is to use the line width as a probe for electron delocalization, while at the same time providing some resolution of the problem of the apparent lack of detectable electron spin density. To proceed we must make two assumptions. Firstly, we assume that the width of the solid state spectrum arises solely from unresolved isotropic hyperfine coupling (i.e. we can ignore anisotropic and residual line width effects). It is particularly easy to relax this assumption should any evidence suggest that it may be necessary, but to date all the evidence suggests that other line width contributions are too small to justify the introduction of a new variable parameter. Secondly, we make a fundamental assumption that the total electron (spin) density is a constant, no matter how many nuclei it is spread over. This is quite a reasonable assumption when the hyperfine structure obviously accounts for the majority of the unpaired electron spin (e.g. a total coupling to nitrogen in ammonia solutions of ~ 110 G), but may not be so reliable when considering line widths dominated by proton interactions. Under these conditions the width of the envelope, ΔH_{ms} is given by

$$\Delta H_{\mathrm{ms}} = \frac{[2I(I+1)]^{1/2}}{(3\ln 2)} \frac{a_{\mathrm{t}}}{n^{1/2}} \tag{10}$$

or

$$\frac{1}{[I(I+1)]^{1/2}} \frac{\Delta H_{\mathrm{ms}}}{a_{\mathrm{t}}} = 0.9805\, n^{-1/2} \tag{11}$$

where the total coupling constant to equivalent nuclei of spin I is a_{t}. A plot of this function is given in Fig. 5. For four systems we have reasonable estimates of the total hyperfine coupling to nitrogen nuclei: ammonia (100 G), ethylamine (70–73 G), HMPA (21.4 G), and pyrrolidine (22.4 G). But, we have narrower solid state line widths for electrons in HMPA (4.2 ± 0.2 G), triethylamine (4.0 G), ethylamine (22 ± 5 G), and butylamine (23 ± 2 G) whilst the maximum line width of ~35 G at ~100 K observed for frozen solutions of potassium in ammonia may possibly represent a value for a localized electron in solid ammonia. This signal is currently unexplained, but is certainly not colloidal metal. Some evidence was also found for a line of this width underlying the very sharp line in frozen sodium or potassium solutions containing alkali halides.

Using Fig. 5 we estimate for HMPA $n = 49 \pm 5$ or an approximate stoichiometry of $e^-(\mathrm{HMPA})_{16\pm2}$. For ethylamine we find $e^-(\mathrm{EtNH_2})_{20\pm4}$, and for triethylamine (using a_{t} for pyrrolidine) $e^-(\mathrm{Et_3N})_{60}$. A line width of 35 G for ammonia yields $e^-(\mathrm{NH_3})_{19}$; the close comparison with the stoichiometry of the solvated electron in ethylamine supports our assignment of the 35 G line to isolated solvated electrons.

An alternative way in which this approach can be exploited is in a direct comparison of fluid and solid state electron resonance line widths. Thus if we make the

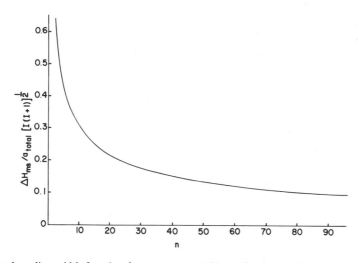

Fig. 5.   Envelope line width function for a constant total hyperfine interaction spread over a variable number, n, of equivalent nuclei. The spin packet width is just sufficient to prevent resolution of hyperfine structure.

additional assumption that the coordination number of the electron is the same in the fluid and solid state (for which we have no supporting evidence and little conflicting evidence[43]) then the fluid solution width can be expressed as

$$\Delta H_{ms}(\text{fluid}) = 1.353 \times 10^7\, I(I+1)\, a_t^2\, \tau c/n \tag{12}$$

whilst the solid state line width is given by[104]

$$\Delta H_{ms}(\text{solid}) = 0.9805\, [I(I+1)]^{1/2}\, a_t/n^{1/2} \tag{13}$$

so that

$$\tau_c = 7.10 \times 10^{-8}\, \Delta H_{ms}(\text{fluid})/[\Delta H_{ms}(\text{solid})]^2 \tag{14}$$

In this case the magnitude of the total coupling is not required and comparing fluid potassium in THF with γ-irradiated MTHF we get

$$\tau_c = 2.1 \times 10^{-10}$$

To complete this section we highlight the importance of the disordered (vitreous) state of the matrix. Solvents such as water, ammonia, and HMPA which are known to be good solvators of electrons in the fluid state, all show a marked tendency to crystallize when frozen: the production of amorphous solids of these materials is extremely difficult. Irradiation of solid water[105], and HMPA[107,108] yield extremely low concentrations of trapped electrons although, in general, the addition of structure breakers[109] (such as electrolytes or mixtures of solvents—see Sect. II.B.4) leads to relatively high yields. Similar effects have been observed in crystalline and solid alcohols[82,83,85], and an attempt has been made[111] to correlate free space volume in amorphous ethanol with yields of trapped electrons. This correlation is not too successful, electrons can only be trapped in a small fraction of the free space in the matrix. This lack of correlation has been attributed to the presence of elongated, channel-like vacancies and supposing that electron trapping is most likely for approximately spherical cavities.

(b) Anion vacancy trapping An important crystalline analogue for the solvated electron is the F-center in ionic crystals. In those systems the electron trapping site is an anion vacancy in the regular crystal lattice. One important case of electron solvation in disordered media bears a very close resemblance to the F-center analogue. The trapping of radiation-induced electrons in aqueous media is extremely inefficient in pure, solid water[112], although yields can be improved considerably by the addition of electrolytes such as alkali halides[113] or covalent structure breakers such as sugars[114]. One important additive in which the relation of the anion to the solvent merits special attention is the alkali metal hydroxide[115–141].

Aqueous solutions of alkali hydroxides ($\gtrsim 1$ M) glassify readily and turn deep blue when exposed to ionizing radiation at 77 K. Electronic spectra comprise a broad, asymmetric band in the visible ($\nu_{max} \sim 17000$ cm^{-1}) closely resembling that recorded for solvated electrons in fluid water. The electron resonance spectra contain signals from O$^-$ ions, hydroxyl radicals (particularly at low hydroxide concentration) and a singlet, attributed to solvated electrons, which can be destroyed by irradiation with visible light.

The mechanism proposed for the production of these solvated electrons illustrates the close analogy with F-centers, whilst providing a means for radiation-induced cavity formation which can account for the high yields of trapped electrons[118]. The primary radiation processes in solutions containing appreciable concentrations of hydroxide ion are the ionization of water and hydroxide.

$$H_2O \xrightarrow{\gamma} H_2O^+ + e_m^- \tag{15}$$

$$OH^- \xrightarrow{\gamma} OH^. + e_m^- \tag{16}$$

The solvent cations H_2O^+ react very rapidly either with hydroxide ion or with another water molecule to yield OH radicals

$$H_2O^+ + H_2O \rightarrow H_3O^+ + OH^. \tag{17}$$
$$H_2O^+ + OH^- \rightarrow H_2O + OH^. \tag{18}$$

In more dilute solutions (< 5 M) hydroxyl radicals are trapped, whilst in more concentrated solutions, or by thermal annealing of the dilute solutions, the hydroxyl radicals disappear with the formation of O^- radical anions. This suggests that the dominant process for the production of O^- is

$$OH^. + OH^- \rightarrow O^- + H_2O \tag{19}$$

whilst the reaction of hydroxyl radicals with water is less favoured

$$OH^. + H_2O \rightarrow O^- + H_3O^+ \tag{20}$$

Any hydronium ions formed by eqn. (17) and, or, eqn. (20) are of course highly mobile in ice and will react with hydroxide ion to produce anion vacancies

$$H_3O^+ + OH^- \rightarrow 2H_2O + \text{anion vacancy} \tag{21}$$

The critical point here is that when the hydroxide solutions are frozen, the water around the hydroxide ions is polarized so as to accommodate a negatively charged ion. If the negative charge is destroyed by eqn. (21), the resulting water molecule is no longer held at the center of the polarization field and can move into the surrounding shell of water molecules leaving a molecular-sized vacancy with a polarization field suitable for electron trapping—a close analogy to the anion vacancy in an ionic crystal.

II.B.4 Mixed solvents

For reasons outlined above, irradiation of pure water, ammonia, and HMPA in the solid phase results in very low concentrations of solvated electrons. The destruction of the crystalline state by structure-breaking additives significantly increased the yields of trapped electrons. In this section we describe how the crystallization process can be inhibited simply by mixing the pure solvents.

(a) Ammonia–water mixtures[142]   Electron solvation in ammonia–water mixtures in the fluid state has been conclusively demonstrated[143] over the entire composition range. Over a wide composition range the mixtures froze[142] to give excellent glasses

without crystallization of either solvent. After short periods of high energy irradiation at 77 K the ammonia–water glasses were coloured blue as would be expected for high concentrations of trapped electrons. Irradiation of the blue glasses with visible light caused photobleaching of the colour exactly as in the aqueous hydroxide glasses, and we must conclude that high yields of solvated electrons had been produced. When the electron resonance spectra were examined there was no trace of any signal attributable to solvated electrons underlying the hydroxyl or amino radical signals, and no significant change in signal when the blue colour was destroyed by photobleaching. It appears, therefore, that the solvated electron in ammonia–water mixtures does not give rise to a detectable electron resonance signal. Since the spectrum of an electron trapped in an aqueous environment gives rise to a prominent 14 G line, it is clear that electron trapping in these mixtures does not take place in a predominantly aqueous environment. The observed total coupling to nitrogen in fluid ammonia systems (~ 110 G) would give rise to a solid-state envelope of ~ 35 G width which should have been detectable under the experimental conditions. The problem remains unsolved.

(b) HMPA–water mixtures[107,108] When water and HMPA are mixed there is considerable evolution of heat indicating strong solvent–solvent interactions. The mixed solvent freezes to give clear glasses over a fairly wide composition range. When irradiated at 77 K the glasses are coloured deep blue and exhibit a central feature ($\Delta H_{ms} = 14$ G, $g = 2.0023$) which can be destroyed along with the blue colour by irradiation with visible light. When HMPA–D_2O mixtures are used the line narrows[107] to ~ 5 G. These results, which are closely similar to those observed in the aqueous hydroxide glasses suggest that the electron trap is very similar to that in the aqueous medium, and the HMPA plays only a minor role in electron solvation in HMPA–water mixtures. Similar results were obtained for HMPA–CD_3OD mixtures[108].

(c) Ethylamine–diglyme mixtures[62] This system is of unique importance in the study of electron solvation as it is the only system for which we have good electron resonance information from both the fluid and solid metal solutions and also from the irradiated solid solvent. This statement must be tempered slightly by the knowledge that the alkali metal cation is also involved in the species responsible for the electron resonance signal from the fluid metal solution, but comparisons of spectra from solutions of ^6Li and ^7Li isotopes have indicated that magnetic interactions between the unpaired electron and the metal cation are extremely weak.

Electron resonance spectra from fluid solutions of lithium in pure ethylamine[50,144] close to the freezing point (~ 180–250 K) show clearly resolved hyperfine structure from four equivalent nitrogen nuclei superimposed upon the narrow singlet characteristic of solvated electrons (Fig. 6). The species responsible for the structured resonance has been identified[50] as an ion-pair of a solvated electron and a solvated lithium ion in which the lifetimes of the lithium–solvent bonds and of the unpaired electron–$Li(EtNH_2)_4^+$ interactions are long ($> 10^{-6}$ s) whilst other electron–solvent interactions have the extremely short lifetime ($\sim 10^{-12}$ s) typical of isolated solvated electrons in fluid. Below the freezing point this structure persists for some degrees, but eventually the sharp singlet and the hyperfine components broaden and merge

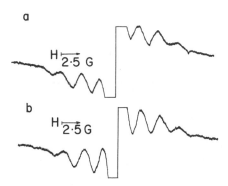

Fig. 6. Electron spin resonance spectra of solutions of lithium in ethylamine at 186.7 K[50]. (a) ^{7}Li/EtNH$_2$; (b) ^{6}Li/EtNH$_2$.

into a broad Gaussian envelope of width ~23 G. Broadening of the hyperfine components alone would yield an envelope width of ~6 G when the resolution is first lost. Accordingly, we conclude that the overall 23 G observed encompasses all electron–solvent interactions (corresponding to an overall coupling to nitrogen of around 70 G if the electron is delocalized over ~20 molecules (see Fig. 5).

Very dilute solutions of potassium in diglyme[145] (CH$_3$ OCH$_2$ CH$_2$ O CH$_2$ CH$_2$ O CH$_3$) have a singlet electron resonance signal of width ~0.15 G, characteristic of rapidly modulated interactions. When these solutions are frozen, a signal with very low signal to noise ratio was observed with width ~4 G. Fluid solutions of lithium in the mixed solvents[62] (up to 56% diglyme) show similar hyperfine structure from four equivalent nitrogen nuclei (the hyperfine splitting constant is dependent upon solvent composition and temperature). On freezing (170–180 K) the lines broaden and merge to a singlet of width ~23 G.

Pure ethylamine crystallizes on freezing and irradiation yields no solvated electrons. In contrast, pure diglyme and ethylamine–diglyme mixtures glassify easily and ionizing radiation produces the characteristic blue colour and central singlet electron resonance signal, both of which can be destroyed by visible light[62]. Widths of the electron resonance lines in diglyme-rich (5 G) and in ethylamine-rich mixtures (19 ±2 G) are close to those observed in frozen solutions of metal in the pure solvents. At intermediate concentrations the two extremes are superimposed, probably indicative of some phase separation in the mixed solvent glasses. It appears that in this case we have the only strong evidence for the structural similarity of the solvated electrons resulting from both alkali metal ionization and radiation-induced solvent ionization.

II.B.5 Photolytically-induced solvated electrons

In this section we make brief reference to the techniques of producing solvated electrons without the influence of either a cation with high surface charge density or of the gross damage produced by high energy ionizing radiation. Suitable solutes can undergo photolytic ionization, the ejected electrons becoming ultimately solvated. Photolysis ($v \sim 30\,000$ cm^{-1}) of dilute solutions of TMPD in ethers causes ionization, leaving behind the paramagnetic cation TMPD$^+$ and generating high yields of solvated electrons[146]. Electron resonance absorption by TMPD$^+$ obscures the resonance from solvated electrons.

Alkali metal reduction of many aromatic hydrocarbons produced relatively stable anionic free radicals (AFR)[147]. Solutions of AFR in ethers glassify and photolysis in the bands of the AFR causes ionization, resulting in the diamagnetic parent hydrocarbon and solvated electrons. The method lends itself to the preparation and study of solvated electrons under "clean" conditions and is worth considerable extension.

Solvated electrons are produced in an alkaline aqueous medium by photolysis of ferrocyanide ion with a low pressure mercury resonance lamp (254 nm)[148,149]. Using this technique again produces a "clean" signal suitable for electron resonance studies. Since the relaxation characteristics of solvated electrons in this system were indistinguishable from those in the presence of O$^-$ radicals, the magnetic isolation of the species seems to be firmly established.

II.C Conclusions

In 1908 Kraus[150] described the solvated electron as "... a new species of anion ... an electron surrounded by an envelope of ammonia molecules ...". A few years later he introduced the idea of a cavity to explain the remarkable volume expansion[151] that accompanies electron solvation (equivalent to a vacancy of ~ 3.5 Å radius for each electron in solution). Kraus' physical picture is still the best we have available, although we now have some idea of how big the envelope is, how long a particular electron experiences a particular environment, of the energetics of solvation[152], and of some details of a wave function describing the state of the electron. Nevertheless, it is true to say that the sum total of almost seventy years of experimentation and theorizing about solvated electrons has enabled us to put some very imprecise numbers to the qualitative ideas of Kraus.

Some very general comments can be made. Firstly, from apparent solvation numbers in water (4), in methanol (5), in HMPA (~ 16), in ammonia and ethylamine (19), and in triethylamine (~ 60), we can conclude that localization is favoured by high polarity of the medium. Secondly, we can find no evidence to support the existence of molecular anions[153] such as NH$_3^-$ in any of the systems that have been examined.

III. Interactions between solvated electrons and other solvated ions

III.A Electron–monopositive cation interactions

Electron spin resonance spectra of solutions of alkali metals in ammonia are almost identical for solutions of lithium, sodium, and potassium, but marked increases in line width are found for solutions of rubidium and cesium[47,158]. The line width increments have been attributed[47] largely to time-averaged electron–cation hyperfine interactions.

$$M^+_{solv.}\, e^-_{solv.} \rightleftharpoons M^+_{solv.} + e^-_{solv.} \qquad (22)$$

Nuclear magnetic resonance studies[35,36] of alkali metal nuclei have been used to probe equilibrium represented in eqn. (22), although sensitivity requirements have restricted the investigations to concentrations greater than normally acceptable for ion-pairing models and in solutions where the electron–electron interactions (Sect. IV) are significant. The metal nuclear resonance is shifted down-field from its expected position, the magnitude of the Knight shift being given by

$$K(M) = \gamma D E_M\, n_M\, a_M\, \alpha / RT \qquad (23)$$

where $E_M = g_e \mu_B / g_M \mu_n$, n_M is the effective number of the cations interacting with each unpaired electron at any given instant (generally taken to be unity) and γ is the fraction of unpaired solvated electrons which are ion-paired to solvated cations. From this point there are two possible approaches to the analysis of the concentration dependence of the metal Knight shift. Firstly, O'Reilly[36] treated the equilibrium constant and the hyperfine coupling constant a_M as independent variables and obtained a "best fit" solution. The ion-pair dissociation constant determined in this way was an order of magnitude higher than the value from analysis of conductivity[157], but this in itself is not too important; differences in ion-pairing constants from spectroscopic and conductivity data are the rule rather than the exception. Secondly,[39] we can utilize the equilibrium constant obtained from conductivity studies and attempt to fit the sodium Knight shift data using a single adjustable parameter, a_M. Coupling constants and spin densities from both approaches are given in Table 3, together with some other estimates.

TABLE 3

Unpaired electron spin densities at alkali metal nuclei in ion-pairs of solvated electrons and solvated cations

	$A_M/g_e\beta_e$	(G)	$10^{-23}\|\Psi(0)\|^2_M$	% Metal ns
(i) By extrapolation of A_{iso} in $EtNH_2/NH_3$ mixtures	^{23}Na	7	1.12	2.2
	^{39}K	2	1.81	2.4
	^{85}Rb	13	5.70	3.6
	^{133}Cs	30	9.68	3.7
(ii) Fitting K(M) using K_1 from conductivity, and adjusting $\|\Psi(0)\|^2_M$	^{23}Na	1.63	0.26	0.5
(iii) Fitting K(M) adjusting both K_1 and $\|\Psi(0)\|^2_M$	^{7}Li	0.62	0.067	0.43
	^{23}Na	3.38	0.54	1.07
	^{85}Rb	7.72	3.37	2.13
	^{133}Cs	21.15	6.75	2.55
(iv) Reducing coupling in (i) by ratio of ^{23}Na values in (i) and (ii) Li value from ratio of ^{7}Li and ^{23}Na values in (iii)	^{7}Li	0.30	0.032	0.21
	^{23}Na	1.63	0.26	0.51
	^{39}K	0.46	0.42	0.56
	^{85}Rb	3.02	1.32	0.84
	^{133}Cs	6.96	2.25	0.85

All approaches predict a very small delocalization of the unpaired electron onto the cation ($\lesssim 1\%$) if we can assume that the electron density at metal nuclei is distributed over a proportion of nuclei similar to that predicted by ion-pairing theory. We cannot, however, rule out the possibility that the shift arises from a very small concentration of species having much higher electron density at the metal nucleus. The metal nuclear relaxation rate allows us to estimate the lifetime of the electron–cation interaction, of the order of 10^{-12} s on an ion-pairing model, but again a few species with high electron density and longer lifetime could give rise to the same average relaxation rate. We can put an upper limit to the lifetime $\sim 10^{-6}$ s since no resolved metal hyperfine structure is observed. An expression for the lifetime τ_M can be obtained from the metal Knight shift

$$K_M = DE_M a_M \gamma / T \tag{24}$$

where $E_M = g_e \mu_B / g_M \mu_n$ and that part of the electron relaxation which arises from rapidly modulated cation hyperfine interactions

$$T_{le}^{-1}(M) = A_M a_M^2 \tau_M \gamma / \alpha \tag{25}$$

For a rubidium or cesium solution of given concentration we get

$$\tau_M = \frac{D^2 E_M^2}{A_M} \frac{T_{le}^{-1}(M)}{K_M} \alpha \gamma T^2 \tag{26}$$

in which, although the magnitude of the coupling constant has been eliminated, we are left with $\tau_m \propto \gamma$, the fraction of cations experiencing a hyperfine interaction with an unpaired electron. Similarly, from a comparison[39] of metal nuclear relaxation rate and Knight shift we can obtain only the product $\tau_m a_m^2$ by eliminating γ.

Thus we must conclude that although metal nuclear resonance spectra in ammonia solutions provide definitive evidence for electron–metal nucleus interactions, and that these interactions must have a lifetime shorter than $\sim 10^{-6}$ s, we cannot conclude that they are weak ion-pair-like interactions.

As mentioned above, solutions of lithium in ethylamine[50,62,144], butylamine[50], and ethylamine–diglyme mixtures[144] have electron resonance spectra (190–250 K) showing clearly resolved hyperfine coupling to four equivalent nitrogen nuclei. The substitution of the ^{6}Li isotope results in only a fractional narrowing[50] of the hyperfine components (Fig. 6) and the signal has been attributed to ion-pairs between solvated electrons and tetracoordinated lithium cations[50]. It is suggested that at a given instant, only one or two of the ethylamine molecules coordinated to lithium take part in the solvation shell of the electron, but on the time scale of an electron resonance experiment the interaction is spread equally over all four molecules. Other ethylamine molecules in the electron's solvation shell have much shorter lifetimes with respect to solvent exchange and do not contribute to the hyperfine structure, although they do contribute to the overall line width of the solid state spectrum.

In solid solutions the evidence for electron–cation interactions is unclear. Presumably the solvated lithium cation in ethylamine is still close to the solvated electron after freezing, but no change in the overall line width could be detected on using $^6Li^{62}$. The expected change was considerably less than experimental error.

Bennet et al.[72,73] have studied deposits of all the alkali metals in water and find no trace of cation interactions even for cesium metal. This implies that during the deposition process the metal atoms are completely ionized, and the electron localized at a distance of at least several water molecules.

The observed line widths and g-factors found for solvated electrons in γ-irradiated alkaline ices have been shown to depend slightly upon the nature of the alkali metal as well as markedly upon the particular hydrogen isotope used. Kevan[160] has applied the simple formalism of Kip et al.[104] to estimate the line width contribution from the alkali metal nucleus. Basically, the technique is similar to that described earlier in the analysis of isolated solvated electrons in solid-state media. In view of the arbitrary estimate of residual line width, and the dependence of line widths upon microwave power[126], we feel that little significance can be attached to the conclusion that cation interactions are involved.

In summary, we have found conclusive evidence for electron–cation interactions in the fluid state (although we cannot estimate the magnitude of the coupling in an electron–cation unit). If any interaction occurs in the solid it is weak and would be consistent with an ion-pair model. Other evidence for ion-pairing between solvated electrons and solvated cations lies outside the scope of this chapter.

III.B Electron–dipositive cation interactions

Although alkaline earth metals are readily soluble in liquid ammonia, and optical spectra[161] quantitatively suggest two solvated electrons per metal atom in solution, there is an almost total lack of magnetic resonance studies on these solutions. Cutler and Powles[14] have shown that electron relaxation in very dilute calcium solutions is effectively identical with that in alkali metal solutions.

Metallic europium shows marked chemical similarities with the alkaline earth metals, including solubility in liquid ammonia[162]. These metal solutions, however, are unique in that the cation itself is paramagnetic and gives rise to an easily recognizable electron resonance spectrum. We thus have the possibility of probing electron–cation interactions by observing the electron resonance signal from the solvated cation Eu^{2+}. If any close approach of solvated electrons and cations occurs in solution this must result in a cross-relaxation process broadening both spectra. If the solvated electron interacts strongly with the solvated cation to give appreciable spin density at the cation (compare the discussion of metal Knight shifts in ammonia, Sect. III.A), then the resulting species should approximate closely to the Eu^+ ion, whilst the addition of two electrons would yield the corresponding europium atom. The ground states of both the europium atom and the dipositive ion are spherically symmetric $^8S_{7/2}$ ($4f^7 5s^2 5p^6 6s^2$ and $4f^7 5s^2 5p^6$, respectively), but that of the monopositive ion ($4f^8 5s^2 5p^6$) contains considerable orbital angular momentum in the f-shell. Accordingly, we expect clear resolution of the hyperfine structure in the electron resonance spectra of Eu^0 and Eu^{2+} but strong broadening to the point of undetectability for Eu^{1+}.

76

Electron spin resonance spectra of solutions of europium(II) iodide in liquid ammonia[158] comprise two six-line patterns characteristic of the two naturally occurring isotopes of europium. Electron spin resonance parameters are closely similar to those observed for Eu(II) ions in crystalline media. Solutions of europium in liquid ammonia have spectra which are strongly dependent upon both metal concentration and temperature. For dilute solutions the spectra approximate to a superposition of the spectrum of Eu^{2+} ions (Fig. 7) and a narrow singlet characteristic of solvated electrons. With increasing concentration both signals broaden, indicating cross-relaxation[163], but under no conditions were signals observed which could be attributed to appreciable penetration of the f-shell by the solvated electron.

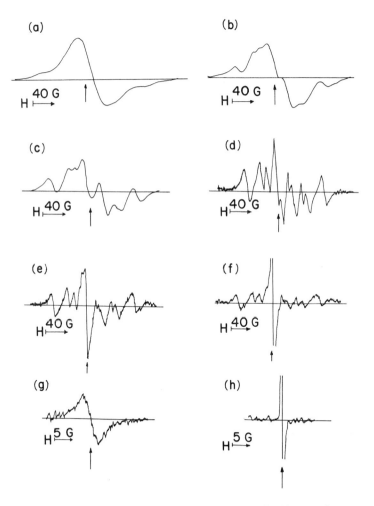

Fig. 7. Electron spin resonance spectra of solutions of europium in liquid ammonia at room temperature[158]. (a) 0.1 M; (b) 0.001 M; (c) through (f) progressive dilution to (g) $\sim 5 \times 10^{-5}$ M; (h) $\sim 10^{-7}$ M. Vertical arrow indicates $g = 2.0004$.

Hasen and co-workers[164] have repeated the work at lower temperature, observing quantitative Eu^{2+} formation at all concentrations but finding an additional broadening of the singlet above the unexpected cross-relaxation process. This additional width is so far unexplained but has been confirmed[39] and could well be related to the spin-pairing process. In this instance we have strong evidence[165] for weak interactions between solvated electrons and solvated cations, and strong evidence against any appreciable concentrations of species exhibiting strong electron–cation interactions.

III.C Electron–anion interactions

We have discussed the representation of interionic interactions in solution of solvated electrons in terms of an ion-pairing approximation in which interactions between ions of like charge are ignored. One way to probe an ion-pairing equilibrium (eqn. (22)) is to constrain it by adding an electrolyte with a common ion; in this case the alkali halides are an obvious choice (we mentioned their use earlier as structure-breakers during freezing experiments. In a detailed study of the effect of added electrolytes on electron spin resonance spectra of dilute solutions of alkali metals in ammonia[32,166,167] a surprising result emerged. Spectra showed negligible dependence upon the nature of the alkali metal cations added but a most marked dependence upon the nature of the anion. The results are illustrated in Fig. 8 for the system $K–KI–NH_3$, and a strong line broadening considerably in excess of viscosity increases is paralleled by a large shift of the resonance signal to higher magnetic fields (lower g-factor). Both these effects are absent for solutions containing added chlorides and amides, and much less marked for solutions containing bromides. Closely similar effects were found for sodium, rubidium and cesium solutions. Iodine NMR studies[39,168] reveal a considerable Knight shift of the iodine resonance for both potassium[168] and cesium[39] solutions confirming the presence of unpaired electron spin density on iodine. These results can only be accommodated by postulating a

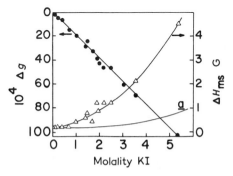

Fig. 8.   Changes in electron spin resonance parameters for solution of potassium in ammonia containing added KI (195K)[39]. $\bullet:\Delta g = g(K–NH_3) - g(K–KI–NH_3)$; $\Delta:\Delta H_{ms}$. Line a calculated for $\Delta H_{ms} \propto T_{2e}^{-1} \propto \eta$ normalized at zero salt concentration.

strong interaction between solvated electrons and solvated anions, the effect being most marked for iodine because of its nuclear properties. An interaction within ions triples[167]

$$M^+_{solv.} \; e^-_{solv.} + I^-_{solv.} \rightleftharpoons e^-_{solv.} \; M^+_{solv.} \; I^-_{solv.} \qquad (27)$$

was preferred to a simple electron–anion interaction

$$e^-_{solv.} + I^-_{solv.} \rightleftharpoons e^-_{solv.} \; I^-_{solv.} \text{ or } I^{2-}_{solv.} \qquad (28)$$

for reasons of charge–charge repulsion and expected ratios of free and ion-paired iodide anions. The ionic quadruple $(e^- M^+ I^- M^+)$ would be an equally valid model.

The presence of iodide anions in frozen aqueous alkali metal hydroxide glasses suppresses the yield of trapped electrons to some extent, but the detection of a new, broader signal with a lower g-factor has been reported[118], suggesting that similar electron–anion interactions occur in the solid state systems produced by γ-radiolysis.

The electron–anion interaction can be rationalized in terms of electron delocalization and the structure of the solvation shells of anions and cations. Delocalization reduces charge–charge repulsion, whilst solvation shells polarized around anions form better electron traps than cation solvation shells.

III.D Impurity states in solid alkali metal–HMPA systems

In this section we take up the case of solvated electron–cation interactions which are too strong to be described as ion-pair-like. We have hinted at the possibility of such states in the analysis of metal Knight shifts in metal–ammonia solutions, but the results for solutions of europium in ammonia appear to demonstrate conclusively that such species are unimportant in ammonia although early work indicated otherwise[169–174]. We turn now to solutions of alkali metals in HMPA which in many ways are closely similar to those in ammonia with the exception of a relatively minor but metal-dependent band in the electronic absorption spectra. Fluid solutions, as we saw in Sect.II.A.2, yield only singlet electron resonance spectra, whilst we have mentioned the 6 G line attributed to isolated solvated electrons in very dilute, frozen solutions of lithium in HMPA. When we examine frozen solutions of the higher alkali metals in HMPA, we find that the dominant feature of the spectra is a metal dependent hyperfine multiplet indicating strong electron–cation interactions.

III.D.1 General features and analysis of spectra[175]

The magnitude of hyperfine splitting constants observed is high, and for sodium, rubidium and cesium solutions indicates such a strong coupling that separate electron and nuclear spin quantum numbers are no longer a valid description of the coupled system. A full treatment of the total spin angular momenta is required in the analysis of these spectra and an analytical expression for line positions is only possible for an experiment in which the magnetic field is held constant whilst the microwave radiation is swept through the resonance condition. Unfortunately, the only technically practicable experiment involves keeping the irradiating frequency constant whilst sweeping the magnetic field through resonance. As a result, analysis of the positions of the hyperfine lines is an iterative process and can lead to such apparent anomalies as the same line occurring at two different fields in a single recording of a spectrum. A correct analysis involves use of the Breit–Rabi expression[176]

Analysis of the dominant multiplets yields electron spin densities at the cation equivalent to 50% occupation (Na) to 70% occupation (Cs) of the outer metal s-orbital. These values are far in excess of anything that could be described as ion-pair-like, and are independent of temperature over a wide range in the solid state. Although the a-factors are approaching those of the free (gaseous) alkali metal atoms, the g-factors deviate markedly from free atom values, indicating that the matrix interaction responsible for some 30–40% of the unpaired electron spin density is also contributing considerable orbital angular momentum to the system; whilst the dependence of the g-factors upon the nature of the alkali metal identifies the states having orbital angular momentum with excited p-states on the metal. Another obvious feature of the spectra is the marked variations in line width of the various hyperfine components, and a consideration of the gradients of the functions suggest that a very reasonable origin of this effect lies in a variation of a-factor from one species to another. In fact, we have been able to derive a distribution function for the variation of the a-factor from this information.

Since we are clearly not studying weak ion-pair interactions in these systems, we base our approach upon the solvated atom of Becker et al.[169], i.e. an orbital comprising appreciable metal s-orbital character but delocalized over several shells of solvent molecules surrounding the metal center. In line with current nomenclature in solid state physics we describe these states as impurity states and note in passing that neither the tight-binding nor the loose-binding (Mott–Wannier) descriptions are appropriate to 30–40% delocalization.

Closer examination of the spectra reveals the presence of a multitude of weaker lines. Careful analysis has revealed the presence of at least six impurity states, all centered on alkali metal nuclei, with s-electron densities from 30 to 80% free atom character. In general, g-factors tend away from the free atom value as the electron density in the metal outer s-orbital tends towards the free atom value. Again, line width variations are generally observed for different hyperfine components. The detailed results for all the states so far identified are presented in Table 4. Although spectra are obviously dominated by one of the impurity states, it should be stressed, firstly that there is no more than an order of magnitude variation in the concentrations of all the impurity states recorded in Table 4, and secondly that the total concentration of all these states represents only a small fraction ($<1\%$) of the total metal in solution. In addition, all spectra have two central singlet resonance absorptions.

III.D.2 A distribution function for the a-factor

Signal-to-noise ratios for spectra of impurity states are, in general, too high for very detailed line shape analysis, but by assuming distribution functions for the variation of the a-factors we are able to calculate electron resonance functions by a numerical convolution process with parameter optimization using a weighted least-

TABLE 4

Electron spin resonance parameters for species showing resolved metal hyperfine splitting in frozen alkali metal–HMPA solutions

Isotope	A_{iso} (G)	% Atomic character	$\lvert\Psi(0)\rvert^2_M \times 10^{-24}$ (electrons cm^{-3})[b]	g-factor (± 0.0004)	$\Delta g \times 10^4$ ($g_{\text{free atom}} - g_{\text{obs}}$)	Sample[a]
^{23}Na	196.7	62.2	3.15[c]	2.0018	5	Na-3, P
	150.7	47.7	2.41	2.0016	7	Na-2, P
^{39}K	54.9	66.7	4.98[d]	2.0005	18	K-4, P
	30.1	36.5	2.73	2.0005	18	K-4, P
^{85}Rb	277.3	76.8	12.15	1.9982	42	Rb-1, P
	251.9	69.8	11.04[e]	1.9985	39	
	233.1	64.5	10.22	1.9992	32	
	210.0	58.2	9.22	1.9985	39	
	179.1	49.6	7.85	1.9998	26	
	253.0	70.1	11.09	1.9982	42	Rb-2, Q
	212.8	59.0	9.33	1.9988	36	
^{87}Rb	849.0	69.6	10.98[f]	1.9980	44	Rb-1, P
	706.3	57.9	9.13	1.9986	38	
	857	70.3	11.08	1.9992	44	Rb-2, Q
	718.4	59.0	9.30	1.9972	52	
^{133}Cs	605.7	73.8	19.54[g]	1.9856	170	Cs-3, Q
	560	68.3	18.07	1.9856	170	
	493.6	60.2	15.90	1.9836	190	
	~655	80	21.1	~1.986	–	
	605	73	19.5	~1.986	–	
	595	72	19.2	~1.986	–	
	555	68	17.9	~1.986	–	

[a] P, Pyrex; Q, Quartz cell; [b] Typical error estimates; [c] ± 0.03; [d] ± 0.02; [e] ± 0.05; [f] ± 0.05; [g] ± 0.07.

squares approach. We find that for a Gaussian distribution of a-factors, we have a four parameter model (a, g, and δa, the variance of the a-factor and ΔH(residual), a line width contribution constant for all hyperfine components) capable of a very precise simulation of the observed electron spectra. A comparison of observed and calculated spectra is given in Fig. 9. Furthermore, we find values of δa which, when expressed as a variance of percent atomic character, are almost independent of metal or atomic character, whilst the residual line width is indistinguishable from that observed for electron–solvent interactions in solid state spectra of isolated electrons in HMPA. It should be stressed that this analysis is particularly well conditioned and that the solution vector represents a deep, clearly defined minimum in the least-squares surface. Although the distribution functions for particular impurity states are considerably greater than observed for normal radical species, the clear resolution between different states indicates that there is a set of very clearly defined energy minima for the atom–HMPA system which must indicate strictly

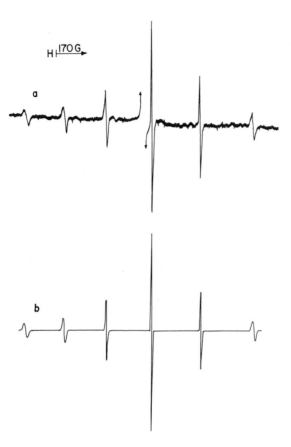

Fig. 9. Comparison of (a) observed and (b) simulated electron spin resonance spectra for the dominant species in frozen rubidium–HMPA solution at 77 K.

preferred values of a geometry coordinate describing the solvent polarization around the trapped atom.

Throughout this discussion no mention has been made of the possibility of a distribution of g-factors. In the regions with which we are concerned, all hyperfine components shift almost equally and linearly to lower field with increasing g-factor and hence a g-distribution cannot make any contribution to line width variations.

III.D.3 Spin lattice relaxation

The microwave power-dependence of the hyperfine components of the solvated atom spectra show only minor differences despite large variations in line width (i.e. for a given metal), but vary strongly from one metal to another. This implies that the dominant mechanism whereby electrons raised to the higher spin state by absorbing energy from the radiation field can return to the lower spin state is via a spin-orbit coupling to the alkali atom. The theory of microwave power saturation for inhomogeneously broadened lines was developed by Portis[177] and by Castner[178] and is summarized by the following relation

$$V_R = \frac{X}{(1 + X^2)^{1/2}} \exp(a^2 X^2) \frac{\{1 - \mathrm{erf}(a(1 + X^2)^{1/2})\}}{[1 \quad \mathrm{erf}(a)]}$$

where V_R is the signal amplitude at a reduced microwave field, $X = H_1/H_{1/2}$, H_1 is the amplitude of the magnetic component of the applied radiation field and $H_{1/2}$ is the microwave field corresponding to a saturation parameter of one half, and a is the ratio of spin packet width to envelope width. Curves calculated using this relation agree well with saturation curves observed for different hyperfine components. The magnitude of the spin lattice relaxation times calculated for the different alkali atoms for a spin-orbit mechanism are in close agreement with those observed.

III.D.4 Theoretical description for intermediate binding states

The species described above involve an almost equal distribution of the unpaired electron between the impurity donor atom and the matrix. Obviously this situation cannot be treated as perturbations of states characteristic of either the donor or the matrix.

A similar situation is encountered in models for P-centers in silicon crystals, but in this case the periodicity of the silicon lattice greatly simplifies the inclusion of large numbers of silicon atoms in the calculation. In the alkali metal–HMPA system, however, we are dealing with a disordered medium lacking in any long range order. The closest approximation so far is probably the description of the monomer in solution of alkali metals in liquid ammonia developed by O'Reilly[171,179] and employed in later descriptions of solvated electrons (see Chap. 1). In this approximation the center is treated as a microcrystal (e.g. $Na(NH_3)_6$) embedded in a matrix characterized by high and low frequency dielectric constants. In this respect the model is in accord with current descriptions of ideal glasses, high short range order combined with negligible long range order. However, since bonding between HMPA molecules and almost-free alkali atoms may not be very strong, the replacement of a perfect microcrystal by less ordered central regions (e.g. steric effects and solvent–solvent dipole repulsions overcoming radial charge dipole interactions) is indicated.

All the impurity states described above involve states in which the electron is approximately equally distributed between the donor impurity atom and the matrix. The results imply a considerable solvation interaction between neutral alkali atoms and the polar solvents, HMPA. It is of interest to compare these interactions with those observed in a non-polar matrix such as a solid rare gas. The alkali atom matrix interactions in rare gas solids are much weaker, and there appears to be little direct bonding of the rare gas atoms to the central metal atom[180–182].

III.E Mott–Wannier states in solid alkali metal–HMPA systems

All electron resonance spectra of solid solutions of sodium, potassium, rubidium and cesium in HMPA have, in addition to the hyperfine multiplets discussed above, a central region comprising two singlet resonances. One of these is very narrow and

TABLE 5

Electron spin resonance parameters for Wannier–Mott impurity ground states in frozen solutions of potassium, rubidium and cesium in HMPA at 77 K

Isotope	Nuclear spin (I)	ΔH_{ms} (G)	g-factor	A_{iso} (M)[a] (G)	% Atomic character	$|\Psi(0)|^2_M \times 10^{24}$
^{39}K	3/2	4.9 ± 0.4	2.0018 (± 0.0004)	0.8	0.97	0.073 (± 0.002)
^{85}Rb	5/2	9.4 ± 0.8	2.0008	1.48	0.47	0.065 (± 0.006)
^{133}Cs	7/2	14.5 ± 1.0	2.0000	1.85	0.23	0.060 (± 0.008)

[a]Obtained from computer simulation of spectra.

independent of metal, and is discussed below. The other singlet is much broader, of Gaussian shape, and has a line width dependent upon the alkali metal nucleus. Assuming that this singlet is an unresolved envelope of hyperfine interactions with a single alkali nucleus, and using the electron–solvent residual width (~ 4.2 G) found previously for both isolated solvated electrons and solvated atoms allows a second moment line shape analysis which yields ultimately the hyperfine coupling constant to the alkali metal nucleus and hence the unpaired electron density at the alkali metal donor (Table 5). A simple chemical picture predicts an approximately constant atomic character for the various alkali nuclei, but the results show that it is the unpaired electron spin density that is constant. In other words, the value of the square of the wave function at the origin of the donor atom is very small and independent of the nature of the donor atom. We must conclude that this impurity state is primarily a characteristic of the matrix with only a minor perturbation resulting from orbitals on the donor impurity atom. These states are thus the closest ground state approximations to Mott–Wannier loose-binding states that have been studied to date.

The lack of any significant parentage in the states of the impurity donor leads to speculation about the degree of delocalization of the electron in these states. Presumably the electron must be smeared out over a relatively large volume of the matrix.

IV. Interactions between excess electrons

On several occasions we have noted an electron spin-pairing process that is operative in metal solutions. In this section we first of all document the extent of these processes—this data was used extensively in Sect. II as the spin-unpaired fraction α at a concentration R. Secondly, we consider the spin-pairing process in its own right and review models for the spin-pairing mechanism. Finally, we consider mutual interactions among large numbers of solvated electrons and consider the formation of narrow impurity bands which contribute significantly to magnetic resonance properties at concentrations as low as 0.01 M.

IV.A Electron spin-pairing of solvated electrons

The spin-pairing process shows up as a drop in the molar paramagnetic susceptibility below the value expected for a metal dissociation process leading to one
unpaired electron per alkali metal atom in solution. Measurements of magnetic
susceptibility have been made by both static methods (Gouy balance) and by
integration of areas under electron resonance spectra.

Solutions of potassium in ammonia[11,30,183] are the only systems for which we have
extensive and reliable data: the spin-unpaired fraction at room temperature is
given in Fig. 10. Data for sodium solutions are more fragmentary and also to some
extent contradictory, but the most recent results[184] indicate susceptibilities close
to those of potassium solutions. Data for other metals are almost non-existent
although it is clear that the spin-pairing process sets in at similar concentrations
for barium[183] and at much lower concentrations in calcium[183] and europium[164]
solutions.

In all cases studied the spin-pairing process is strongly temperature-dependent,
pairing being favoured by low temperatures. Spin-unpaired fractions in HMPA[51]
are almost an order of magnitude lower, whilst values in primary aliphatic amines
and THF are again orders of magnitude lower[52].

Electronic absorption spectra of solutions of alkali metals in liquid ammonia
show only a very slight shift[156] to low energy throughout the spin-pairing concentration region. To a very good approximation[155] the electronic transition is unperturbed
by the spin-pairing process. This observation led Gold et al.[155] to propose that
pairing of electron spins occurred within ionic clusters such as the quadruple
$(M^+ e^-)_2$.

Solutions of europium metal in ammonia[158], however, manage to pair the solvated
electron spins very extensively, presumably in ionic triples $e^- Eu^{2+} e^-$ without any
appreciable interaction with the *f*-shell of the cation. This observation led to a
resurrection of the concept of solvated electron pairs[185], two electrons of opposite
spin trapped in the same solvation well. Neither model is universally accepted and

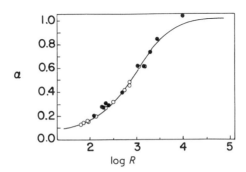

Fig. 10.   Fraction α of unpaired electron spins at 298 K in potassium–ammonia solutions.

indeed there are stong arguments[154] against any spin-paired species in a thermo-dynamic sense. One important point[185] is that any spin-pairing process

$$e_\alpha^- + e_\beta^- \rightleftharpoons e_2^= \tag{30}$$

must necessarily limit the lifetime of unpaired spin states although electron exchange processes such as

$$e_\alpha^- + e_2^= \rightleftharpoons e_2^= + e_\alpha^- \tag{31}$$

will have no observable effect. From the electron relaxation rate and relative con-centrations of metal solute and unpaired electrons we can estimate a lower limit to the lifetime of the spin-paired species of about 10^{-5} s. This contrasts markedly with the lifetime of 10^{-12} s found for the ion-pair $M^+ e^-$.

There is a considerable and growing body of information and concensus of opinion that the metal-dependent electron absorption band in solutions of alkali metals in amines and ethers arises from solvated alkali anions. The evidence for these species is certainly stronger than that for any particular model for the spin-paired species in liquid ammonia. Indeed, Dye et al.[186] have recently isolated crystalline com-pounds containing alkali cations solvated by polycyclic ethers and unsolvated alkali anions. Although the metallic nature of these compounds suggests that the valence shell electrons are perhaps better described as conduction electrons, nuclear resonance studies have confirmed beyond reasonable doubt the existence of Na^-[187].

When alkaline ices are irradiated[118,188-190], the concentration of trapped electron centres first grows linearly with dose, then levels off and finally decreases again to near zero for very high radiation doses. Similar observations have been recorded in hydrocarbon glasses[191]. When trapped electrons in alkaline ices are photolyzed with visible light, the signal again disappears. In both cases, warming to ~ 100 K causes a return of the blue colour and the electron resonance signal from solvated electrons. These changes have been interpreted in terms of the formation and thermal ionization of trapped electron pairs.

IV.B Impurity band formation

A problem we have repeatedly by-passed in earlier sections has been the persis-tance of motional narrowing of electron spin resonance spectra of solvated electrons into the solid state. Solutions of alkali metals in both ammonia[64,65] and HMPA[66] show this property, although the crystallization problem has restricted its observa-tion in ammonia to relatively high metal concentrations (0.1–0.4 M) in the presence of added electrolytes. The situation of HMPA is much clearer; a time-averaged electron resonance signal is observed down to concentrations of $\sim 10^{-2}$ M. Below this concentration the averaging stops and we are left with spectra attributed to isolated solvated electrons. Since the solvent is effectively rigid some 200 degrees below its freezing point, we must associate the time-dependence of hyperfine interactions with electronic motion. It is proposed[66] that the solvated electron centers overlap sufficiently at 10^{-2} M to be best described as a narrow impurity band. Electronic motion within this narrow band is not thermally activated, but mobilities are several orders of magnitude lower than in the much wider bands characteristic of metallic states. Only at concentrations an order of magnitude higher ($\sim$0.4 M) does the mobility of the electrons approach the value required to yield electron resonance spectra characteristic of rapid spin diffusion in the relatively narrow skin depth of microwave penetration[192], whilst a full merger of the impurity band with the conduction band does not occur[193] until concentrations greater than $\sim$1 M.

V. Concluding remarks

Perhaps the most fundamental distinction between models for solvated electrons is the degree of confinement of the electron to a small volume element of space. A confined model employing a crude square well formalism has the advantage of extreme simplicity and describes some of the optical properties adequately[106]. Such a model is inappropriate to any discussion of magnetic resonance spectra, and one aim of this chapter has been to establish the degree of confinement of the solvated electron. Although data are extremely scanty for serious analysis, we were able to draw some conclusions (Sect. II.C).

In addition, we have shown that solvated electrons can undergo ionic aggregation interactions typical of normal solvated ions, but that they are also capable of more specific interactions to yield impurity states. Throughout the chapter we have attempted to stress the close similarities between excess electrons in fluid or solid media, whether produced by ionization of alkali metals or by radiolysis methods, and the essential unity of the subject is amply demonstrated.

Acknowledgements

I would like to record my debt to many colleagues and mentors, particularly Professors M. C. R. Symons and J. L. Dye, whose stimulating arguments have helped to develop many of the ideas here. However, all mistakes and misconceptions are mine alone. Finally I would like to thank my colleague Dr. P. P. Edwards for innumerable discussions on these topics over the last three years.

References

1 A. Carrington and A. D. McLachlan, Introduction to Magnetic Resonance, Harper and Row, New York, 1957.
2 M. A. Bonin, K. Tsuji and F. Williams, Nature, (London), **218** (1968) 946.
3 R. J. Egland and M. C. R. Symons, J. Chem. Soc., A, (1970) 1326.
4 B. S. Gourary and F. J. Adrian, Solid State Phys., **10** (1960) 127.
5 J. J. Markham, F-Centers in Alkali Halides, Academic Press, New York, 1966.
6 J. H. Schulman and W. D. Compton, Color Centers in Solids, Pergamon, London, 1963.
7 G. Feher, Phys. Rev., **114** (1959) 1219.
8 A. G. Milnes, Deep Impurities in Semi-conductors, Wiley, New York, 1973.
9 G. Baldini and R. S. Knox, Phys. Rev. Lett., **11** (1963) 127.
10 W. Kohn, Solid State Phys., **5** (1957) 257; B. Raz and J. Jortner, Proc. Roy. Soc., Ser. *A*, **317** (1971) 113.

11 C. A. Hutchison, Jr. and R. C. Pastor, J. Chem. Phys., **21** (1953) 1959.

12 V. L. Pollak and R. E. Norberg, Bull. Amer. Phys. Soc., **1** (1956) 397; V. L. Pollak, J. Chem. Phys., **34** (1961) 864.

13 D. Cutler and J. G. Powles, Proc. XI Colloq. Ampere, (1962) 147.

14 D. Cutler and J. G. Powles, Proc. Phys. Soc., **80** (1962) 130.

15 D. Cutler and J. G. Powles, Proc. Phys. Soc., **82** (1963) 1.

16 C. A. Hutchison, Jr. and R. C. Pastor, Phys. Rev., **81** (1951) 282.

17 C. A. Hutchison, Jr. and R. C. Pastor, Rev. Mod. Phys., **25** (1953) 285.

18 C. A. Hutchison, Jr., J. Phys. Chem., **57** (1953) 546.

19 G. Feher and R. A. Levy, Phys. Rev., **98** (1955) 264.

20 R. A. Levy, Phys. Rev., **102** (1956) 31.

21 W. Wysocyanski and J. A. Cohen, Bull. Amer. Phys. Soc., **2** (1957) 318.

22 A. A. Galkin, Ia. L. Shamfarov and A. V. Stefanishina, Zh. Eksp. Teo. Fiz., **32** (1957) 1581, translation in J. Exp. Theor. Phys. **6** (1958) 1291.

23 A. Charru, Compt. Rend., **246** (1958) 3445; **247** (1958) 195; Ann. Phys. (Paris), **5** (1960) 1449.

24 M. A. Garstens and A. H. Ryan, Phys. Rev., **81** (1951) 888.

25 E. C. Levinthal, E. H. Rogers and R. A. Ogg, Jr., Phys. Rev., **83** (1953) 182.

26 T. R. Carver and C. P. Slichter, Phys. Rev., **102** (1956) 975.

27 R. Beeler, Arch. Sci., **10** (1957) 186.

28 R. J. Blume, Bull. Amer. Phys. Soc., **1** (1956) 397; Phys. Rev., **109** (1958) 1867.

29 G. Ebert, Z. Anorg. Allg. Chem., **294** (1958) 129.

30 C. A. Hutchison, Jr. and D. E. O'Reilly, J. Chem. Phys., **34** (1961) 1279.

31 O. Stirand, Czech. J. Phys., Sect. **B11** (1961) 72; Sect. **B12** (1962) 207.

32 R. Catterall and M. C. R. Symons, J. Chem. Soc., (1964) 4342.

33 C. Lambert, J. Chem. Phys., **48** (1968) 2389.

34 H. M. McConnell and C. H. Hohn, J. Chem. Phys., **26** (1957) 1517.

35 J. Acrivos and K. S. Pitzer, J. Phys. Chem., **66** (1962) 1693.

36 D. E. O'Reilly, J. Chem. Phys., **41** (1964) 3729.

37 T. R. Hughes, Jr., J. Chem. Phys., **38** (1963) 202.

38 R. W. Coutant, Ph.D. thesis, Purdue University, 1963.

39 R. Catterall, in J. J. Lagowski and M. J. Sienko (Eds.), Metal–Ammonia Solutions, Butterworths, London, 1970, p. 105.

40 R. A. Newmark, J. C. Stephenson and J. S. Waugh, J. Chem. Phys., **46** (1967) 3514.

41 J. Stoh and T. Takeda, Bull. Phys. Soc. Jap., **18** (1963) 1560.

42 A. L. Cederquist, Ph.D. thesis, Washington University, 1963.

43 R. A. Pinkowitz and T. J. Swift, J. Chem. Phys., **54** (1971) 2858.

44 C. Lambert, in J. Jortner and N. R. Kestner (Eds.), Electrons in Fluids, Springer-Verlag, Berlin, 1973, p. 57.

45 R. Catterall, Nature Phys. Sci., **229** (1971) 10.

46 D. W. G. Smith and J. G. Powles, Mol. Phys., **10** (1966) 451.

47 D. E. O'Reilly, J. Chem. Phys., **50** (1969) 4743.

48 D. E. O'Reilly, Phys. Rev. Lett., **11** (1963) 545.

49 Y. Nakamura, M. Yamamoto, S. Shimokawa and M. Shimoji, Bull. Chem. Soc. Jap., **44** (1971) 3212.

50 R. Catterall, M. C. R. Symons and J. W. Tipping, in J. J. Lagowski and M. J. Sienko (Eds.), Metal–Ammonia Solutions, Butterworths, London, 1970, p. 317.

51 R. Catterall, L. P. Stodulski and M. C. R. Symons, in J. J. Lagowski and M. J. Sienko (Eds.), Metal–Ammonia Solutions, Butterworths, London, 1970, p. 151.

52 J. Slater, Ph. D. thesis, University of Leicester, 1970.

53 R. Catterall, L. P. Stodulski and M. C. R. Symons, J. Chem. Soc., A, (1967) 437.

54 L. P. Stodulski, Ph.D. Thesis, University of Leicester, 1969.

55 M. Clark, A. Horsfield and M. C. R. Symons, J. Chem. Soc., (1959) 2478.

56 R. Catterall and M. C. R. Symons, J. Chem. Soc., (1964) 4357.

57 R. Catterall, unpublished results.

58 R. Catterall, J. W. Tipping and M. C. R. Symons, J. Chem. Soc., A, (1966) 1529.

88

59 R. Catterall, J. Slater and M. C. R. Symons, in J. J. Lagowski and M. J. Sienko (Eds.), Metal–Ammonia Solutions, Butterworths, London, 1970, p. 329.
60 H. Linschitz, M. G. Berry and D. Schweitzer, J. Amer. Chem. Soc., 76 (1954) 5833.
61 R. Catterall, J. W. Tipping and M. C. R. Symons, unpublished results.
62 R. Catterall, I. Hurley and M. C. R. Symons, J. Chem. Soc. Dalton Trans., (1972) 139.
63 R. Catterall and P. P. Edwards, J. Chem. Soc. Chem. Commun., (1975) 96.
64 R. Catterall, Phil. Mag., 22 (1970) 779.
65 R. Catterall, W. T. Cronenwett, R. J. Eglard and M. C. R. Symons, J. Chem. Soc., A, (1971) 2396.
66 R. Catterall and P. P. Edwards, J. Phys. Chem., (1975) Dec.; Chem. Commun., (1975) 96.
67 R. Catterall and P. P. Edwards, Advan. Mol. Relaxation Processes, 7 (1975) 87.
68 D. Schulte-Frohlinde and K. Eiben, Z. Naturforsch. A, 17 (1962) 445.
69 R. Catterall, unpublished results.
70 E. Bosch, Z. Phys., 137, (1954) 89.
71 G. Beuermann, Z. Phys. 236 (1970) 70.
72 J. E. Bennett, B. Mile and A. Thomas, Nature (London), 201 (1964) 919.
73 J. E. Bennett, B. Mile and A. Thomas, J. Chem. Soc., A, (1967) 1393.
74 R. Marz, S. Leach and M. Horani, J. Chim. Phys., 60 (1963) 726.
75 D. R. Smith and J. J. Pieroni, Can. J. Chem., 45 (1967) 2723.
76 C. Chachaty, J. Chim. Phys., 64 (1967) 614.
77 H. Hase, M. Noda and T. Higashimura, J. Chem. Phys., 54 (1971) 2975.
78 J. H. Baxendale and P. Wardman, Chem. Commun., (1971) 429.
79 J. T. Richards and J. K. Thomas, J. Chem. Phys., 53 (1970) 218.
80 N. V. Klassen, H. A. Gillis and D. C. Walker, J. Chem. Phys., 55 (1971) 1971.
81 C. Chachaty and E. Hayon, Nature (London), 200 (1963) 59.
82 C. Chachaty, Compt. Rend., 259 (1964) 2219.
83 C. Chachaty and E. Hayon, J. Chim. Phys., 61 (1964) 1115.
84 L. Shields, J. Phys. Chem., 69 (1965) 3186.
85 M. J. Blandamer, L. Shields and M. C. R. Symons, J. Chem. Soc., London (1965) 1127.
86 K. Tsuji and F. Williams, J. Phys. Chem., 72 (1968) 3884.
87 A. Ekstrom and J. E. Willard, J. Phys. Chem., 72 (1968) 4599.
88 D. R. Smith and J. J. Pieroni, Can. J. Chem., 42 (1964) 2209.
89 D. R. Smith and J. J. Pieroni, Can. J. Chem., 43 (1965) 876.
90 F. S. Dainton and G. A. Salmon, Proc. Roy. Soc., A, 285 (1965) 319.
91 K. Fueki, Z. Kuri and H. Goto, Kogyo Kagaku Zasshi, 68 (1965) 1544.
92 K. Tsuji, H. Yoshida and K. Hayashi, J. Chem. Phys., 46 (1967) 810.
93 J. Lin, K. Tsuji and F. Williams, J. Amer. Chem. Soc., 90 (1968) 2766.
94 M. Gric, K. Hayashi, S. Okamura and H. Yoshida, J. Chem. Phys., 48 (1968) 922.
95 D. R. Smith and J. J. Pieroni, J. Phys. Chem., 70 (1966) 2379.
96 D. R. Smith, F. Okenka and J. J. Pieroni, Can. J. Chem., 45 (1967) 833.
97 K. Tsuji and F. Williams, J. Amer. Chem. Soc., 89 (1967) 1526.
98 J. Lin, K. Tsuji and F. Williams, J. Chem. Phys., 46 (1967) 4982.
99 J. Lin, K. Tsuji and F. Williams, Chem. Phys. Lett., 1 (1967) 66.
100 H. Hase, J. Phys. Soc. Jap., 24 (1968) 589.
101 F. S. Dainton, G. A. Salmon and J. Teply, Proc. Roy. Soc., Ser. A, 286 (1965) 27.
102 W. T. Cronenwett and M. C. R. Symons, J. Chem. Soc., A, (1968) 2991.
103 H. Tsjuikawa, K. Fueki and Z. Kuri, J. Chem. Phys., 47 (1967) 256.
104 A. F. Kip, C. Kittel, R. A. Levy and A. M. Portis, Phys. Rev., 91 (1953) 1066.
105 K. Eiben and I. A. Taub, Nature (London), 216 (1967) 782.
106 M. J. Blandamer, R. Catterall, L. Shields and M. C. R. Symons, J. Chem. Soc., (1964) 4357.
107 K. V. S. Rao and M. C. R. Symons, J. Chem. Soc. Faraday II, 68 (1972) 2081.
108 R. Catterall and P. P. Edwards, unpublished work.
109 B. G. Ershov and A. K. Pikaev, Radiat. Res. Rev., 2 (1969) 1.

110 M. C. R. Symons, in J. J. Lagowski and M. J. Sienko (Eds.), Metal–Ammonia Solutions, Butterworths, London, 1970, p. 309.

111 M. C. R. Symons, unpublished work.

112 V. N. Shubin, V. A. Zhigunov, V. I. Zolotarevsky and P. I. Dolin, Nature (London), 212 (1966) 1002.

113 B. G. Ershov and A. K. Pikaev, Khim. Vys. Energ., 1 (1967) 29.

114 W. R. Elliott, Science, 157 (1967) 558.

115 J. Jortner and B. Scharf, J. Chem. Phys., 37 (1962) 2506.

116 M. J. Blandamer, L. Shields and M. C. R. Symons, Nature (London), 199 (1963) 902.

117 B. G. Ershov, A. K. Pikaev, P. Ya. Glazunov and V. I. Spitsyn, Dokl. Akad. Nauk SSSR, 149 (1963) 363.

118 M. J. Blandamer, L. Shields and M. C. R. Symons, J. Chem. Soc., (1964) 4352.

119 B. G. Ershov, A. K. Pikaev, P. Ya. Glazunov and V. I. Spitsyn, Izvest. Akad. Nauk S.S.R., Ser. Khim., 10 (1964) 1755.

120 T. Henriksen, Radiat. Res., 23 (1964) 63.

121 P. N. Moorthy and J. J. Weiss, Phil. Mag., 10 (1964) 659.

122 K. Eiben and D. Schulte-Frohlinde, Z. Phys. Chem., 45 (1965) 20.

123 L. Kevan, J. Phys. Chem., 69 (1965) 1081.

124 L. Kevan, J. Amer. Chem. Soc., 87 (1965) 1481.

125 P. N. Moorthy and J. J. Weiss, Adv. Chem., 50 (1965) 180.

126 J. Zimbrick and L. Kevan, J. Amer. Chem. Soc., 88 (1966) 3678.

127 W. A. Seddon and D. R. Smith, Can. J. Chem., 45 (1967) 3083.

128 J. Zimbrick and L. Kevan, J. Chem. Phys., 47 (1967) 2346.

129 B. G. Ershov, O. Ya. Grindberg and Ya. S. Lebedev, Zh. Strukt. Khim., 9 (1968) 694.

130 G. B. Ershov and A. K. Pikaev, Adv. Chem., 81 (1968) 1.

131 O. F. Khodzhaev, B. G. Ershov and A. K. Pikaev, Izvest. Akad. Nauk SSSR, Ser. Khim., 10 (1967) 2253.

132 B. G. Ershov, O. F. Khodzhaev and A. K. Pikaev, Dokl. Akad. Nauk SSSR, 179 (1968) 911.

133 B. G. Ershov and A. K. Pikaev, Izvest. Akad. Nauk SSSR, Ser. Khim., 9 (1966) 1637.

134 H. Brazynski and D. Schulte-Frohlinde, Z. Naturforsch. A, 22 (1967) 2131.

135 B. G. Ershov, O. F. Khodzhaev and A. K. Pikaev, Khim. Vys. Energ., 2 (1968) 42.

136 B. G. Ershov and A. K. Pikaev, Khim. Vys. Energ., 1 (1967) 29.

137 B. G. Ershov, O. F. Khodzhaev and A. K. Pikaev, Khim. Vys. Energ., 2 (1968) 35.

138 L. T. Bugaenko and V. N. Belenskii, Zh. Fiz. Khim., 39 (1965) 2958.

139 B. G. Ershov and A. K. Pikaev, Izvest. Akad. Nauk SSSR, Ser. Khim., 2 (1966) 386.

140 L. T. Bugaenko and O. S. Povolotskaya, Dokl. Akad. Nauk SSSR, 174 (1967) 378.

141 B. G. Ershov and A. K. Pikaev, Zh. Fiz. Khim., 41 (1967) 2573.

142 R. J. Egland and M. C. R. Symons, private communication.

143 J. L. Dye, M. G. DeBacker and L. M. Dorfman, J. Chem. Phys., 52 (1970) 6251.

144 K. Bar-Eli and T. R. Tuttle, Jr., J. Chem. Phys., 40 (1964) 2508.

145 R. Catterall, J. Slater and M. C. R. Symons, J. Chem. Soc., in press.

146 W. H. Hamill, in L. Kevan and E. T. Kaiser (Eds.), Radical Ions, Wiley-Interscience, New York, 1973, p. 321.

147 J. D. W. Van Voorst and G. J. Hoijtink, J. Chem. Phys., 42 (1965) 3995.

148 P. B. Ayscough, R. G. Collins and F. S. Dainton, Nature (London), 205 (1965) 965.

149 P. Wardman and W. A. Seddan, Can. J. Chem., 47 (1969) 2155.

150 C. A. Kraus, J. Amer. Chem. Soc., 30 (1908) 1323.

151 R. Catterall, in G. Lepoutre and M. J. Sienko (Eds.), Metal–Ammonia Solutions, Benjamin, New York, 1964, p. 41.

152 See Chapter 1.

153 T. R. Tuttle, Jr. and P. Graceffa, J. Phys. Chem., 75 (1971) 843; M. C. R. Symons, J. Phys. Chem., 75 (1971) 3904; T. R. Tuttle, Jr. and P. Graceffa, J. Phys. Chem., 75 (1971) 3905.

154 J. L. Dye, in J. J. Lagowski and M. J. Sienko (Eds.), Metal–Ammonia Solutions, Butterworths, London, 1970, p. 1.

155 M. Gold, W. L. Jolly and K. S. Pitzer, J. Amer. Chem. Soc., **84** (1962) 2264; M. Gold and W. L. Joly, Inorg. Chem., **1** (1962) 818.
156 I. Hurley, T. R. Tuttle, Jr. and S. Golden, in J. J. Lagowski and M. J. Sienko (Eds.), Metal–Ammonia Solutions, Butterworths, London, 1970, p. 503.
157 E. C. Evers and P. W. Frank, Jr., J. Chem. Phys., **30** (1959) 61.
158 R. Catterall and M. C. R. Symons, J. Chem. Soc., (1965) 3763.
159 J. E. Bennett, B. Mile and A. Thomas, J. Chem. Soc., A, (1969) 1502.
160 L. Kevan, J. Amer. Chem. Soc., **87** (1965) 1481.
161 C. Hallada and W. L. Jolly, Inorg. Chem., **2** (1963) 1076.
162 J. C. Warf and W. L. Korst, J. Phys. Chem., **60** (1956) 1590.
163 R. G. Pearson and T. Buch, J. Chem. Phys., **36** (1962) 1277.
164 D. S. Thompson, E. E. Hasen and J. S. Waugh, J. Chem. Phys., **44** (1966) 2954.
165 R. Catterall and M. C. R. Symons, J. Chem. Phys., **42** (1965) 1466.
166 R. Catterall, J. Corset and M. C. R. Symons, J. Chem. Phys., **38** (1963) 272.
167 R. Catterall and M. C. R. Symons, in G. Lepoutre and M. J. Sienko (Eds.), Solutions Metal–Ammonia, Benjamin, New York, 1964, p. 277.
168 D. E. O'Reilly, J. Chem. Phys., **50** (1969) 4320.
169 E. Becker, R. H. Lindquist and B. J. Alder, J. Chem. Phys., **25** (1956) 971.
170 W. E. Blumberg and T. P. Das, J. Chem. Phys., **30** (1959) 251.
171 D. E. O'Reilly, J. Chem. Phys., **41** (1964) 3736.
172 K. D. Vos and J. L. Dye, J. Chem. Phys., **38** (1963) 2033.
173 R. Catterall and M. C. R. Symons, J. Chem. Soc., (1965) 6656.
174 D. E. O'Reilly and T. Tsang, J. Chem. Phys., **42** (1965) 3333.
175 R. Catterall and P. P. Edwards, J. Phys. Chem., (1975).
176 G. Breit and I. I. Rabi, Phys. Rev., **38** (1931) 2082.
177 A. M. Portis, Phys. Rev., **104** (1956) 584.
178 T. G. Castner, Phys. Rev., **115** (1959) 1506.
179 R. H. Land and D. E. O'Reilly, J. Chem. Phys., **46** (1967) 4496.
180 C. K. Jen, V. A. Bowers, E. L. Cochran and S. N. Foner, Phys. Rev., **126** (1962) 1749.
181 J. P. Goldsborough and T. R. Kochler, Phys. Rev., **133** (1964) 135.
182 S. L. Kuffermann and F. M. Pipkin, Phys. Rev., **166** (1968) 207.
183 S. Freed and N. Sugarman, J. Chem. Phys., **11** (1943) 354.
184 A. Demortier, M. DeBacker, and G. Lepoutre, J. Chim. Phys., **69** (1972) 380.
185 R. Catterall and M. C. R. Symons, J. Chem. Soc., A, (1966) 13.
186 J. L. Dye, J. M. Ceraso, Mei Tak Lok, B. L. Barnett and F. J. Tehan, J. Amer. Chem. Soc., **96** (1974) 608.
187 J. L. Dye and J. M. Ceraso, J. Chem. Phys., **61** (1974) 1585.
188 J. Zimbrick and L. Kevan, J. Amer. Chem. Soc., **89** (1967) 2483.
189 O. F. Khodzhaev, B. G. Ershov and A. K. Pikaev, Bull. Acad. Sci. USSR, Div. Chem. Sci., **8** (1967) 1816.
190 L. Kevan, D. R. Renneke and R. J. Friauf, Solid State Commun., **6** (1968) 469.
191 M. Shirom and J. E. Willard, J. Amer. Chem. Soc., **90** (1968) 2184.
192 R. Catterall, J. Chem. Phys., **43** (1965) 2262.
193 R. Catterall and N. F. Mott, Advan. Phys., **18** (1969) 665.

Chapter 3

EXCITED STATES OF EXCESS ELECTRONS STUDIED BY FAST PHOTOABSORPTION METHODS

D. C. WALKER

I. Introduction

1.A The familiar problem

It was Henry Adams' contention[1] that by continually recasting a problem one gets closer to embracing its meaning. Perhaps that alone justifies this chapter. Photobleaching is a field in which ambiguity seems to lurk in the interpretation of nearly every new experiment, so that definitive conclusions and rationalisations are rare. However, one certainly should learn a lot about the nature of electron solvation— a central theme of this book—from photobleaching studies, since they are concerned with the origin of the absorption band and the character of the excited state.

Every time the problem is recast the familiar old questions are posed. For instance, in a review[2] which was written nine years ago the following points were discussed, or alluded to as questions, and consequently these must have been raging issues amongst solvated electron zealots of the time, which was three years after Hart and Boag[3] reported the hydrated electron's absorption band, one year after Jortner[4] gave a quantum-mechanical rationalisation for it, and concurrent with the measurement by Schmidt and Buck[5] of its diffusion coefficient:

(i) Does a single bound–bound optical transition account for the width and asymmetry of the solvated electron's absorption band?

(ii) Does a bound-free continuum (ionisation efficiency profile) contribute significantly to the band?

(iii) Does the medium provide a variety of trap-depths for electron localisation, thereby giving rise to a broad structureless absorption band merely as the envelope of individual sub-structure bands?

(iv) Will photobleaching experiments be able to decide between these possibilities?

(v) Can any of these pictures account for nearly identical spectra with a given medium in liquid, solid and highly concentrated solution, at low and high temperatures?

(vi) Is the electron stabilised at indigenous, pre-existing, sites ("trapped") or is most of the dipolar orientation induced after localisation ("solvated"), and is diffusive motion best described as tunnelling between suitable sites?

These same questions form the essence of this chapter. They are still not satisfactorily answered despite considerable efforts to seek solutions through photobleaching, photoconductivity and photoelectron emission studies. It is likely that many answers to each question will have to be accepted, different ones for electrons solvated in different classes of media. The time-resolved bleaching studies to be reviewed here force one, for the moment, towards this viewpoint, because the ramifications which would be required to encompass data from all media in a single model seem too prodigious.

I.B Ambiguities inherent in "steady-state" experiments

Several types of photoinduced effects have been studied in attempts to elucidate the origin of the optical absorption band of solvated electrons. Experiments involving low intensity illumination will be referred to as "steady-state" experiments when the illumination time is much longer than the mean lifetime of the individual chemical entities involved in the observed process. Measurements have been made of: photobleaching—loss of light-absorbing species; photoshuffling[6]—shifting of the absorption band, apparently due to exchange or relocation between absorbing states; photoconductivity—production of highly mobile charge states; photoelectron emission[7]—loss of electrons into the gas phase from illuminated surfaces.

Positive observations of any of these effects would be consistent with the optical transition being described as "bound-free", where the photoexcited state is a quasi-free (continuum) state (represented by a continuous wave function). However, each of these effects is also possible following a "bound–bound" transition, where the bound excited state implies that following excitation the electron is still localised in the same region of space. Bound excited states can have greatly enhanced "reactivity" and "mobility", relative to the ground state, for several reasons:

(i) the charge distribution of the excited state is predicted to be much wider than that of the ground state[4];

(ii) quantum-mechanical tunnelling is facilitated from an excited state because the potential barrier is lowered[8];

(iii) electrons may be thermally promoted to the conduction band;

(iv) autoionisation may occur, or

(v) the electron's binding trap may be "weakened" by the thermal shock resulting from de-excitation following photoexcitation.

For these reasons, interpretations of data in terms of "bound" or "free" excited states are subject to ambiguity. For instance, even photoshuffling in low temperature glasses can be qualitatively explained without invoking any form of electron migration, in terms of point (v). A bound–bound excitation, followed by de-excitation, causes a significant thermal shock to the molecules of the electron's trap due to dissipation of the photon energy. Suppose that such shocks can cause either deepening of shallow traps or weakening of deep traps—in analogous manner to the proposed variation with temperature of irradiation of trap-depth distributions[8]. When a sample of trapped electrons (which are assumed to be distributed in a range of trap-depths) is illuminated by light corresponding to the red end of the absorption band, this light will be absorbed predominantly by shallow traps so that the net effect must be a deepening of shallow traps and hence a blue-shift will be observed. Conversely, when blue light is used, deep traps dominate the absorption and the overall net effect can only be a weakening of deep traps and a red-shift. This issue is really beyond the scope of the present discussion since it is connected with the whole question of whether electrons are basically "diggers" or "seekers" with regard to spectral shifts. However, arguments have been presented, based on the influence of scavengers, that spontaneous (thermal) spectral shifts indicate a digging process[9].

Ambiguity concerning the electron's release does seem to have been abrogated, in at least one case, by detailed study of photoelectron emission[7]. Delahay shows how an analysis of the energy distribution of the photoelectrons emitted from the surface of an illuminated solution of electrons in hexamethyl phosphoric triamide (HMPA) reveals that the high energy region corresponds to photoionisation. This work and other aspects of photobleaching are reviewed in Chap. 4[7].

Photobleaching studies are particularly equivocal. Bleaching is registered in steady-state experiments when there is a net loss of absorbing species, or, succinctly, when photochemical decomposition occurs. Loss may be caused by chemical reactions of electrons with themselves, with "hole" centers (parent ions, etc.) or solutes (scavengers or impurities). However, optical excitation to bound, relaxed, relatively stable excited states may also be featured, and here there is no net loss and no need for electron release. The quantum yield for bleaching, ϕ_B, measures the efficiency of loss. It changes with wavelength, λ, generally increasing with decreasing λ, particularly on the high energy tail of the absorption band[10,11]. This could be indicative of: (a) an autoionising excited state, (b) a progressively increasing fraction of the extinction coefficient being from the photoionisation continuum[10,11], or (c) simply arising from the fact that electrons which carry off more excess kinetic energy from the absorbed photon have a greater chance of finding a reaction partner before being retrapped[12]. This latter point is a curious one because it has been argued[13] that in solvents of low dielectric constant at low temperatures the trapped electrons are geminate and consequently can only escape recombination (to become retrapped elsewhere) if they are released with excess kinetic energy. In this case ϕ_B should decrease with decreasing λ. Since this is not observed for α-methyltetrahydrofuran (MTHF)[10] it implies that the wavelength dependence of ϕ_B is not dominated by the excess kinetic energy factor.

Further complications can be anticipated when attempting to make grand correlations of photobleaching studies, because it is reported that ϕ_B, and the spectral shifts, are different at different low temperatures in several glasses[8,14,15]. Generally, trap-deepening processes are slowed down, weaker traps become more stable and the conduction band is less accessible, the lower the temperature and the harder the glass. Several of these uncertainties are circumvented by time-resolved studies using short, high intensity light pulses for photobleaching studies.

I.C Solvated electrons and their absorption bands

When placed in an unreactive, polarisable medium an excess electron can acquire a variety of possible energy states. Unfortunately, the working nomenclature is disorganised, for even "trapped" and "solvated" have widely differing meanings to different authors (compare, for instance, the common usage with those suggested in refs. 2 and 16). Most commonly, these adjectives simply distinguish mobility—a trapped electron being an immobile solvated electron. Disorder has led to the adaptation of colloquialisms for clarity (though Alice would disapprove[17]). Thus we have dry, damp and wet, ranging, respectively, from unsolvated (quasi-free) through different degrees of dampness (described variously as weakly solvated, shallowly trapped or IR-absorbing) to wet, the thoroughly solvated, thermal equilibrium state. In addition, of course, there are frozen (chilled in glasses), iced (in crystals) and, presumably, "aerosolled" (gaseous cluster[18]) electrons. (One sometimes wishes electron states in ethanol had been named first so that a more colourful and descriptive imagery would have emerged:) All these states, except dry, are localised (bound) states and hence are, to some extent, solvated. Thus "solvated electron" has to be used here occasionally as a *generic* term.

Some typical absorption bands are shown in Fig. 1. This figure includes electrons in KI crystals, liquid water, in both the wet and damp state for ethanol, in a hydrocarbon and in the aprotic dipolar solvent HMPA, at the temperatures given in the legend. There is evidently no correspondence between the photon energy of the band maximum ($E_{\lambda\text{max}}$) and the static dielectric constant of the medium[21]. In fact the striking correlation is between the magnitude of $E_{\lambda\text{max}}$ and the solvation energy

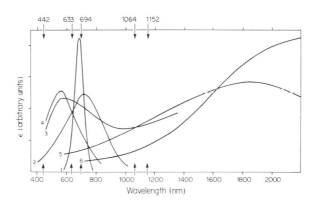

Fig. 1. Various "solvated" electron spectra. 1, F-centers in KBr crystals at 295 K[13]; 2, hydrated electrons in water at 295 K[13]; 3, "damp" electrons in liquid ethanol at 160 K measured at the end of a 5 ns high energy electron pulse[19] (contains significant contributions from "wet" electrons); 4, "wet" electrons in liquid ethanol at 160 K 20 ns after formation[19]; 5, trapped electron in 3 methylhexane glass at 76 K observed 8 ms after 40 ns pulse[26]; 6, solvated electron in liquid hexamethylphosphoric triamide at 295 K[20]. The extinction coefficients, ε, are on arbitrary and different scales for each media, so the relative magnitudes are not drawn to scale. Arrows indicate the positions of the laser lines utilised in laser-bleaching experiments.

(ΔH_s) for small negative ions (as indicated by parameters such as Y, a, Z and E_T^{13}). This was demonstrated in a study of dimethylsulphoxide which has a high dielectric constant of 46 but $E_{\lambda max}$ resembles that of a saturated hydrocarbon[21], as does its solvating power for anions. Thus the different values of $E_{\lambda max}$ for the wet and damp electrons presumably simply reflect the different solvation energies involved.

The progressive blue-shift of the spectrum accompanying the second step in the solvation process depicted by eqn. (1)

$$e^-_{(dry)} \rightarrow e^-_{(damp)} \rightarrow e^-_{(wet)} \tag{1}$$

is slowed down by reducing the temperature. For ethanol the lifetime of $e^-_{(damp)}$ is of the order of 10^{-11}, 10^{-8}, 10^{-6} and $>10^3$ s at 295 K[22], 160 K[19], 77 K[23] and 4 K[24], respectively. Spontaneous changes depicting this second step on various time scales have been observed in other media (see, for example, refs. 25 and 26).

Solvated electron absorption bands generally have oscillator strengths of about unity and a common shape, neither of which alters significantly on changing the phase from liquid to solid, nor the temperature, nor by the addition of inert solutes. In fact, the product of the extinction coefficient at the band maximum (ϵ_{max}) multiplied by the band half-width is nearly a constant despite variations in both individually[13]. Furthermore, mixtures of two solvents invariably show a single absorption band with $E_{\lambda max}$ intermediate between that of the pure solvents. When the mixtures are imperfect one solvent tends to dominate the spectrum, but for nearly perfect mixtures an approximately linear variation of $E_{\lambda max}$ with volume fraction seems to exist[20]. (Volume fraction generally seems to show a better correlation than mole fraction[27], as would be expected for perfect mixtures.) Vannikov and Marevtsev[28] have noted that solvated electrons in pure monoethanolamine have the same $E_{\lambda max}$ as those in a 50/50 mixture of ethanol and monobutylamine, a mixture with approximately the same overall composition of functional groups. It is interesting to note that the spectrum of the latter is much broader, perhaps reflecting a wider distribution of trap-depths due to there being a greater variety of polarised regions. Thus it may be revealing to study photobleaching in this solvent mixture.

Solvated electrons, for bleaching studies, have been produced by radiolysis, photolysis, or by dissolving an alkali metal in the solvent. They are often stable indefinitely in glassy media at 77 K or lower temperatures; but are reactive and short-lived in the liquid phase (except for some metal solutions when the equilibria involved can maintain a relatively high concentration for long periods of time).

II. Flash-photobleaching methods

II.A Advantages of flash methods

Just as pulse radiolysis enables one to identify intermediate states and directly measure the rates of the intervening reactions of radiation chemistry, so flash-photolysis methods contribute insight to the evolution of photobleaching processes. From the overall effects of steady-state photolysis, and the influence of scavengers, temperature changes, variation in light intensity, etc., on these, valid deductions have been made about the mechanisms involved. But such studies will commonly fail to reveal effects such as unproductive excitation processes, rapidly repairing events, transitory "hole-burning" which may be masked by subsequent relaxations and spectral shifts, or purely thermal effects proceeding quickly after photolysis. Some of these events can be deciphered by time-resolved bleaching studies. To do this one needs to inject, and have absorbed, a sufficient number of photons to significantly alter the number of species present in a time period which is short compared to the time scale of the reactions. This turns out to be rather difficult for solvated electrons because of the exceedingly short excited state lifetimes that they apparently have. It is a particularly tricky task when attempted with hydrated electrons at room temperature because the ground state species themselves are quite short-lived.

II.B Flash-photobleaching of hydrated electrons

This problem was attacked several years ago in a series of three types of experiments[29]. In each case photobleaching was to be identified as either a permanent or temporary decrease in the absorbance of a sample of hydrated electrons, or as a significantly enhanced rate of reaction with scavengers, upon exposure to an intense flash of light at a wavelength within the hydrated electron's absorption band. Scavengers were assembled which were thought to be more reactive towards quasi-free than towards fully solvated electrons[30].

Conventional flash-photolysis methods were used in the first attempt at hydrated electron bleaching[31]. A dilute aqueous solution of OH^- and H_2 was contained in a 50 cm-long reaction vessel placed between two flash-photolysis lamps. One lamp was separated from the solution by a space containing light filters which allowed transmission of visible light only. The lamps could be fired independently with any pre-selected delay time. Hydrated electrons were produced from OH^- in the first flash, by UV light at $\lambda < 240$ nm, and their concentration (10^{-7} to 10^{-6} M) was monitored at 633 nm using a continuous wave (cw) He/Ne laser. At delay times ranging from 10 to 200 μs the second flash, of 20 μs duration and 1000 J energy, was fired, thereby photolysing the hydrated electrons (e_{aq}^-) with light in various parts of their absorption band. No decrease in the absorbance of e_{aq}^- was observed; in fact there was an increase for an entirely different reason[31].

The second attempt utilised very short, intense (probably super-radiant) light flashes created from plastic scintillators by a 7 J, 3 ns pulse of high energy electrons from a 600 kV Febetron accelerator. Figure 2 shows a sketch of the apparatus. Most of the electron beam from the accelerator was used to create a light pulse in the scintillator. This was in the form of a hemisphere with highly polished surfaces which were completely silvered, except for the central part of a small groove in the flat face, which was therefore the only region from which the trapped scintillation light could emerge. Exactly where the light emerged there was a thin-walled glass capillary cell containing water in which hydrated electrons were produced by a part of the Febetron's pulse of high energy electrons. The concentration of e_{aq}^- was monitored at 633 nm by a laser beam. On firing the Febetron, hydrated electrons were produced and immediately photolysed by the flash ($\tau \sim 5$ ns) from the scintillator.

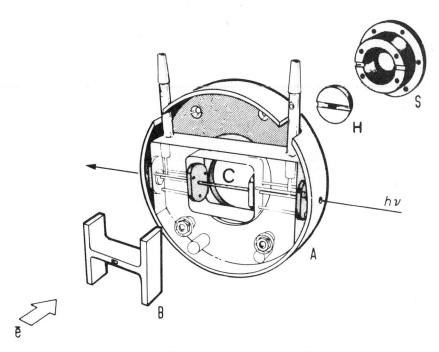

Fig. 2. Exploded view of pulsed scintillator apparatus. C is the capilliary glass reaction cell; A is the cell holder which attaches to the face-plate of the Febetron accelerator, whose electron beam approaches in the direction shown for e⁻. The cell holder also permits solutions to be flowed through C; H is a hemispherical scintillator silvered on the entire outer-surface except for the central 5 mm of the groove opposite the slot in B; S is an aluminium cup which supports H in such a position that C lies in the groove across the front of H; $h\nu$ represents the analysing laser light beam which monitors the concentration of hydrated electrons in C; B is an electron beam restricter, which allows electrons to strike only the central 5 mm of capilliary C and the scintillator H.

Three scintillators were used having broad bands centred at ∼420, 490 and 575 nm. Comparisons were made of the absorbance and decay rates of e_{aq}^- in the presence and absence of the light flash, but no differences could be detected. It looked as if de-excitation and repair occurred within a nanosecond or so[12].

From the third attempt it was possible to place a limit of $k > 1.8 \times 10^{11} \mathrm{s}^{-1}$ on the de-excitation rate and therefore $\tau < 6$ ps.on the excited state lifetime of e_{aq}^-[12,32]. This conclusion was reached from the failure to observe induced transparency by a Q-switched pulse at 100 MW power from a ruby laser. The laser method, and treatment of data, will be discussed shortly.

II.C The time scales and techniques involved

If the rate at which the ground state is repopulated in the case of e_{aq}^- is typical for solvated electrons in general, then it is clear that severe demands will be placed on available flash techniques to usefully study flash photobleaching. In any event, one knows from the position and width of the absorption band that the radiative lifetime is of the order of 10^{-9} s, and therefore that the natural lifetime is probably less than this; but the possibility exists of rapid transfer to non-radiative "relaxed excited states". Such relaxed states could result from relatively minor adjustments in the nuclear configuration of the solvent molecules in response to the new charge distribution of the excess electron in the excited state, and these might be relatively long-lived. This evidently does not happen with e_{aq}^- and May's data[13] suggests that in fact long-lived excited states for solvated electrons are very rare, F-centers at 77 K being an exception[33,34]. Direct observation of de-excitation and bleaching processes in most cases will consequently require picosecond, or sub-picosecond pulses. Furthermore, in order to detect significant fractional changes in the ground state population during the excited state lifetime, such pulses must be intense, for one requires about 10^{-3} J cm^{-2} of light within a reasonably narrow wavelength range. (This estimate of 10^{-3} J cm^{-2} is based on 10 % bleaching of a sample having an absorbance of 0.5 with an extinction coefficient of $1.5 \times 10^4 \, \text{M}^{-1}$ cm^{-1} in a 10 mm optical path length.) These requirements seem to be outside the scope of conventional flash lamps. However, single pulses from mode-locked solid state lasers at 1064 and 694 nm can meet these conditions. Also, from these laser lines other wavelengths are available as simple harmonics, or the pulses may be used as pumping sources for dye lasers or as picosecond continuum sources[35].

Q-switched ruby and neodymium-doped glass (or YAG) lasers produce single pulses at wavelengths of 694 and 1064 nm, respectively which are ~ 20 ns duration and have an energy per pulse of 1–2 J cm^{-2}. The pulse is bell-shaped and over the central 3 ns or so it is of nearly constant intensity at 300 to 1000 Einstein cm^{-2}s^{-1}. An outline is given in Sect. III of the use of such pulses to probe the picosecond time scale, by considering the photostationary condition which exists during the Q-switched pulse peak when the lifetime for de-excitation is very much less than the pulse duration. This represents a steady state treatment over the 3 ns of the pulse peak. The light flux is so large that the lifetimes probed may be one thousand-fold shorter than the period of the measurements. In addition, the absorbance can be monitored using ordinary spectrophotometric methods so that the post-pulse processes occurring on the time scale of 10^{-9} s and longer can be measured directly at the same time.

It transpires that most of the significant post-pulse bleaching processes of solvated electrons are complete in less than one microsecond[13], so that spectrophotometric measurements are required primarily for the period 10^{-9}–10^{-6} s. This poses a technical problem because of light scattered into the detection system from the giant pulse laser beam, which is typically $\sim 10^{10}$ times more intense than a normal monitoring light beam. Either exceptionally good optical arrangements, or ultra-fast electronic switching devices for the detector, are required for successful use of a white light analysing system. These difficulties are readily alleviated however by using cw lasers for monitoring purposes. In this case scattered light is removed by taking advantage of the highly collimated and monochromatic character of these probe beams[34]. Monitoring can thus be achieved in a direction which is almost co-linear with the photolysing pulse and it can proceed during the pulse, as well as afterwards. The principal limitation of course is that wavelengths are restricted to those for which cw lasers are available. Continuously tunable dye lasers have changed that from a technical limitation to one of cost.

Figure 1 indicates the wavelengths of photolysis and monitoring lasers in May's experiments[13,36]. Though very restricted in number these lines are at quite useful wavelengths for solvated electron studies. When using the ruby laser at 694 nm for bleaching purposes, the absorbance can be monitored concurrently at an adjacent wavelength 633 nm and at a considerable separation on each side, 442 and 1152 nm.

Two of the principal questions on the origin of the absorption band that these studies are attempting to answer—whether the band is homogeneous and the excited state bound—require comparisons to be made of bleaching in different parts of the band. This has been achieved by May for stable trapped electrons in two ways. First, one selects for study a trapped electron system in which the band maximum appears directly between the two photolysing wavelengths of 694 and 1064 nm and compares the bleaching effects of the two lasers[36]. Second, one chooses a series of trapped electron systems so that a particular bleaching laser line falls in different regions of the spectrum for different systems[13,36].

Short-lived solvated electrons, such as e_{aq}^-, were produced by conventional flash-photolysis methods in situ and in these cases the bleaching laser pulse was triggered at an appropriate time (typically $\sim 10^{-4}$ s) thereafter. Stable trapped electrons were produced by γ-radiolysis at the appropriate temperature, stored in the dark, then bleached at leisure. Details of these experimental arrangements are given elsewhere[13,32,35].

III. Analysis of transient photobleaching data

III.A Transient effects

There are two distinct aspects of photobleaching. One concerns the net effects which are manifest as permanent decreases in the absorbance of the sample. (In the case of compensatory spectral shifts, there can be concurrent small increases in the absorbance in other regions of the spectrum[6].) Permanent changes are monitored and treated in accordance with standard spectrophotometric procedures. Generally, they will be ordinary photolytic processes in which photoexcitation leads to release, or tunnelling, and thence to reaction. The other aspect concerns transient bleaching, which occurs during the flash but recovers thereafter and does not lead to permanent loss.

From the point of view of seeking the origin of the absorption bands, transient effects are probably more instructive. They are also the dominant effects sometimes. For instance, in the case of F-center electrons at $\sim$295 K, a pulse of ruby laser light causes complete transparency during the pulse, but $\sim 98\%$ of the original absorbance recovers within 1 μs and permanent bleaching is a trivial side-effect, with a quantum yield < 0.01[34].

This section will outline the analysis of data for: Sect. III.B, transient bleaching at the photolysing wavelength; Sect. III.C, transient bleaching at an adjacent monitoring wavelength; Sect.III.D, differences in transient effects arising from homogeneously and heterogeneously broadened absorption bands; and Sect. III.E, the magnitude of extinction coefficients when the bands are heterogeneous. All of these have been considered in greater detail in a recent report[13]. In Sect. III.F the concentration gradients caused by photobleaching are noted.

III.B Transient bleaching at the wavelength of photolysis

In a typical spectrophotometric measurement, absorption of the light beam causes a negligible change in the concentration of absorbing species, even when the absorption is large. Thus the absorbance measured is independent of the intensity of the light used. In effect, the number of absorbing species per unit area (cl) greatly exceeds the photon flux, I, multiplied by the life-time for de-excitation, τ. Thus c is independent of I so the Beer and Lambert laws (eqn. (2))

$$-dI/dl = Ic\varepsilon(2.303)$$ (2)

can be integrated to give the familiar

$$\log I_0/I_t = A = \varepsilon cl$$ (3)

Throughout, ε will represent the molar decadic extinction coefficient in $M^{-1}cm^{-1}$, I the light flux in Einstein $cm^{-2}s^{-1}$, c the concentration in M and A the absorbance.

Equation (3) is valid only in the limit of $I_0 \rightarrow 0$ and it becomes noticeably inaccurate when either I_0 is very large or τ very long. Under such conditions c, or dc/dt, must be expressed as a function of I prior to integration of eqn. (2). In order to do this it is necessary to specify an energy level scheme corresponding to absorption and de-excitation for the system under study. General procedures for this have been discussed[37] and applied, particularly to the non-linear behaviour of bleachable dyes used for Q-switching[38]. It has been pointed out[39] that there are no general analytical solutions to the differential equations involved so that one must resort to various approximations.

Two important and valid simplifications immediately alleviate these mathematical difficulties when one is specifically considering the treatment of experimental data on photo bleaching of solvated electrons[32]. First, when the de-excitation processes occur on a time scale very much less than the period of photolysis (the 3 ns pulse peak), as evidenced by the experimental result itself, then a photostationary condition exists and steady state kinetic procedures eliminate the time variable. Second, the energy level scheme invoked is relatively simple for solvated electrons since there are obviously no secondary manifolds for excited states of different multiplicity, nor does one expect significant absorption by any state other than the ground state. Therefore, an energy level scheme is used in which there is only one absorbing state.

An energy level diagram is given in Fig. 3 which should be sufficiently comprehensive to cover all the possibilities for solvated electrons. Absorption and stimulated emission occur between the ground state (1) and excited state (2). State (2) can undergo various spontaneous degradation processes directly to state (1), all of which

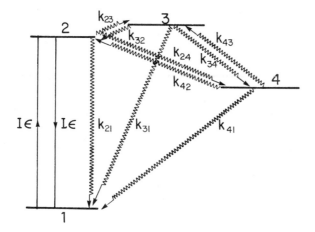

Fig. 3. Energy level scheme for photoexcitation of solvated electrons. (1) is the ground state (the only absorbing state); (2) is the photoexcited state from which stimulated emission can occur; states (3) and (4) represent a pair of excited or intermediate states which could be involved in the de-excitation from state (2) to (1). k_{21}, k_{23}, k_{24}, k_{31}, k_{32}, k_{34}, k_{41}, k_{42} and k_{43} signify pseudo-first-order rate constants which represent all the processes by which energy transfer occurs between the levels specified in the subscripts.

are represented by a rate constant k_{21}. Alternatively, the excitation may be dissipated in a variety of indirect routes involving transfer to intermediate states represented by levels (3) or (4). States (2), (3) and (4) encompass "quasi-free", "weakly trapped", "relaxed excited states", or "autoionising" situations as used in the discussion. All four states are interconnected by processes designated as apparent first-order rate constants where k_{ij} represents conversion from state i to state j.

Adopting the procedures used previously[32] to integrate eqn. (2) by expressing c_1 and c_2 as functions of I, noting that $c_1 + c_2 + c_3 + c_4 = c_0$, the total initial concentration (since only transient bleaching is being considered) one obtains eqn. (4).

$$A^0 - A = \varepsilon I(1 - 10^{-A})10^3\left[(2 + K'/k^*)/(k_{21} + k_{23} + k_{24} - K''/k^*)\right] \qquad (4)$$

where

$$K' = k_{23}(k_{41} + k_{42} + k_{43}) + k_{24}(k_{31} + k_{32} + k_{34}) + (k_{24}k_{43} + k_{23}k_{34})$$
$$K'' = k_{23}k_{32}(k_{41} + k_{42} + k_{43}) + k_{24}k_{42}(k_{31} + k_{32} + k_{34}) + k_{24}k_{32}k_{43} + k_{23}k_{34}k_{42}$$
$$k^* = (k_{31} + k_{32} + k_{34})(k_{41} + k_{42} + k_{43}) - k_{34}k_{43}$$

Thus it is seen that the absorbance at low light levels A^0 ($= \varepsilon c_0 l$) differs from that, A, at high intensity (I) by the parameters on the right-hand side of eqn. (4). These include ε, I and the term in square brackets, where the latter is related to the mean lifetime to repopulate the ground state, τ_g.

Equation (4) is more general than those presented previously for solvated electrons[13,32]. The term in square brackets will normally simplify very considerably because one particular decay mechanism will be predominant in a given system so that only one excited state is significantly populated. Thus when k_{23} and k_{24} are much smaller than k_{21}, direct de-excitation (2) → (1) dominates—as in the simple two-level scheme which seems appropriate for hydrated electrons[32]—and the term in square brackets reduces to $2/k_{21}$. For this case $(A^0 - A)$ is seen to be simply proportional to the mean lifetime of the photoexcited state. Alternatively, when the de-excitation pathway is principally via state (4), $(2) \xrightarrow{k_{24}} (4) \xrightarrow{k_{41}} (1)$, then the term in square brackets approximates to $[2/k_{24} + 1/k_{41}]$ as before[13]. If k_{41} happens to be much smaller than k_{24}, $(A^0 - A)$ becomes directly proportional to the mean lifetime of state (4). In this example it is state (4) which is populous and in the photostationary condition almost all electrons exist in that state. This situation probably pertains to F-centers at 77 K where the mean lifetime of the long-lived excited state in KI crystals is evidently 1.4 μs[34]. For F-centers at room temperature, it can be argued that k_{23} is significant because electrons seem to be released from a bound excited state (2) to the quasi-free state (3), from whence they are trapped by F-centers to form F^1-centers. In the case of hydrated electrons a distinction cannot be made as to whether state (2) is bound or free (or autoionising) since there are reasons for believing the resolution times in hydroxylic solvents are of the order of 10^{-12} s at room temperature[22,40].

Thus one sees that when measuring concentrations by light absorption, the measurement process itself alters that concentration and hence the absorption. In fact, the essence of the transient photobleaching method is to measure the absorbance difference at low and high intensities $(A^0 - A)$ and from this to evaluate the lifetime to repopulate the ground state.

At what light levels does significant bleaching occur? Suppose $\tau_g \sim 10^{-9}$ s (the radiative lifetime for a strong broad band in the visible or near-IR) and $\epsilon = 10^4$ $M^{-1} cm^{-1}$ for a sample having $A^0 = 0.5$. Equation (4) shows that A will be 10% less than A^0 when $I = 4$ Einstein $cm^{-2} s^{-1}$. This is not excessive. Typical flash-photolysis lamps of a few hundred joules energy used "end-on" as spectroflash sources can give such intensities over fairly broad wavelength bands. Q-switched ruby and Nd lasers give 20 ns pulses of monochromatic light up to 200 times this intensity. Individual mode-locked pulse intensities are much higher still and would cause significant bleaching of strong bands in species with sub-picosecond lifetimes. For instance, even the interrogating light beam could have been on the brink of bleaching in the studies by Rentzepis et al.[40] on the picosecond photoformation of hydrated electrons.

III.C Transient bleaching at a monitoring wavelength in a homogeneous band

The transmittance by an absorbing sample increases with light intensity because of depletion of absorbing species and because stimulated emission occurs when there is a significant concentration of excited states. On the other hand, the absorption process itself, which creates this population disturbance, leads to a decrease in transmission. These effects are specified collectively in eqn. (4). Only the first of these effects, the depletion of ground state molecules, affects the transmittance at other wavelengths within the same absorption band. The absorbance at a monitoring wavelength, α, within a homogeneous band, will be reduced due to intense illumination within that band at wavelength β, but, when the light intensity at wavelength α is low, absorption there does not itself contribute significantly to depletion of ground state species. Similarly, stimulated emission is unimportant at α because the excited state corresponding to absorption at α will not generally be populated by illumination at β (unless, by chance, it happens to be a relatively long-lived state involved in the de-excitation process following excitation at β). It follows then, that the bleaching at α must normally differ from that at β.

On applying the laws of Beer and Lambert to wavelength α we have eqn. (5)

$$-dI_\alpha/dl = I_\alpha \epsilon_\alpha c_1 (2.303) \tag{5}$$

in which c_1 is the same function of I_β as in Sect. III.B for the scheme of Fig. 3 and is independent of I_α. Since I_α and I_β are independent of each other, eqn. (5) is readily integrated, after applying the photostationary condition[13] to c_1, to give eqn. (6).

$$A_\alpha^0 - A_\alpha = A_\alpha^0 I_\beta \epsilon_\beta 10^3 / (k I_\beta \epsilon_\beta 10^3 + k') \tag{6}$$

where $k = (2 + K'/k^*)/(1 + K'/k^*)$ and $k' = (k_{21} + k_{23} + k_{24} + K''/k^*)/(1 + K'/k^*)$ with K', K'' and k^* having been defined for eqn. (4).

Again, the term containing the rate constants will invariably simplify very considerably since only one excited state will prevail in importance in each particular system. For the case of de-excitation via the sequence (2) → (4) → (1) the equation given previously is obtained[13].

Two limiting situations are worth noting;

(i) when the spontaneous direct de-excitation (2) → (1) dominates the decay so that state (2) is the only significantly populated excited state, then $k \to 2$ and $k' \to k_{21}$. This means that, even when I_β is so large that there is complete transparency due to saturation of the transition at β, the fractional bleaching at α, $(A_\alpha^0 - A_\alpha)/A_\alpha^0$ cannot exceed 0.5. If $(A_\alpha^0 - A_\alpha)/A_\alpha^0$ does exceed 0.5, then a simple two-level scheme can evidently be discounted;

(ii) when state (4) is the most populated state in the photostationary condition as a result of k_{24} being relatively large and k_{41} small, then $k \to 1$ and $k' \to k_{41}$. Now 100% bleaching is possible at α.

The ratio of fractional bleaching at β compared to that at α for the two limiting cases above is given by eqn. (7) for direct de-excitation (2) → (1) and by eqn. (8) for decay via long-lived state (4)

$$[(A_\beta^0 - A_\beta)/A_\beta^0]/[(A_\alpha^0 - A_\alpha)/A_\alpha^0] = (1 - 10^{-A\beta})(4I_\beta \epsilon_\beta 10^3/k_{21} + 2)/A_\beta^0 \tag{7}$$
$$[(A_\beta^0 - A_\beta)/A_\beta^0]/[(A_\alpha^0 - A_\alpha)/A_\alpha^0] = (1 - 10^{-A\beta})(I_\beta \epsilon_\beta 10^3/k_{41} + 1)/A_\beta^0 \tag{8}$$

It is clear that this ratio is never unity (except by coincidence under one set of values of A_β, A_β^0, I_β, ϵ_β and k_{21} or k_{41}). Thus the extent of bleaching at a monitoring wavelength never equals that at the photolysing wavelength—even though the band is homogeneous. However, the fractional bleaching at α depends only on factors related to β, such as I_β, ϵ_β and the de-excitation processes of the excited state reached by wavelength β. Consequently, a fingerprint of a homogeneously broadened band is that the transient fractional bleaching be the same at all monitoring wavelengths, though different to that at the photolysing wavelength.

III.D Transient bleaching at a monitoring wavelength in a heterogeneous band

The above considerations pertain when a band simply displays the variation in extinction coefficient with wavelength for absorption by one type of ground state species—the homogeneous band. The photobleaching situation changes, however, when the observed spectral band is an envelope of several, or many, individual bands each of which represents a distinct ground state energy level. Now the band is certainly inhomogeneous. This is commonly referred to as a heterogeneously broadened band and can be thought of as "vertically stacked"[13], at least in part. Spectral-shifts have been observed[6,41] to result from bleaching of trapped electrons by exposure to low intensity illumination in specific wavelength ranges (This was sometimes totally compensated by increased absorption in other spectral regions and consequently termed photoshuffling[6] or photoshuttling[41b].) Such effects have been found in a variety of low temperature glasses and interpreted as evidence for heterogeneous absorption bands. This evidence is convincing, for it is difficult to envisage spectral-shifts without presupposing the existence of a variety of electron traps which each contribute absorption predominantly to one spectral region, the weaker traps to the red end, the deeper traps to the blue.

One would expect the extent of transient bleaching at a monitoring wavelength to change with wavelength for a heterogeneous band. In the limit, of course, if there are no species which contribute to absorption at both α and β then the absorbance at α is unaffected by intense illumination at β. As overlap of sub-structure bands at α and β increases so will the influence of illumination at β change the absorbance at α. An analysis has been given[13] for the fractional bleaching in terms of the extinction coefficients at α and β of species contributing to absorption at both wavelengths and their relative population, for a relatively simple energy level scheme. The steepness with which $(A_\alpha^0 - A_\alpha)/A_\alpha^0$ falls off as α is moved away from β is correlated with the widths of the underlying bands and, conceivably, such a sub-structure could be deconvoluted if enough data were available.

Since a sub-structure arises from different ground state energies, there could be relatively rapid exchange and redistribution of species between these states. This can confuse the interpretation of spectral-shift data, especially with regard to photo-excitations which lead to electron release and retrapping. Furthermore, there are unexplained thermal bleaching processes at least in pulsed laser bleaching studies[13], which can further aggravate the interpretation of all spectral-shift data.

III.E   Extinction coefficients in heterogeneous bands

A further consideration that has been discussed recently[13] concerns the actual magnitude of trapped electron extinction coefficients. Different values for ε will be evaluated from knowledge that the band is heterogeneous rather than homogeneous, the former being the larger. Their actual magnitude cannot be determined, however, without details of the individual sub-structure bands.

A broadened homogeneous band can arise from a common ground state with many unresolvable excited states or simply from rapid dynamic fluctuations in fine structure levels of ground and excited states. In any event, the band stems from the presence of only one type of species, each species contributing equally to absorption throughout the spectrum. Only the extinction coefficient changes with wavelength. Thus measurements of the concentration, c, and absorbance, A_α, yield the extinction coefficient of the species at wavelength α from eq. (3), $\epsilon^\alpha_{hom} = A_\alpha/cl$. For solvated electrons the extinction coefficients at their band maxima are typically $15000-50000$ M^{-1} cm^{-1}.

A heterogeneous band, in the sense of this discussion, is composed of discrete underlying bands each characterising distinct electron states which do not exchange rapidly on the time scale of the photolysis. Since there is some vertical stacking, the absorbance at a particular wavelength is composed of different contributions from species in different energy states. Electrons in some types of traps may not contribute at all to certain regions of the overall spectrum. The effective concentration of species contributing to absorbance is consequently less than the total concentration of species (c), as measured chemically. If, for the moment, we assume perfect vertical stacking (i.e. no overlapping of adjacent underlying bands) then we can suppose that only a certain fraction, x, of the total concentration of species contributes to absorption at α. Thus we deduce that $\epsilon^\alpha_{het} = A_\alpha/xcl$. It is seen in this case that ϵ_{het} exceeds ϵ_{hom} by a factor of $1/x$.

Since the individual discrete bands in the heterogeneous case each have an oscillator strength numerically equal to that evaluated for the whole band on the basis of it being homogeneous[13], which for solvated electrons is generally about unity, it follows that the shape of à heterogeneous band partly displays the population distribution. If the underlying bands are very narrow and non-overlapping then the overall band envelope is simply representative of the number of species contributing to absorption in different regions of the spectrum. When the underlying bands are broad then the absorbance is, of course, a complex composite summing populations and individual extinction coefficients.

In order to evaluate ϵ_{het}, the heterogeneous band must be deconvoluted. This could possibly be achieved by bleaching methods or by measuring the separate concentrations in individual electron states. Perhaps some selective chemical process could be devised which progressively eats through the absorption band from red to blue due to a reaction rate constant which is quite sensitive to the energy state of the trapped electron. The only indications at the present time on the widths of individual bands come, firstly, from steady-state spectral-shift measurements (which suggests there are only a few, rather broad bands in aqueous media[41]) and secondly, from transient laser bleaching data (which indicates very narrow isolated bands on the high energy tail in MTHF[36,42]). Even if the band consists of only ten discrete species, ϵ_{het} could reach values of the order of 150,000 to 500,000 M^{-1} cm^{-1}. This latter figure corresponds to a cross-section of ~ 20 Å^2, which is comparable to the physical area displayed by the major part of the charge distribution ascribed to a hydrated electron.

A heterogeneous band will doubtless be a composite of sub-bands which are broad in some regions, narrow in others, and it will be difficult to decipher. Furthermore, the origin of differences giving rise to various trap-depths is not clear. One view that is favoured by Kestner is that trap distortions cause spectral broadening[43]. Nevertheless, it is necessary to know the real magnitude of ϵ_{het} in order to evaluate excited state lifetimes from transient bleaching data. Hitherto it has been common practice to use ϵ_{nom} in all experiments. Whereas this is adequate when evaluating G from $G\epsilon$ pulse radiolysis data since both G and ε refer to the indiscriminately determined total concentration[13], it is not appropriate for bleaching studies.

III.F Concentration gradients arising from photobleaching

Unless a sample is "optically thin", photobleaching will create a concentration gradient down the length of the cell. After a period of illumination the concentration will be lower at the front of the cell than at the rear, and for typical trapped electron studies the ratio of front-to-rear concentrations can approach 2 or more. This obviously introduces severe complications for quantitative measurements of bleaching when the experimental arrangement has the analysing light beam traversing the sample perpendicular to the photolysing beam. The analysis is somewhat simpler when a co-linear monitoring light system is used. Occurrence of a concentration gradient is also a problem in cases where repair of "hole-burning" or spectral relaxations take place, since the redistribution rates presumably change with population gradients and these will vary down the length of the cell. However, such gradients as do exist will normally not invalidate most of the general inferences drawn from qualitative bleaching observations.

It is possible to solve the appropriate equations in order to see how the concentration varies with distance through the sample and with total illumination. An analysis which applies to permanent bleaching, or to transient bleaching during a photo-stationary state, is analogous to that developed for an unstirred photochemical reaction system[44]. For the case of permanent bleaching with a quantum yield ϕ_B, using incident light intensity I for time t, the concentration at depth l in the cell (c_l) is given by eqn. (9)

$$c_l = c_0 / \{1 + \exp(-\epsilon'c_0 l)[\exp(\epsilon'\phi_B I t) - 1]\} \tag{9}$$

where c_0 is the initial (uniform) concentration. Interestingly enough, eqn. (9) indicates that the concentration varies nearly linearly with distance through the cell for up to 80% bleaching. A linear variation in concentration is, of course, very convenient and enables one to make corrections rather simply.

IV. Time-resolved laser bleaching effects

IV.A General conditions employed

Lasers have been used in cw mode[45], or as time-averaged light sources when repetitively pulsed[14], for overall, steady-state, bleaching studies of trapped electrons. Pulsed lasers have been used to generate solvated electrons[40,46] and to study the time scale of the solvation process[40]; but, to the writer's knowledge, no studies have been published on time-resolved bleaching other than those of Kenney-Wallace[32] and May[13,34,42] utilising Q-switched, giant pulse ruby and Nd lasers, and those of Mahr on F-centers at 77 K which were limited to microsecond response times[47].

It has been known for several years that the quantum yield for photobleaching of trapped electrons at low temperatures often decreases markedly with progressive bleaching of a sample[10,11]. In the pulsed laser studies under discussion the same effect is evident for permanent bleaching. This means that under high intensity pulses, ϕ_B is less than it is with low intensity pulses, but the actual amount of bleaching per pulse increases monotonically with pulse intensity. Both the pulsed and steady-state observations are consistent with the notion that a sample of trapped electrons has a wide range of bleaching efficiencies, the range stemming from a wide distribution of distances which separate trapped electrons from potential reaction sites. In fact, upon initial formation by radiolysis, there is even spontaneous thermal decay occurring at $\sim$77 K[26], whose efficiency falls off as more of it proceeds, again indicating a distribution of efficiencies with which electrons may be lost by "tunnelling" processes at low temperatures.

For these reasons the quantum yield was not a parameter of concern in these studies. Thus, in order to optimise the observation of bleaching during and after a laser pulse, the maximum pulse intensity was generally used. Permanent bleaching efficiencies were often < 0.01 under these conditions though the same effects could sometimes be discerned with values of 0.1–1 at lower intensities.

Basically, two types of process were explored. First, the depletion of the ground state during the pulse, as registered by induced-transparency, and the recovery afterwards without reaction was studied. Second, the time-profile of electron loss processes which created permanent bleaching was measured. These were studied at a few wavelengths within each solvated electron absorption band. All post-pulse processes proceeded in the dark in these studies, whereas, in the analogous steady-state bleaching studies, there was continual illumination whence photostationary conditions would exist.

Transient bleachings of organic dyes were used as control experiments to demonstrate that the laser-bleaching procedures worked as expected[32,34]. Thus, 10^{-5} M solutions of vanadyl phthalocyanine showed saturation of the optical transition to the point where the fraction bleached was $\sim$0.8 at the maximum laser intensity. Methylene Blue solutions were bleached and their recovery recorded microseconds after the pulse. A 2×10^{-6} M solution of cryptocyanine showed a four-fold decrease in absorbance during the pulse at the highest intensities[32] and this solute is known to have excited state lifetimes of only 40 ps. On the basis of these demonstrations it was possible to draw definitive conclusions from failure to see induced transparency in some cases for solvated electrons, and thereby place upper limits of a few ps on lifetimes of the excited states involved.

IV.B F-centers at $\sim$295 K

These colour centers in alkali halide crystals are well characterised as electrons trapped at anion vacancies[48]. They can be created by radiolysis or by injection of an excess of alkali metal into the crystal lattice. In most alkali halide crystals, but not all, F-centers are stable in the dark at room temperature. In some respects they can be regarded as the simplest example of trapped electrons. Their absorption bands are similar in shape to other trapped electrons and they too have an oscillator strength of about unity. As can be seen in Fig. 1, the band, though substantially narrower than the others, is remarkably broad—despite the polarization field being generated by a thoroughly ordered crystal lattice. Absorption in the F-band has been assigned to the bound–bound ($2p \leftarrow 1s$) transition[48], with the $2p$ level some 0.1–0.2 eV below the conduction band. A K-band, attributed to ($3p \leftarrow 1s$), is resolvable from the F-band and lies above the bottom of the conduction band.

In May's experiments[34] stable F-centers were produced in KBr and KI single crystals by γ-irradiation or exposure to K metal vapour. These were subjected to 20 ns pulses of 694 nm light at intensities up to 350 Einstein cm^{-2} s^{-1}. Transmission during the pulse was measured at 694 nm and the absorbance was monitored at 633 and 1152 nm.

At the highest intensities the samples became completely transparent during the pulse but almost all of the original absorbance recovered within microseconds. Permanent bleaching occurred with $\phi_B < 10^{-2}$. Figure 4 shows the transient bleaching (as fractional bleaching—in order to normalise the 694 and 633 nm data) as a function of laser pulse intensity. The latter is also proportional to total light per pulse. Some 30% of the absorbance recovered too rapidly to measure its rate (essentially with the pulse, $k \sim 10^9$ s^{-1}) and the remaining 70% absorbance redeveloped in a first-order manner with rate constants of 1.7×10^6 s^{-1} and 0.9×10^5 s^{-1} for KBr and KI, respectively[34]. Absorbance was created at 1152 nm by the laser pulse. This decayed concurrently with the redevelopment of absorbance at 633 nm.

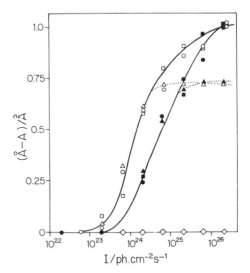

Fig. 4. Fractional bleaching, $(A^0-A)/A^0$, as a function of pulse peak light intensity, I, for F-centers in alkali halide crystals at 295 K[34]. Open data points refer to KI, closed ones to KBr; ● and ○, 694 nm during ruby pulse peak; ■ and □, 633 nm during ruby pulse peak; ▲ and △, 633 nm about 10 ns after ruby pulse peak; ◇ (for both KI and KBr), 2 ms or 2 mins. after ruby laser pulse. Uncertainties in the data were typically ±0.1 (see ref. 34). (Note how bleaching sets in at a lower light intensity for KI consistent with its higher extinction coefficient at 694 nm.)

Several notable features pertinent to time-resolved studies are exemplified by the F-center data. The following inferences were made[34].

(i) Bleaching processes were identical throughout the absorption band. Bleaching in this case arises from loss from the absorbing manifold, therefore this result implies the band is homogeneous—as already purported[48].

(ii) Infrared absorption signalled the formation of F^1-centers—presumably formed by reaction of quasi-free, or conduction band, electrons with other F-centers. Thus, although photoexcitation gives a bound excited state this, nevertheless, leads to electron release.

(iii) 100% bleaching during the pulse probably stems from complete saturation of the optical transitions of both F- and F^1-centers. This conclusion seems to be preferred to the possibility that by the pulse peak essentially all electrons occupied non-absorbing relaxed excited states, because at the pulse end there are approximately equal numbers of F and F^1-centers (noting that each F'-center decomposes to two F'-centers).

(iv) Equation (4) cannot be applied to transient bleaching at the pulse peak in this case because the photostationary condition cannot hold, and the pulse shape is distorted, due to the extensive and accumulative bleaching in forming F^1-centers. Nevertheless, from the intensity required to produce 5% bleaching in KI, one estimates a lifetime for the excited state of $\sim 10^{-9}$ s. The relaxed excited state seems to have a lifetime of $\sim 10^{-8}$ s and the F'-center $\sim 10^{-6}$ s.

(v) Permanent bleaching is seen to be an incidental perturbation of predominantly transient effects and probably occurs when a conduction band electron chances upon a hole reactive center or impurity.

IV.C F-Centers at 77 K

Bleaching of F-centers at low temperatures has been studied extensively by various methods, including ruby laser bleaching[47]. As there was no permanent bleaching induced by a laser pulse[34], F^1-centers apparently were not formed, since they are known to be stable at 77 K. This means that electrons are not thermally promoted from the excited state to the conduction band at this temperature. Extensive bleaching is observed at 694 and 633 nm during the pulse but the absorbance at 633 nm in KI crystals recovers in a first-order manner with a rate constant of $5 \times 10^5 \, \text{s}^{-1}$ which is the known decay rate of the relaxed excited state of the F-center at 77 K[34]. F-centers at 77 K represent cases in which transient bleaching occurred because of the accumulative stacking of photoexcited electrons in a non-absorbing excited state which was long-lived relative to the 20 ns laser pulse.

IV.D Hydrated electrons at ~295 K

At the other extreme, where no type of bleaching was observed, lies the mobile hydrated electron in very dilute solution at room temperature. Even at the highest ruby laser pulse intensities, there was no self-induced transparency at 694 nm[32], nor a change in absorbance at 633 nm[36], nor was the decay rate ($\tau \sim 100 \, \mu s$) altered measureably[36]. On applying eqn. (4) to the transmission data at 694 nm, an upper limit of 5.5 ps was evaluated for the excited state lifetime, this limit being set by the experimental error on the transmission data arising from random fluctuations in pulse intensity. Failure to observe bleaching at 633 nm indicates that this upper limit is ≤3 ps through the application of eqn. (6), which presupposes that the band is homogeneous between 633 and 694 nm. It was also calculated using ε_{hom}. If either of these assumptions were invalid the estimated upper limit to the lifetime would be even smaller, or if the simple two level scheme invoked were an over-simplification this too would lower the lifetime still further.

Unfortunately, this result does not resolve the question of whether the excited state reached at 694 nm is bound or free. The result is consistent with it being free because a solvation time of electrons in liquid water of 2–4 ps seems to be indicated[39], so that the electron would become resolvated within that time. Whether an excited state lifetime of <3 ps is inconsistent with a bound state awaits theoretical demonstration. In any event, the excited state apparently does not transform rapidly to a comparatively long-lived IR absorbing species as happens to the trapped electron in aqueous glasses at 77 K[49].

IV.E Trapped electrons in glasses at 77 K

Many solvents produce stable trapped electrons when γ-irradiated as amorphous solids at 77 K. Laser bleaching studies have been conducted on hydrocarbons (3 methylhexane and 3 methylpentane), an ether methyltetrahydrofuran (MTHF), an amine (triphenylamine), various alcohols (including n-propanol and 2-butanol) plus aqueous glasses made from concentrated solutions of eight ionic solutes[13,36]. Absorption bands in some of these media are given in Fig. 1 to show their relationship to the two bleaching and three monitoring wavelengths used. It should be noted that these bands seem to be somewhat blue-shifted and slightly narrower than in the same medium as a liquid at room temperature, but are otherwise very similar.

Results are given eleswhere though many have been summarised in a recent comprehensive report[13]. In this latter publication[13] the diagram given in Fig. 5 was utilised to show various classes of bleaching characteristics in schematic form on a logarithmic time scale. From the extensive compilation of data available the following salient features emerge:

(i) Bleaching by ruby laser pulses of trapped electrons in some aqueous glasses creates transient absorption in the IR (observed at 1152 nm) and is particularly strong in aqueous sodium formate and sucrose glasses[49]. This is only one of several bleaching processes to occur in these glasses. It manifests another example of the ambiguity confounding deductions drawn from photobleaching data, for it is impossible at the present time to decide whether or not this implies electron release. It could arise from a relaxed excited state, or from dielectron formation (analogous to the F'-center) or it could signify a shallowly trapped state as an intermediate in the retrapping process but having a very low efficiency[49].

(ii) Despite similarities in the absorption band of trapped electrons in aqueous glasses made from a variety of solutes, the photobleaching characteristics vary in

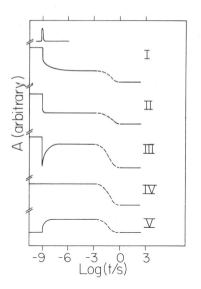

Fig. 5. A schematic display of the several types of bleaching behavior found for trapped electron absorption in a wide-range of glasses at 77 K. The figure indicates, in a qualitative way, changes in absorbance as a function of time (logarithmic scale) following photoexcitation by a ruby laser pulse. The laser pulse is shown in the uppermost row. Type (I), (II), (III) or (IV) bleaching behavior was observed at 633 and 442 nm in systems in which trapped electrons initially absorbed at these wavelengths, whereas type (V) behavior was observed at 1152 nm in systems when there was negligible absorption there initially. (Figure taken from ref. 13.)

magnitude quite profoundly, as do the radiation yields, level of impurities, solvation energies and thermal properties of the glasses.

(iii) In all glasses studied, the quantum yield for permanent bleaching decreases with increasing intensity. At the highest intensities permanent bleaching is a rare event for the photoexcited species, as most return to the ground state within 10^{-9} s. Self-induced transparency at 694 nm is evident in aqueous K_2CO_3 and $NaClO_4$ glasses for instance[13]. Consequently, for electrons in a wide variety of media, de-excitation or retrapping is extremely efficient.

(iv) Much of the permanent bleaching sets in surprisingly slowly. There are three general time scales involved—immediate (<10 ns), fast (occurring largely over the first 10^{-6} s) and slow (taking up to 10^{-2}–10^{-1} s for completion) as indicated in Fig. 5. Relative proportions of these vary from one solvent, or solute, to another and are summarised elsewhere[13].

(v) All of the post-pulse bleaching of electrons (i.e. the "fast" and "slow" fractions) must originate from secondary photochemical processes. It is presumed throughout this discussion that solvated, or trapped, electrons are the only species absorbing the photolysis light at 694 or 1064 nm, so that photoexcitation, followed by de-excitation, seems to have initiated processes which subsequently destroyed trapped electrons. Two possibilities have been proposed[13]. First, photoexcited electrons may react with other trapped electrons. Perhaps such reactions are relatively slow at 77 K and involve tunnelling rather than promotion to a quasi-free state. Second, the many excitations and de-excitations to which each trapped electron is committed during an intense laser pulse, results in the local deposition of considerable energy (up to 200 photons of ~2 eV each). Dissipation of this energy will cause gross thermal shocks, and possibly melting, in the locality of the traps. Temperature fronts emanating from these centers could cause trapped electron migration (and hence possibly reaction—perhaps during a period of microseconds) and eventually, after overlapping, result in bulk structural settling or relaxation, at which time (perhaps the millisecond region) further trapped electron loss takes place.

(vi) In only one case so far, that of MTHF[36,42], is there direct evidence of a heterogeneous absorption band from these pulsed-laser experiments. For trapped electrons in MTHF immediate bleaching was observed at 694 and 1152 nm but not at 633 and 442 nm when subjected to a ruby laser pulse. Only post-pulse bleaching occurred at 633 and 442 nm, type IV behaviour (Fig. 5), in this particular glass. Although 694 nm also occurs well out on the high energy tail of the band in hydrocarbons like 3-methylhexane there was no similar selective bleaching observed in them.

(vii) Aqueous glasses made from 5 M LiCl showed strong transient bleaching of type III, which had a comparable recovery time to that of F-centers at 77 K. There was also IR absorption produced which decayed on the same time scale as the trapped electron recovery[13,49]. These facts suggest the occurrence of relaxed excited states with lifetimes of $\sim 10^{-6}$ s, and indicate that this particular solute causes quite different bleaching behaviour to that observed in other aqueous glasses.

(viii) Experiments were performed to measure the circular dichroism shown by solvated electron absorption bands in a chiral solvent, in the hope that the magnitude of the optical activity might be used to distinguish bound–bound (specifically $p \leftarrow s$) from bound-free transitions. None was observed. Several types of measurement were made, but the most sensitive involved trapped electrons in d-2-butanol being subjected to circularly polarised light pulses from the ruby laser whilst the absorbance was monitored with crossed polarisers[36,50]. In another type of experiment electrons solvated in chiral solvents (2-butanol or aqueous sugar glasses) failed to show or to create a sufficient concentration difference for circular dichroism to be detectable[51].

V. Concluding remarks

Perhaps the most intriguing general points to emerge from these time-resolved bleaching studies concern (i) the high rate of de-excitation and (ii) the occurrence of slow photobleaching processes. On the first point, only F-centers and aqueous LiCl glasses at 77 K gave evidence of excited states which were sufficiently long-lived for the repair process to be observed at 633 nm. In all other cases the ground state was totally repopulated within a few picoseconds, except for the permanent bleaching which took place. There must be either super-fast internal conversions from bound states, or complete and rapid retrapping of photo-released electrons. The latter seems plausible for liquid systems, but in pulse radiolysis studies at low temperatures the majority of electrons are trapped rather slowly and pass through relatively long-lived, IR-absorbing states[19,23,26]. Whereas some IR absorption was produced in these laser photolysis studies[49], the quantum yield of formation of such states was small—otherwise the visible band would have been totally bleached part of the way through the pulse.

A large fraction of the overall net bleaching, following an intense laser pulse, was observed to occur in microseconds and milliseconds after the pulse. There seems little doubt that this portion of the bleaching was not a direct photochemical loss, but rather set in as a result of secondary reactions or thermal relaxations. Such data certainly introduces further ponderables for the interpretation of steady-state bleaching effects.

It is often very convenient for practical reasons if one can effect a "time dilation" of processes by means of temperature reduction. This seems to work, at least qualitatively, for the progressive electron stabilisation sequence depicted in eqn. (1), but a temperature reduction is not expected to have a consequential effect upon the photophysical processes of light absorption and it evidently has not had a serious impact on the curtailment of de-excitation, in most cases. Thus, more time-resolved studies are needed, preferably with shorter times and greater resolution. Were one able to study induced transparency at sub-picosecond times for all photolysing and monitoring wavelengths then surely one could resolve the major issue of what systems yield bound or free excited states and homogeneous or heterogeneous absorption bands.

Acknowledgements

The studies described in this chapter are the result of collaborations with Dr. Roger May and Dr. Geraldine Kenney-Wallace, to both of whom the author is greatly indebted. Financial support was gratefully received from the National Research Council and Defence Research Board of Canada.

References

1 G. R. Young, M. A. Thesis, University of North Carolina, (1962).
2 D. C. Walker, Quart. Rev., 21 (1967) 79.
3 E. J. Hart and J. W. Boag, J. Amer. Chem. Soc., 84 (1962) 4090.
4 J. Jortner, Radiat. Res. Suppl., 4 (1964) 24.
5 K. H. Schmidt and W. L. Buck, Science, 151 (1966) 70.
6 G. V. Buxton, F. S. Dainton, T. E. Lantz and F. P. Sargent, Trans. Faraday Soc., 66 (1970) 2962.
7 P. Delahay, this volume Chap. 4, p. 115 and references therein.
8 J. Kroh and A. Plonka, Int. J. Radiat. Phys. Chem., 6 (1974) 211.
9 L. Kevan, J. Chem. Phys., 56 (1972) 838.
10 P. J. Dyne and O. A. Miller, Can. J. Chem., 43 (1965) 2696.
11 A. Bernas, D. Grand and C. Chachaty, Chem. Commun., (1970) 1667.
12 G. A. Kenney-Wallace and D. C. Walker, Ber. Bunsenges. Phys. Chem., 75 (1971) 634.
13 D. C. Walker and R. May, Int. J. Radiat, Phys. Chem., 6 (1974) 345.
14 (a) S. L. Hager and J. E. Willard, Chem. Phys. Lett., 24 (1974) 102; (b) G. C. Dismukes, S. L. Hager, G. H. Morine and J. E. Willard, J. Chem. Phys., 61 (1974) 426.

114

15 P. Hamlet and L. Kevan, J. Amer. Chem. Soc., **93** (1971) 1102.

16 T. Higashimura, Ann. Rep. Res. React. Inst. Kyoto Univ., **6** (1973) 38.

17 G. A. Kenney and D. C. Walker, in A. J. Bard (Ed.), Electro-analytical Chemistry, Vol. 5, Marcel Dekker, New York, 1971, p. 1 (Authors' Preface),

18 E. J. Hart, B. D. Michael and K. H. Schmidt, J. Phys. Chem., **75** (1971) 2798.

19 (a) J. H. Baxendale and P. Wardman, Nature (London), **230** (1971) 449; (b) J. H. Baxendale and P. Wardman, J. Chem. Soc. Faraday Trans. I, (1973) 584.

20 E. A. Shaede, L. M. Dorfman, G. J. Flynn and D. C. Walker, Can. J. Chem., **51** (1973) 3905.

21 D. C. Walker, N. V. Klassen and H. A. Gillis, Chem. Phys. Lett., **10** (1971) 636.

22 L. Gilles, J. E. Aldrich and J. W. Hunt, Nature (London), **243** (1973) 70.

23 J. T. Richards and J. K. Thomas, J. Chem. Phys., **53** (1970) 218.

24 H. Hase, M. Noda and T. Higashimura, J. Chem. Phys., **54** (1971) 2975.

25 L. Kevan, J. Advan. Radiat. Chem., **4** (1974) 181.

26 N. V. Klassen, H. A. Gillis and G. G. Teather, J. Phys. Chem., **76** (1972) 3847.

27 F. Y. Jou and L. M. Dorfman, J. Chem. Phys., **58** (1973) 4715.

28 A. V. Vannikov and V. S. Marevtsev, Int. J. Radiat. Phys. Chem., **5** (1973) 4533.

29 G. A. Kenney-Wallace, M.Sc. thesis (1968), and Ph.D. thesis (1970), University of British Columbia.

30 R. K. Wolff, M. J. Bronskill and J. W. Hunt, J. Chem. Phys., **53** (1970) 4211.

31 N. Basco, G. A. Kenney-Wallace, S. K. Vidyarthi and D. C. Walker, Can. J. Chem., **50** (1972) 2059.

32 G. Kenney-Wallace and D. C. Walker, J. Chem. Phys., **55** (1971) 447.

33 R. K. Swank and F. C. Brown, Phys. Rev., **130** (1963) 34.

34 R. May and D. C. Walker, Can. J. Chem., **51** (1973) 2306.

35 (a) G. E. Busch, R. P. Jones and P. M. Rentzepis, Chem, Phys. Lett., **18** (1973) 178; (b) S. C. Wallace and G. A. Kenney-Wallace, Int. J. Radiat. Phys. Chem., **6** (1975) 345.

36 R. May, unpublished data.

37 (a) R. W. Keyes, I. B. M. J. Res. Develop., **7** (1963) 334; (b) F. Gires, I.E.E.E. J. Quantum Electron, Q.E.2, (1966) 624.

38 L. Huff and L. G. De Shazer, J. Opt. Soc. Amer., **60** (1970) 157.

39 A. Zunger and K. Bar-Eli, J. Chem. Phys., **57** (1972) 3558.

40 (a) P. M. Rentzepis, R. P. Jones and J. Jortner, Chem. Phys. Lett., **15** (1972) 480; (b) P. M. Rentzepis, R. P. Jones and J. Jortner, J. Chem. Phys., **59** (1973) 766.

41 (a) F. S. Dainton, G. A. Salmon and U. F. Zucker, Chem. Commun., **8** (1968) 1172; (b) F. S. Dainton, Ber. Bunsenges. Phys. Chem. **75** (1971) 608; (c) O. F. Khodzhaev. B. G. Ershov and A. K. Pikaev, Dokl. Akad. Nauk SSSR, **179** (1968) 911; (d) H. Hase and L. Kevan, J. Chem. Phys., **54** (1971) 908; (e) S. Noda, K. Fueki and Z. Kuri, Can. J. Chem., **50** (1972) 2699; (f) J. R. Miller and J. E. Willard, J. Phys. Chem., **76** (1972) 2341; (g) H. B. Steen and J. Moan, J. Phys. Chem., **76** (1972) 3366.

42 R. May and D. C. Walker, J. Chem. Soc. Chem. Commun., (1972) 1064.

43 N. R. Kestner, Radiat. Res., **59** (1974) 255, (Abstract).

44 B. Rabinovich, Photochem. and Photobiol., **17** (1973) 479.

45 K. K. Ho and L. Kevan, Int. J. Radiat. Phys. Chem., **3** (1971) 193.

46 (a) M. Andorn and K. H. Bar-Eli, J. Chem. Phys., **55** (1971) 5017; (b) P. M. Rentzepis and M. R. Topp, in D. L. Horrocks and C. T. Peng (Eds.), Organic Scintillators, Academic Press, New York, 1971, p. 91.

47 H. Mahr, in W. B. Fowler (Ed.), Physics of Color Centers, Academic Press, New York, 1968, p. 243.

48 W. B. Fowler, in W. B. Fowler (Ed.), Physics of Color Centers, Academic Press, New York, 1968, p. 53.

49 R. May and D. C. Walker, Chem. Phys. Lett., **30** (1974) 69.

50 R. May and D. C. Walker, Radiat. Res., **59** (1974) 198, (Abstract).

51 (a) M. Ulrich and D. C. Walker, in press; (b) M. Ulrich, Ph.D. thesis, University of British Columbia (1974).

Chapter 4

CONTINUUM TRANSITIONS OF EXCESS ELECTRONS STUDIED BY PHOTOELECTRON EMISSION SPECTROSCOPY

P. DELAHAY

I. Introduction

Irradiation ($\lambda \leq 1000$ nm) of the surface of a solvated electron solution causes emission of electrons into the solvent vapor above the solution. Quantitative work on solvated electrons has been done so far with only two solvents (liquid ammonia and hexamethyl phosphoric triamide (HMPA)), but the phenomenon of photoelectron emission (PEE) should be a general one for solvated electrons and other emitters in solution (inorganic ions, organic anion radicals, etc.) or even for pure liquids.

Two new and significant results have already been obtained by PEE spectroscopy of solutions: (i) photoionization spectra of species in solution were determined, and (ii) mechanisms of photoionization via autoionization of excited bound states were established. In general, these new results could not have been obtained by other methods, at least for solutions of high ionic conductivity. Such methods (e.g. electron spin resonance) do provide valuable information on solvated electron solutions, but generally not the type of information obtained by PEE spectroscopy. These methods are generally not suitable, in this respect, for a variety of reasons. For instance, changes in conductivity caused by the production of quasi-free electrons upon irradiation (photoconductivity) are too small to be detected when the conductivity is determined by an overwhelming majority of other charge carriers in solution. Photobleaching does not discriminate between bound–bound and bound–continuum transitions.

Photoelectron emission by the solution can be detected by means of an apparatus resembling, at least in principle, a vacuum photodiode: a cup containing the emitting solvated electron solution faces a collector electrode located in the gas phase; the diode circuit is completed by a current detector and a power supply. Two types of experimental information can be obtained whenever proper experimental requirements are met.

(i) The spectral response for emission quantum yield[1] (Fig. 1). It should be stressed that the quantum yield in the work reviewed in this chapter is expressed as the number of electrons emitted per incident photon.

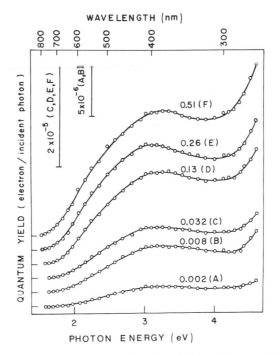

Fig. 1. Plots of quantum yield as electron emitted per incident photon against photon energy for solutions of potassium of varying molar concentration in liquid ammonia at $-60\,°C$. Quantum yields are measured at 0.0685 V m^{-1} torr^{-1} and are not corrected for backscattering of electrons in the gas phase[1].

(ii) The number distribution of emitted electrons according to their kinetic energy for monochromatic irradiation. The distribution is obtained by a retarding potential method (Sect. V.A). It is displayed[2] (Fig. 2) as an energy distribution curve (EDC) at a given photon energy. An EDC is a plot of the derivative of the quantum yield, at a given photon energy, with respect to retarding potential, against the retarding potential. The retarding potential in the present discussion is defined as the voltage being applied between the collector electrode and the metallic electrode making electric contact with the solution being irradiated. The kinetic energy of emitted electrons at any given retarding potential is equal to the electronic charge multiplied by the sum of the retarding potential and some constant depending on the contact potentials in the cell circuit. This constant is generally not known, but the value of the retarding potential corresponding to zero kinetic energy can be assigned approximately by procedures mentioned in the captions to the various EDC's in this chapter.

Energy distribution curves yield the functional dependence of the photoionization cross-section on photon energy, that is, the photoionization spectrum (Sects. V and VI). Spectral response curves provide, in conjunction with absorption spectra, criteria for establishing the occurrence of photoionization via autoionization of excited bound states (Sect. VII). These two topics are central to this chapter, but

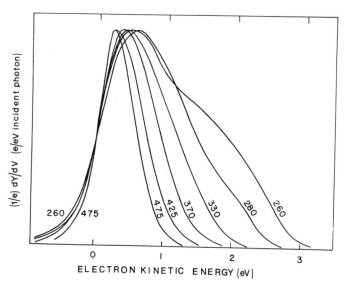

Fig. 2. Plots of EDC's at different wavelengths (in nm) for 6×10^{-2} M anthracene monovalent anion radical in HMPA at room temperature and for a collector pressure of 0.4×10^{-3} torr. The ordinate is the absolute value of $(1/e)dY/dV$, where e is the electronic charge, Y the quantum yield as electron emitted per incident photon, and V the difference of potential applied between the emitting liquid and the collector electrode. Zero kinetic energy is set by convention at the inflection point in the ascending branch of EDC's. Curves are normalized to the same height; ordinates at the maxima from 475 to 260 nm, respectively, in 10^{-5} electron per incident photon and per electron volt are 0.09, 1.4, 8.2, 9.0, 2.2, and 28.0. Segments of EDC's to the left of zero abscissa correspond to electrons emitted with negligible kinetic energy but accelerated in the emitter-collector field[2].

some preliminary material must be covered first: limitations imposed by the use of liquid emitting surfaces (Sect. II), some experimental aspects (Sect. III), and processes involved in PEE by liquids and solutions (Sect. IV).

We conclude these introductory remarks with a brief historical survey. A more extensive bibliography of early work is given in ref. 3. The first observation of PEE by solutions was reported by Stoletow[4] in 1888, and several investigations of this phenomenon appeared at the turn of the century. Emitters in solution were generally organic dyes, and complications arose from spurious photochemical reactions and the attending formation of a film at the surface of the solution. In 1940, Häsing[5] reported the first investigation of PEE by solutions of solvated electrons. He obtained the quantum yield spectral response for solutions of alkali metals in liquid ammonia. His work was not followed up until the writer and co-workers initiated work on the PEE spectroscopy of solutions in 1968[3]. The photoionization spectrum of solvated electrons (in HMPA) was obtained only recently[6] by this method. The development of PEE spectroscopy of liquid systems beyond the stage reached by Häsing was influenced by PEE studies in solid-state physics during the last fifteen years and particularly by the work of Spicer[7] and co-workers.

118

II. Experimental problems posed by a liquid emitting surface

The scattering of electrons by solvent molecules in the vapor phase in the space between the emitting solution and the collector electrode has two effects.

(i) PEE quantum yields are smaller than their actual values because of backscattering (Thomson scattering) of electrons to the emitting surface. The extent of backscattering increases with kinetic energy of electrons at emission, and consequently the spectral response for quantum yield is somewhat distorted.

(ii) Electrons lose kinetic energy in inelastic collisions with solvent molecules in the gas phase, and consequently EDC's are distorted. There is enrichment in collected low energy electrons at the expense of more energetic electrons (Fig. 3).

Quantum yields can be corrected for backscattering of electrons in the gas phase if collisions are not too markedly inelastic. In contrast, no simple correction for energy loss in inelastic collisions is available. Meaningful EDC measurements are feasible, as of this writing, only with solvents of sufficiently low vapor pressure and, if necessary, with equipment in which electron–molecule collisions in the gas phase are minimized by efficient continuous pumping.

Correction for backscattering in the gas phase has been studied in some detail[3,8 10], and a simple correction, based on the Thomson equation, was developed and tested. Backscattering can be extensive, e.g. approximately 50% of electrons emitted in nitrogen were backscattered to the emitting surface in an experiment under the most favorable conditions[9]. However, the backscattering correction will not be discussed any further for several reasons.

(i) The correction is very tentative for the markedly inelastic electron–molecule

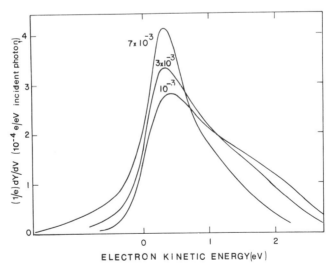

Fig. 3. Plots of EDC's at 260 nm and three pressures at the collector (torr) for the solution of Fig. 2. Zero kinetic energy is set at inflection point in the ascending branch of EDC at 7×10^{-3} torr. The same remark as in Fig. 2 applies about EDC segments to the left of zero abscissa[2].

collisions prevailing in the vapor of most solvents[10]. It turns out that backscattering is less important for such inelastic collisions than for elastic collisions[10].

(ii) The distortion of the shape of spectral response curves caused by backscattering is not so pronounced as to render these curves useless, e.g. in establishing the mechanism of photoionization (Sect. VII). This can be seen from the uncorrected and corrected spectral response curves in ref. 9.

(iii) Distorted EDC's cannot be corrected by this method, and undistorted EDC's provide the prime data needed in the determination of photoionization cross-sections.

Since there appears to be no relatively simple correction for the EDC distortion arising from electron–molecule collisions in the gas phase, experimental conditions must be achieved in which such collisions are relatively rare. The probability of collision is lowered by continuous and efficient pumping to achieve a steep pressure gradient in the gas phase near the solution–gas interface.

The pumping requirements were analyzed in ref. 2, and the probability that emitted electrons be collected without undergoing a single collision was derived. The calculation was based on simple arguments from the kinetic theory of gases and was performed on the simplifying assumption that the emitting surface and the collector are concentric spheres. It was shown that EDC distortion should be minor when the following condition is satisfied

$$\sigma n_0 (r_c - r_0)(p_c/p_0) \leq 0.2 \tag{1}$$

where, σ is the total cross-section for inelastic electron–molecule collisions in cm^2; n_0 the number of molecules per cm^3 corresponding to the equilibrium vapor pressure p_0 of the solvent; r_c and r_0 are the radii of the spherical collector electrode and emitting surface, respectively, in cm; and p_c is the pressure of the collector. Equation (1) holds when $r_0/r_c \ll 1$ and $p_c/p_0 \geq 10^{-2}$, but a more general condition is given in ref. 2. If, for instance, $\sigma = 10^{-16}$ cm^2 and $p_c/p_0 = 10^{-2}$, eqn. (1) implies that $p_0 \leq 2 \times 10^{-2}$ torr for the usual geometry of a PEE cell[2]. Thus, the selection of solvents is severely limited if significant EDC distortion is to be avoided.

Actual conditions may be less strenuous than is implied by the foregoing example for two reasons.

(i) The cross-section σ for inelastic collision often drops off markedly at kinetic energies of a few electron volts[11], that is, in the high energy tail of EDC's (Sect. V.A.).

(ii) It is precisely the tail segment of EDC's that is needed in obtaining the photoionization cross-section (Sect. V.B), and consequently distortion in the low energy segment of EDC's can be tolerated as long as the high energy tail is not distorted. In any event, the absence of EDC distortion in the high energy tail must be verified experimentally. This test can be performed by varying the pumping speed.

III. Some instrumental aspects

III.A Cell

Discussion will be limited to the description and discussion of a convenient cell for air-sensitive solutions and a block diagram of the optics and electronics. The reader is referred to the literature[1,2,6,8–10,12] (refs. 2 and 6 in particular) and detailed, unpublished reports[13–16]. In general, experiments in which only quantum yields are measured are less demanding than those in which EDC's are also determined.

The cell[16] of Fig. 4 is composed of three parts: the main body with the emission cup and the collector electrode, the cryogenic pumping compartment, and the attachment for overflowing and recycling the solution. Overflow of the solution complicates the experiment but is essential for the following reasons:

(i) To avoid the accumulation of impurities at the surface of the emitting solution.

(ii) To avoid formation of a solid film because of solvent evaporation under continuous pumping.

(iii) To prevent the freezing of the solvent, under pumping, in experiments near the solution freezing point (to lower the equilibrium vapor pressure). The operation is as follows:

The solution flows ($\approx$ 18 ml min^{-1}) through a capillary tube from reservoir A (90 ml) into cup B where it is irradiated from the top of the cell. The solution over-

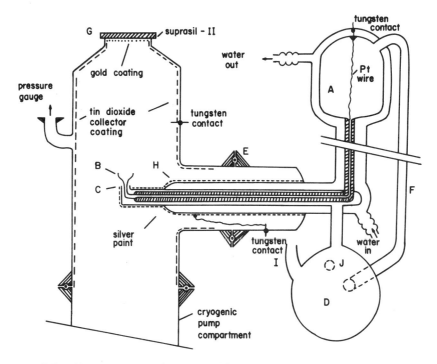

Fig. 4. Cell for the measurement of quantum yields and EDC's with air-sensitive solutions. Some significant dimensions are as follows: diameter of the main body of the cell, 11 cm; distance between the window G and the rim of cup B, 10 cm; outer diameter of cup B, 1 cm; height of the compartment for cryogenic pumping, 20 cm; horizontal distance between the axis of cup B and the flange of O-ring E, 14 cm; vertical distance between the bottom of reservoir A and the rim of cup **B**, 21 cm. Reservoirs A and D have a capacity of 90 ml. The diffusion pump is connected to the cryogenic pump compartment[16].

flowing from cup B into the concentric cup C is collected in reservoir D (90 ml). The attachment with reservoirs A and D is connected to the main body of the cell via O-ring E to allow rotation (plane perpendicular to the paper in Fig. 4) while maintaining the system under vacuum. Once reservoir A is nearly empty (approximately every 5 min), the attachment is slowly rotated by approximately 90°, and the solution collected in D flows back to A via tubing F. After completion of this transfer, the attachment is rotated back to its original position, and measurements are resumed. The overflow system is maintained at constant temperature by a water-jacket shown in Fig. 4. Cleaning procedure, preparation of solution (in reservoir D using inlets I and J) and other practical details are given in ref. 16.

A few important points to be considered in the design of PEE cells for EDC measurements will be noted.

(i) The geometry of the emitting cup-collector electrode assembly must approach or be equivalent to two concentric hemispheres, and the area of the emitting surface must be much smaller than the area of the collector electrode. The geometry in Fig. 4 (cylindric collector) fulfils this requirement for all practical purposes.

(ii) The collector electrode must have a much lower quantum yield than the emitting solution in the range of photon energies being covered. The tin dioxide coating[17] in the cell of Fig. 4 meets this requirement.

(iii) Accumulation of static charge on electrically insulated parts of the cell must be avoided strenuously. EDC's are worthless if this precaution is not taken, and much attention must be paid to this seemingly trivial problem. Three features of the cell of Fig. 4, which are important from this point of view, are noted: the window, G, is covered with a transparent gold layer; the stem, H, is covered with a conducting layer maintained at ground potential; the cup, C, has a smaller outer diameter than cup B, and the rim of C is below[15] the body of cup B.

(iv) Provision must be made for efficient cryogenic pumping in addition to evacuation by a diffusion pump. Details on the cooling element are given in ref. 15.

III.B Optics and electronics

Optics and electronics will be discussed on the basis of the block-diagram of Fig. 5 which shows schematically the equipment utilized in recent work[6,16] on solvated electrons. Alternative equipment used in previous investigations[9,10,12,17] will be noted where appropriate.

The optical train includes a 250-watt lamp (tungsten, xenon, xenon–mercury), two Suprasil II focusing lenses, interchangeable blocking and third-order interference filters, a chopper and a Suprasil II beam splitter. This equipment provides the rather high photon flux required for PEE studies with dilute ($< 10^{-2}$ M) solutions of emitter. A monochromator can conveniently be substituted for the set of interference filters (30 filters in ref. 6) in work limited to higher concentrations of emitters. Requirements for removal of stray light are quite severe, especially at wavelengths below 500 nm, whether a monochromator or a set of interference filters is used. Blocking filters are essential for this purpose.

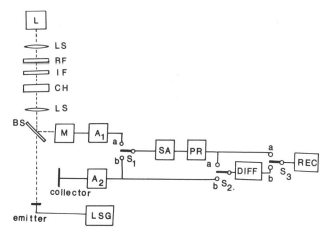

Fig. 5. Schematic diagram of optical train and electronics. L, lamp; LS, Suprasil-II lens; BF, blocking filter; IF, interference filter; CH, chopper; BS, Suprasil-II beam splitter; M, monitor detector; LSG, linear sweep generator; A_1, amplifier; A_2, amplifier; SA, selective amplifier; PR, precision rectifier with filter in the output; DIFF, feedback differentiator; REC, recorder. The positions of the switches are as follows: S_1 (a), S_2 (open) and S_3 (a) in the monitoring of the photon flux; S_1 (b), S_2 (open) and S_3 (a) in the measurement of quantum yields; S_1 (b), S_2 (a) and S_3 (b) in the recording of EDC's for PEE; S_1 (open), S_2 (b) and S_3 (b) in the recording of EDC's for thermionic emission[16].

The electronics of Fig. 5 fulfil the following functions: monitoring of the photon flux, measurement of PEE currents, recording of EDC's for photoelectron or thermionic emission. The modulated output from the monitoring detector or the PEE cell is amplified, and the output square-wave signal from A_1 or A_2 is changed into a sinusoidal signal by SA. The output of SA is read on REC after full rectification by PR. Thus, the photocurrent from either the monitor detector or the PEE cell is read directly on the recorder REC. The voltage applied to the PEE cell is set at a constant value on LSG in measurements of PEE currents. The output of PR is differentiated by operational amplifiers (DIFF) in the recording of EDC's for PEE by the solution, and the voltage applied to the PEE cell is scanned linearly by LSG in this operation. However, SA and PR are by-passed in the recording of EDC's for thermionic emission, and the output of A_2 is directly differentiated by DIFF.

The electronic equipment of Fig. 5 can be utilized for PEE studies in general, but it was especially designed for solutions exhibiting strong thermionic emission (alkali metals[8] in HMPA). Thus, the PEE current is modulated by the chopper, but the thermionic current is not (cf. also Sect. V.A). In general, thermionic emission by solutions is not observed, and it is then simpler to differentiate the photocurrent by superposition of a small (0.1–0.25 V) alternating signal on the voltage applied to the cell. This method of obtaining EDC's is widely used in PEE studies of metals and semiconductors, and some relevant literature is quoted in ref. 17. The beam impinging on the solution is not modulated in this method but is still modulated in the monitoring of the photon flux and the measurement of PEE currents.

IV. Processes involved in photoelectron emission by liquids and solutions

IV.A The three-step model

Spectral response curves for quantum yield and EDC's will be interpreted on the basis of a model[18] in which the PEE process is decomposed into the following three consecutive steps:

(i) Generation of quasi-free electrons in the bulk of the liquid by optical excitation.

(ii) Random walk of quasi-free electrons in the liquid prior to emission into the gas phase. This step also involves transfer of energy between moving quasi-free electrons and the liquid medium.

(iii) Emission of electrons into the gas phase. This last step is affected by the barrier at the solution–gas interface.

This model is inspired from Spicer's three-step model[7] for PEE by solid emitters (metals, semiconductors), but the details of each step are different for liquid and solid emitters. The decomposition of the PEE process into three steps is somewhat artificial[19], but a unified model for PEE by liquid systems seems quite untractable. Improvement, beyond the initial work[18], in the description of each step and in the correlation between these steps should be feasible and indeed has already been achieved for the random walk[20].

Semi-classical treatment of the model considerably simplifies matters. The generation of quasi-free electrons requires a quantum-mechanical description, whereas the random walk and barrier problems can be treated classically to a first approximation. The first two steps are discussed in Sect. IV.B and IV.C, but our treatment of the barrier will be brief. We shall consider the barrier to be classical and of constant height. Thus, we neglect several possible complications: tunneling, reflection of electron waves, details of the interfacial structure, and image potentials. This simplified approach is somewhat questionable, but the ensuing approximation does not affect the determination of photoionization cross-sections. The barrier problem would be a serious one, however, if one were to calculate quantum yields from a theoretical model involving an integration over all kinetic energies. Fortunately, such a theory is not necessary for obtaining photoionization cross-sections.

IV.B Generation of quasi-free electrons by optical excitation

The production of quasi-free electrons in liquids and the emission of electrons in vacuum will be discussed with reference to the energy level diagrams of Fig. 6. These diagrams were prepared on the basis of ideas developed in refs. 21–23. It is assumed that the surface potential at the liquid–vacuum interface is equal to zero, but this simplification will be removed at the end of the discussion. The following energies are defined with reference to the vacuum level, that is, with respect to the zero level corresponding to electrons in vacuum at infinity.

(i) V_0 is the ground state energy of quasi-free electrons in the liquid. The level V_0 thus corresponds to the bottom of the conduction band in the liquid. The value of V_0 or even its sign is generally not known for polar liquids, and consequently two

124

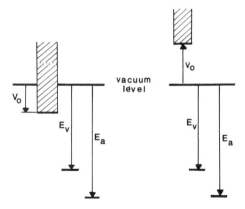

Fig. 6. Energy level diagram. The shaded area indicates the conduction band. Symbols are defined in text.

diagrams are displayed in Fig. 6 according to the sign of V_0. A negative value of V_0 indicates a positive affinity of the liquid for quasi-free electrons and vice versa.

(ii) E_v is the difference of energy between the vacuum level and the ground state of the emitter in solution for a process involving removal of an electron without change in the nuclear coordinates of the system (vertical process). The energy E_v is a negative quantity. Since there is a distribution of configurations of the emitter in solution in the ground state, E_v is defined for the most probable of these configurations. It should be noted, as indicated below, that the minimum photon energy required for photoelectron emission into vacuum always depends on E_v but also depends on V_0 when the latter is positive.

(iii) E_a is defined in the same way as E_v but for a process involving relaxation of the nuclear coordinates to the most probable configuration of the emitter minus its lost electron (adiabatic process). The energy E_a, which is also negative, is relevant to electron emission by thermal excitation. Reorganization of the medium requires some energy, and one expects $|E_a| > |E_v|$.

Several deductions follow directly from Fig. 6.

(i) Quasi-free electrons are generated in the liquid provided that the photon energy E satisfies the condition $E \geqq R_0 - E_v$. This condition holds regardless of the sign of V_0.

(ii) Photoelectron emission requires $E \geqq -E_v$ for $V_0 < 0$ and $E \geqq V_0 - E_v$ for $V_0 > 0$.

(iii) When $V_0 - E_v \leqq E < -E_v$ and $V_0 < 0$, quasi-free electrons are generated at a level below the vacuum level, and consequently no emission into vacuum occurs.

(iv) If $V_0 > 0$, the minimum photon energy $V_0 - E_v$ required for production of quasi-free electrons in the liquid is larger than $-E_v$. Electrons are emitted into vacuum with a kinetic energy equal to V_0 in the absence of energy loss between generation of quasi-free electrons and emission into vacuum.

(v) Likewise, if $E > -E_v$ and $V_0 < 0$, electrons are emitted into vacuum with kinetic energy $E + E_v$ in the absence of energy loss in the liquid and at the interface.

(vi) The energy corresponding to the surface potential at the liquid–vacuum interface must be added to the above values of photon energies when this surface potential hinders emission into vacuum. No correction is required in the opposite case.

IV.C Transport of quasi-free electrons

Quasi-free electrons undergo a net displacement in the liquid during their lifetime, that is, during their random walks between generation and ultimate retrapping (thermalization and possibly scavenging). This net displacement is much smaller than the length of the tortuous path in the random walk. The net displacement varies from one quasi-free electron to another, but the statistics of the process can be characterized by an average net displacement, that is, by the range of the quasi-free electrons. This range depends on the kinetic energy of quasi-free electrons upon generation and the rate at which kinetic energy is transferred to the liquid medium during random walks. Quantitative information on the dynamics of this type of energy transfer in polar liquids is limited at low kinetic energies (a few electron volts at most). Chapter 5 of this volume[24] is germane to this problem.

To develop the three-step model, we introduce a cut-off approximation consistent with the concept of range. We assume that quasi-free electrons emitted into the gas phase originate from a slab of liquid adjacent to the liquid–gas interface (referred to as the *emission slab*). In the absence of an interfacial barrier and in the simplified case of $V_0 = 0$, the emission slab has a thickness equal to the product of the range of quasi-free electrons by a numerical constant accounting for the geometry of the overall process. The emission slab is somewhat thinner for emission hindered by an interfacial barrier, since the maximum kinetic energy available for transfer to the liquid is then reduced for emitted electrons, by an amount equal to the height of the barrier (classical description). Quasi-free electrons generated in the liquid beyond the emission slab are retrapped before they reach the liquid–gas interface. The probability of emission for quasi-free electrons generated in the emission slab is smaller than unity and is determined by the statistics of the random walk and the dynamics of energy transfer to the liquid[18,20].

The photon flux is hardly attenuated over the thickness of the emission slab, except possibly for PEE by pure liquids, and consequently most incident photons are absorbed at greater depths in the liquid. The quantum yield, expressed as the number of emitted electrons per incident photon, therefore is much smaller than unity (Fig. 1) at the usual concentrations of emitter in solution.

One expects from the foregoing analysis that the quantum yield at a given photon energy be proportional, to a first approximation, to the product of the photoionization cross-section and the range (corrected for the interfacial barrier) corresponding to the energy of irradiation. This relationship presupposes only negligible attenuation of the photon flux in solution over a distance equal to the range of quasi-free electrons. Moreover, the photoionization cross-section and the range are functions of photon energy. The corresponding functional dependencies are not known, and these two quantities cannot be extracted from the spectral response curves for quantum yield in the absence of complementary information. Of course, one could assume a known functional-dependence for one of these quantities, but the procedure is rather tentative at this stage[18].

V. Energy distribution curves

V.A Interpretation

The morphology of EDC's will be discussed by considering first plots of quantum yield (electrons emitted per incident photon) against retarding potential at a given photon energy. Such plots will be referred to as *retarding potential curves*. The EDC then corresponds to the first derivative of the retarding potential curve with respect to retarding potential (Sect. I and Fig. 2). For convenience, EDC's in this chapter are plotted against the kinetic energy of emitted electrons, rather than the retarding potential in the presentation and discussion of results. The procedure for conversion of retarding potential to kinetic energy is noted in each caption. The conversion involves some arbitrariness, and the zero of kinetic energies is known only approximately. This uncertainty is of no consequence in the determination of photoionization cross-sections, as we shall see.

The morphology of retarding potential curves is readily understood from simple considerations. A single PEE process is assumed, and backscattering caused by the solvent vapor (Sect. II) is supposed to be negligible. At any given retarding potential and photon energy, only those electrons are collected which have sufficient kinetic energy at emission into the gas phase to overcome the adverse electric field applied between the emitting liquid surface and the collector electrode. The situation is comparable to the throwing of a ball in the vertical direction against the gravitational field of the earth. A change in the retarding potential, at a given photon energy, would correspond to a change in the gravitational field. Practically none of the emitted electrons are collected when the collector is highly negative with respect to the emitting surface. Conversely, practically all emitted electrons are collected when the electric field is equal to zero or accelerates electrons. The saturation photocurrent is then observed and the retarding potential curve exhibits a plateau. The Schottky effect is indeed negligible at the low prevailing accelerating fields[9]. The plateau should be reached at zero field, but in practice it is often observed at accelerating potentials because of imperfect geometry (Sect. III). There is a progressive transition from one extreme situation to the other, and the quantum yield increases in a monotonic fashion from zero to its saturation value in the plateau region.

The sigmoid shape of retarding potential curves is reflected in EDC's (Fig. 2): a peak at low kinetic energies and a drawn-out tail at higher kinetic energies. EDC's become wider as the photon energy is increased, primarily because of a change in the tail region. Widening of EDC's reflects the increase in the maximum kinetic energy available for transfer to the liquid medium upon increase of photon energies[2,17] (Sect. IV). This description of EDC's has been expressed in a detailed quantitative treatment[20]. The initial distribution of quasi-free electrons according to their kinetic energy was taken into account in this treatment.

The ascending part of EDC's and the position of the peak in the scale of kinetic energies are affected only in a minor way by a change in photon energy. The shape of EDC's in this region of low kinetic energies is strongly dependent on the effect of the barrier at the interface, and realistic quantitative analysis may be quite arduous. Fortunately, this region of EDC's is not needed to obtain photoionization cross-sections (Sect. V.B).

The EDC's at 260 and 280 nm in Fig. 2 exhibit a hump in the tail. This hump indicates PEE by two processes having different spectral response curves for quantum yield. It can be shown that the occurrence of two distinct PEE processes does not result in the appearance of two peaks in the EDC[20].

The foregoing features of EDC's are wholly consistent with the model of Sect. IV, and they were observed thus far for PEE by anthracene monovalent anion radical[2] in HMPA and ferrocyanide anion in glycerol[17,25]. EDC's for solvated electrons[6], observed with solutions of sodium metal in HMPA, exhibit an unusual feature, namely a high energy peak which is apparent in the EDC's at 300 and 440 nm in Fig. 7. The solution in this experiment contained solvated electrons and a metal species, presumably[26] Na^-, which also exhibited PEE (the second step in the curve at 10 mM in Fig. 8). However, the EDC's of Fig. 7 were obtained at a sodium

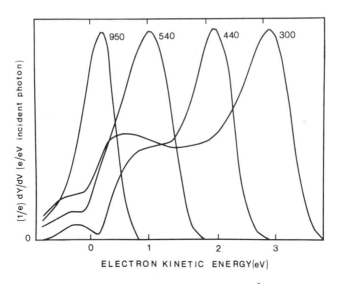

Fig. 7. Plots of EDC's at different wavelengths (nm) for 5×10^{-3} M sodium metal in HMPA at approximately 10 °C. Zero kinetic energy is set by convention at the peak of the EDC for thermionic emission (measured without irradiation). The maximum to the left of zero kinetic energy in the curve at 300 nm results from incomplete suppression of the EDC for thermionic emission (Sect. V.A). This spurious effect distorts the ascending branch of the EDC, and the resulting maximum is not the genuine maximum of the EDC for thermionic emission obtained without irradiation. The zero kinetic energy corresponds approximately to a difference of potential of 3.5 V between the collector electrode and the emitting liquid. Curves are normalized to the same height; ordinates at the maxima from 950 to 300 nm, respectively, in 10^{-6} electron per incident photon and per electron volt are 2.7, 6.4, 5.9, and 4.6. The same remark as in Fig. 2 applies about EDC segments to the left of zero abscissa[6].

128

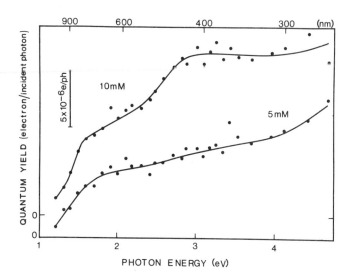

Fig. 8. Plot of quantum yield as electron per incident photon against photon energy for two solutions of sodium metal of different concentrations in HMPA at approximately 10 °C[6].

analytical concentration at which PEE by Na$^-$ is quite negligible, and the high energy peak in the EDC's is not related to the Na$^-$ species. There is also a small peak near zero kinetic energy in two EDC's in Fig. 7. This peak corresponds to an incompletely suppressed EDC for thermionic emission arising from modulation of the temperature at the surface of the solution under intense irradiation. This artifact, discussed further in ref. 6, need not concern us here.

Several interpretations of the high energy peak in Fig. 7 are considered in ref. 6. It was concluded that the peak is probably caused by enrichment of solvated electrons in the double layer at the solution–gas interface. Such a complication was not introduced in the model of Sect. IV, and apparently this simplification was justified for EDC's obtained thus far with systems[2,17,25] other than solvated electrons in HMPA. The argument is developed in detail in ref. 6.

V.B Superposition of energy distribution curves

Photoionization cross-sections are determined by taking advantage of a remarkable property of EDC's observed for all systems studied so far[2,6,25], namely the superposition of EDC's over wide ranges of kinetic and photon energies according to the following procedure:

Consider two EDC's for a given system measured at different photon energies. The treatment of data involves two steps.

(i) The EDC obtained at the higher photon energy is shifted toward lower kinetic energies along the abscissa axis by a constant increment equal, in absolute value, to the difference between the photon energies at which the EDC's were measured.

(ii) The ratio of the ordinate of the shifted EDC to the ordinate of the unshifted EDC, at a given common abscissa, is computed. It was found so far[2,6,25] that this ratio is constant, within experimental errors, over a wide range of abscissae. Hence, the shifted and unshifted EDC's are superimposed over the range of abscissae in which the calculated ratio of ordinates is constant. A whole series of EDC's can be treated in this fashion by using a single EDC as unshifted curve, e.g. the EDC at 300 nm in Fig. 9.

The superposition of EDC's is essentially an experimental fact, the implications of which are analyzed in ref. 2. The analysis, which is too lengthy to be reproduced here, involves two main steps.

(i) A general equation was proposed for EDC's on the basis of the model of Sect. IV. Details of each step of the model were not introduced, and this EDC equation therefore is of general validity, inasmuch as the model of Sect. IV is adequate in its broad features.

(ii) The implications of EDC superposition were then analyzed on the basis of the EDC general equation by using purely mathematical arguments. Thus, it was shown that the ratio of ordinates for a pair of shifted and unshifted EDC's, in the range of superposition, is equal to the ratio of the photoionization cross-sections at the photon energies at which the EDC's were obtained. Hence, variations of this ratio with photon energy, for one common unshifted EDC, represent the functional dependence of the photoionization cross-section on photon energy. The photo-ionization spectrum, thus obtained, represents a bulk property of the solution since

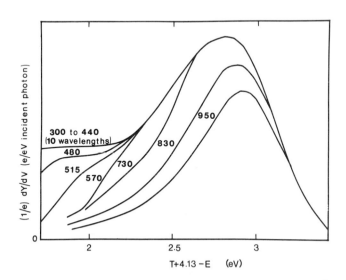

Fig. 9. Plots of superimposed EDC's at different wavelengths (nm) for 5×10^{-3} M sodium metal in HMPA at approximately 10 °C. The scale of electron kinetic energy for each EDC (T) is shifted by 4.13 eV (300 nm) $- E$ (photon energy for each EDC) so as to normalize to the EDC at 300 nm. The ordinates of all EDC's, except those at 830 and 950 nm, are normalized at the maximum. The EDC's at 830 and 950 nm were normalized in the superposition range in the high energy tail[6].

the effect of the interfacial barrier on PEE is removed by the method of analysis of EDC's. Absolute values of photoionization cross-sections are not obtained by this procedure, and cross-sections are expressed in arbitrary units.

The validity of this method was ascertained for PEE by solutions of anthracene monovalent, anion in HMPA[2]. The photoionization spectrum obtained by EDC superposition in this case could be compared with the independent evidence provided by the absorption spectrum.

We consider now EDC superposition for solutions of solvated electrons in HMPA, as reported in ref. 6. Superposition is achieved for typical EDC's in Fig. 9 over a range of kinetic energies covering approximately 2 eV for the EDC's between 300 and 440 nm. The superposition range becomes narrower as the photon energy decreases because the shape of EDC's at low kinetic energies is complicated by the barrier problem (Sect. V.A). The standard deviation for any pair of superimposed EDC's in the experiment of Fig. 9 was below 10% and was compatible with experimental errors in such rather delicate experiments. Further details on superposition are given in ref. 6.

VI. Photoionization spectrum of solvated electrons

The photoionization spectrum of solvated electrons in HMPA was obtained in ref. 6 by EDC superposition for solutions of sodium metal at different concentrations in this solvent. The resulting spectrum (Fig. 10) is independent of sodium concentra-

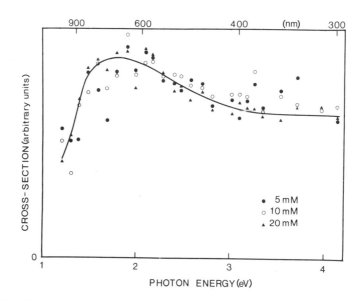

Fig. 10. Plot of photoionization cross-section of solvated electrons against photon energy for three sodium metal concentrations in HMPA at approximately 10 °C. Ordinates for the 5, 10 and 20 mM solutions are normalized in the ratios 1:1.57:2.63[6].

tion, although PEE by the Na⁻ species (Sect. V.A) is significant at the two highest concentrations (Fig. 8). Thus, the effect of PEE by the Na⁻ species was effectively eliminated by EDC superposition. The data in Fig. 10 are representative of a bulk property and not of a cross-section characteristic of the interfacial region. This question is taken up in ref. 6 where it is pointed out that the loss of kinetic energy by quasi-free electrons in the superposition domain in Fig. 9 corresponds to a range of quasi-free electrons much greater than the thickness of the interfacial region.

The photon energy of ~1.7 eV at the broad maximum in Fig. 10 is much higher than the energy of ~0.56 eV at the maximum of the absorption band of solvated electrons in HMPA[27-29]. Comparison of the photoionization and absorption[29] spectra shows that the contribution of photoionization to the oscillator strength does not exceed a few per cent in this particular instance. The relative changes in photoionization cross-section and absorbance[28] between 1.7 and 2.5 eV appear to be the same. Hence, the high energy region in the tail of the absorption spectrum seems to correspond mostly to bound–continuum transitions, as was previously suggested[18]. Unfortunately, comparison is not feasible above 2.5 eV because the absorption spectrum is complicated by spurious species generated by pulse-radiolysis[28,29] or by the Na⁻ species for solutions of sodium in HMPA[27].

There is a definite resemblance between the cross-section curve of Fig. 10 and the spectral response curve for quantum yield in Fig. 8 (curve for 5 mM). The quantum yield curve therefore provides a crude approximation to the photoionization cross-section curve. It should be noted, however, that the two curves diverge progressively with increasing photon energy because of the concomitant increase in the range of quasi-free electrons (Sect. IV). This effect of the range and possibly a small contribution to PEE by Na⁻ at the higher photon energies account for the absence of a maximum in the spectral response curve (5 mM) for quantum yield. The curve at 10 mM in Fig. 8 exhibits a definite step for PEE by Na⁻, and comparison with the photoionization spectrum is limited to low photon energies.

The photoionization spectrum of Fig. 10 and similar spectra for other solvents should prove valuable in the testing of theoretical models of solvated electrons[30]. For instance, the variation of cross-section in the tail of Fig. 10 is too small to be accounted for by the 8/3 power-dependence on photon energy which one would expect for a simple model based on hydrogen-like wave functions[18,31]. Photoionization spectra should provide a more severe and less ambiguous test than absorption spectra since the latter may be composed of several unresolved, overlapping bands[30,32]. In fact, experimental absorption and photoionization spectra supply complementary information.

Two approaches can be followed in theoretical work.

(i) One predicts optical and other properties by starting with some model, as has generally been done[30].

(ii) One starts with the photoionization spectrum and deduces by theory the potential well accounting for the photoionization spectrum. A calculation of this type was recently reported[33], but the resulting potential well may prove quite approximate since the Born approximation was introduced. The second approach, however, appears interesting despite the difficulties inherent to such inverse calculations (from spectrum to potential well) beyond the Born approximation. The curve of Fig. 10 is being analyzed from this point of view at present in the writer's laboratory.

VII. Autoionization of excited bound states in solution

VII.A Anthracene anion radicals

One might think that bound–continuum transitions provide the only mechanism for the photoionization of solutions of solvated electrons since this type of emitter may be regarded as a one-electron system. However, there are indications of photoionization via the autoionization of excited bound states in solutions of alkali metals in two solvents which have been investigated from this point of view, namely HMPA[6] and liquid ammonia[1]. The evidence for autoionization for these systems is not as conclusive as for anthracene anions in solution[2,12], and we shall consider this last case first to show how the occurrence of autoionization processes in solution can be established by PEE studies. This was the case for which such arguments were initially presented[12].

The central argument in favor of photoionization via autoionization involves the comparison of the absorption spectrum of the emitter with the spectral response curve for PEE. Such a comparison is made in Fig. 11 for the anthracene monovalent anion radical A^- in tetrahydrofuran (THF). The absorption spectrum was taken from ref. 34. A similar and just as conclusive comparison is given in ref. 12 for the anthracene divalent anion A^{2-}, but this case will not be considered here.

Absorption and PEE bands can be compared on two counts: (i) the position of the bands in the scale of photon energies and (ii) the relative intensities of each type of band for spectra exhibiting multiple bands (discussed below). The absorption and PEE bands in Fig. 11 are closely matched in the scale of photon energies, and the matching is just as excellent when the absorption spectrum is compared with the photoionization spectrum. The latter, obtained by EDC superposition, is given in ref. 2 for PEE by A^- in HMPA. The matching for A^- is in marked contrast with the case of solvated electrons in HMPA for which the absorption and photoionization spectra exhibit their maxima at very different energies, i.e. 0.56 and ≈ 1.7 eV, respectively (Sect. VI).

The absorption bands in Fig. 11 have been assigned to bound–bound transitions in several detailed studies mentioned in ref. 12, and yet the PEE bands match these absorption bands in the scale of photon energies. This matching and other observations in refs. 2 and 12 are explained by the following mechanisms of photoionization of A^-.

Photoionization via a direct bound–continuum transition is assumed to require photon energy of the order of at least 2.2 eV (approximate experimental threshold; see caption to Fig. 11). The corresponding continuum is marked (I) in Fig. 12. Several bound–continuum transitions probably occur at photon energies above 2.2 eV but their contribution to PEE is quite minor. However, the excited bound states corresponding to the absorption bands at 3.35, 3.8 and 4.8 eV, respectively, autoionize

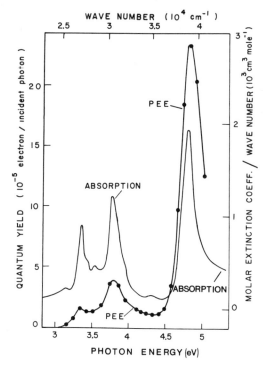

Fig. 11. Comparison of absorption spectrum and spectral-response curve for quantum-yield for anthracene monovalent anion A⁻ in THF. Concentration is $2 \times 10^{-2} M A^- + 3 \times 10^{-2} M$ anthracene in the PEE studies; concentration was not specified in ref. 34 for the absorption spectrum, but it is presumably of the order of 10^{-4} M. Absorption bands below 2.5 eV are not shown. No detectable PEE was observed below approximately 2.2 eV[12].

(Fig. 12). Configurations for these excited states interact with configurations of similar energies in the continuum, and radiationless transitions occur from the bound excited states into the continuum.

Autoionization thus accounts for the matching of bands in the energy scale in Fig. 11. Moreover, each pair of matching bands has nearly the same width since the range of quasi-free electrons does not vary markedly over the narrow interval of photon energy corresponding to the band width. However, the ratios of the intensities are not necessarily the same for the absorption and PEE bands because the efficiency of the autoionization process may vary from one band to the other. Ratios of intensities should really be compared for absorption and photoionization spectra (and not spectral response curves for quantum yield, as in Fig. 11) to eliminate the effect of the varying range of quasi-free electrons (Sect. IV).

The arguments are presented in much greater detail in refs. 2 and 12, and other points, e.g. the continuum (II) in Fig. 12, are also examined. The foregoing outline, however, suffices for our present purpose, which is primarily to consider possible autoionization mechanisms in PEE by solutions of alkali metals.

134

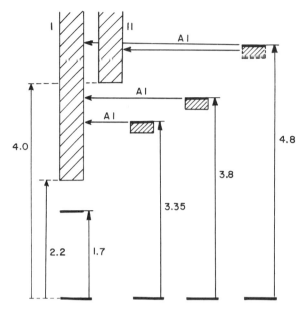

Fig. 12. Energy level diagram for discussion of autoionization of anthracene monovalent anion in THF or HMPA. Ground states are set by convention at a common level for transitions. AI refers to autoionization. Shaded areas adjacent to vertical arrows indicate approximate ranges of autoionizing states as deduced from EDC studies[2].

VII.B   The Na⁻ species in hexamethyl phosphoric triamide

The spectral response curve for quantum yield for 10 mM in Fig. 8 exhibits a second step which was assigned to PEE by the Na⁻ species[6,9]. This assignment is consistent with the observation of two absorption bands[27] in not too dilute solutions of sodium metal in HMPA. The photoionization cross-section of Na⁻ obtained by EDC superposition[6] is displayed in Fig. 13. This curve shows a marked decrease in cross-section at higher photon energy in contrast with the curve for solvated electrons in Fig. 10. Moreover, arguments presented in ref. 6 point toward the occurrence of two maxima in the cross-section curve: the maximum near 3 eV in Fig. 13 and another maximum near 1.7 eV. The latter coincides with the maximum of the absorption band of Na⁻ in HMPA[27]. These and other considerations[6] suggest the following mechanisms for the photoionization of Na⁻ in HMPA.

(i) Photoionization of Na⁻ via direct bound–continuum transitions requires photon energies somewhat below 1.7 eV.

(ii) The bound excited states corresponding to the absorption band at 1.7 eV autoionize. Photoionization is observed, as noted in ref. 6, at photon energies in the absorption band of Na⁻ in solution for several solvents[35,36]. The matching of the absorption spectrum of Na⁻ and the plot of photoionization yield against photon energy[35] strongly suggests an autoionization mechanism. Moreover, it

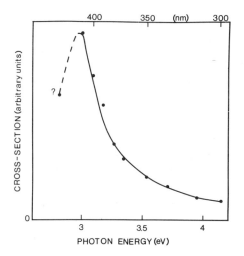

Fig. 13. Plot of photoionization cross-section against photon energy for Na$^-$ species for 96 mM solution of sodium metal in HMPA at approximately 10 °C[6].

was recently reported[37] that negative ions of the alkali metals in the gas phase undergo photodetachment of electrons, via autoionization. Finally, PEE by the Na$^-$ species was inferred in ref. 6 from the analysis of EDC's obtained with a concentrated solution (96 mM) of sodium in HMPA.

(iii) The cross-section curve of Fig. 13 corresponds chiefly to autoionization of excited bound states produced by absorption by Na$^-$ near 3 eV. The case of Na$^-$ is similar, in this respect, to that of the anthracene anion A$^-$ for which one observes multiple absorption bands and their matching PEE bands (Sect. VII.A). This interpretation is consistent with the sharp decrease in cross-section above 3 eV in Fig. 13, in contrast with the slow variation of cross-section in the curve of Fig. 10 for bound–continuum transition. Observation of the absorption band near 3 eV would require reflectance spectroscopy studies of concentrated solutions of sodium in HMPA, and such studies have not been made to the writer's knowledge. In the absence of this piece of evidence and further studies, the proposed interpretation should be regarded as a plausible one.

VII.C Solutions of potassium and sodium in liquid ammonia

The quantum yield curve for solutions of potassium (and also sodium) in liquid ammonia exhibits a band at ~4.6 eV (Fig. 14). The ascending branch of this band is also apparent in the quantum yield curves of Fig. 1. The origin of this band was examined in ref. 1, and an autoionization mechanism was tentatively suggested. Autoionization requires by its very nature a system with at least two electrons, but it cannot be ruled out a priori for solvated electrons. One-electron models are approximations of the solvent–solvated electron entity. It is conceivable, then, that

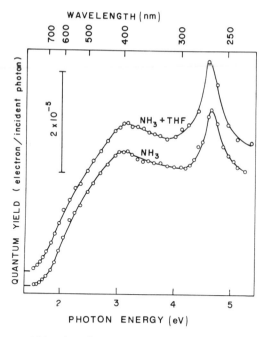

Fig. 14. Plots of quantum yield against photon energy at $-60\,°C$ for 0.5 M potassium in liquid ammonia and in liquid ammonia/THF (75/25 by volume). Quantum yields are measured at $0.0685\ V\ m^{-1}\ torr^{-1}$ and are not corrected for backscattering of electrons in the gas phase[1].

bound excited states of the solvent for the solvent–solvated electron system auto-ionize by coupling with states in the continuum of the solvated electron.

Several arguments favor the foregoing interpretation of the PEE band at 4.6 eV[1]:

(i) Assignment to PEE by amide is ruled out by an experiment with a sodium amide solution (0.01 M) in liquid ammonia in which no PEE was detected.

(ii) The ratio of the quantum yield at the maximum of the PEE band for the solvated electron (Fig. 1) to the quantum yield at any point along the high energy (4.6 eV) PEE band was independent of the nature of the metal (sodium or potassium) and its analytical concentration.

(iii) The narrowness of the PEE band at 4.6 eV points toward autoionization rather than to bound–continuum transitions.

(iv) There are some indications[38,39] of an absorption band above 4 eV for a saturated solution of lithium in liquid ammonia.

(v) The absorption spectrum of hydrated electrons exhibits the shoulder of a high energy band starting at $\sim$210 nm, and consequently the case of solvated electrons in liquid ammonia does not appear to be a unique one. Further work on the absorption spectrum of alkali metal solutions in liquid ammonia in the high photon energy range seems in order. The same comment could be made about solutions of solvated electrons in general, in view of the above considerations on autoionization mechanisms.

VIII. Conclusion

PEE spectroscopy of solutions has provided so far two significant results already mentioned in a general way in Sect. I:

(i) The photoionization spectrum of solvated electrons was obtained for solutions of high ionic conductivity in HMPA.

(ii) Autoionization mechanisms were established for photoionization of solutions of anthracene anions A^- and A^{2-} and were suggested for solutions of alkali metal anions in HMPA and for solvated electrons in liquid ammonia. These results were not available, it would seem, prior to these studies, and PEE spectroscopy appears as a useful tool for the study of photoionization processes in liquids and solutions. Further work extending application to the vacuum UV range is desirable and, in fact, is already being carried out[25]. Novel or improved methods for EDC measurements with liquids of significantly higher vapor pressure than HMPA are desirable. EDC determinations may also prove useful in other problems than those considered here, e.g., energy transfer between quasi-free electrons and liquids, photoionization processes in inorganic and organic glasses.

Acknowledgements

The work from the writer's laboratory, reviewed here, was supported by the National Science Foundation and the Office of Naval Research. Valuable comments were made about the manuscript of this chapter by H. Aulich, L. Chia, L. Nemec and N. Presser.

References

1 H. Aulich, B. Baron, P. Delahay and R. Lugo, J. Chem. Phys., **58** (1973) 4439.

2 H. Aulich, P. Delahay and L. Nemec, J. Chem. Phys., **59** (1973) 2354.

3 B. Baron, P. Chartier, P. Delahay and R. Lugo, J. Chem. Phys., **51** (1969) 2562.

4 M. A. Stoletow, Compt. Rend., **106** (1888) 1593.

5 J. Häsing, Ann. Phys. (Leipzig), **37** (1940) 509.

6 H. Aulich, L. Nemec and P. Delahay, J. Chem. Phys., **61** (1974) 4235.

7 W. E. Spicer, Phys. Rev., **112** (1958) 114. Numerous papers on PEE by metals and semiconductors from this author's group have appeared since 1958.

8 B. Baron, P. Delahay and R. Lugo, J. Chem. Phys., **53** (1970) 1399.

9 B. Baron, P. Delahay and R. Lugo, J. Chem. Phys., **55** (1971) 4180.

10 P. Delahay, P. Chartier and L. Nemec, J. Chem. Phys., **53** (1970) 3126.

11 L. G. Christophorou, Atomic and Molecular Radiation Physics, Wiley-Interscience, New York, 1971, pp. 321–407.

12 H. Aulich, B. Baron and P. Delahay, J. Chem. Phys., **58** (1973) 603.

13 B. Baron, P. Delahay and R. Lugo, Preparation and Handling of Sodium Solutions in Hexamethyl Phosphoric Triamide for Thermionic and Photoelectron Emission Studies, Technical Report No. 13 to the Office of Naval Research, Task No. NR 051-258, 1970.

14 H. Aulich, B. Baron, P. Delahay and R. Lugo, Experimental Methods in Photoelectron Emission Spectroscopy: Solvated Electrons in Liquid Ammonia, Technical Report No. 21 to the Office of Naval Research, Task No. NR 051-258, 1972.

15 H. Aulich, P. Delahay and L. Nemec, Experimental Methods for the Determination of Energy

Distribution Curves for Photoelectrons Emitted by Solutions, Technical Report No. 23 to the Office of Naval Research, Task No. NR 051-258, 1973.

16 H. Aulich, L. Nemec, and P. Delahay, Experimental Methods for the Determination of the Photo-ionization Spectrum of Solvated Electrons, Technical Report No. 26 to the Office of Naval Research, Task No. NR 051-258, 1974.

17 L. Nemec, B. Baron and P. Delahay, Chem. Phys. Lett., **16** (1972) 278.

18 P. Delahay, J. Chem. Phys., **55** (1971) 4188.

19 C. Caroli, D. Lederer-Rozenblatt, B. Roulet and D. Saint-James, Phys. Rev., *B*, **8** (1973) 4552.

20 L. Nemec, J. Chem. Phys., **59** (1973) 6092.

21 J. Jortner, in G. Stein (Ed.), Radiation Chemistry of Aqueous Solutions, Wiley-Interscience, New York, 1968, pp. 91–107.

22 J. Jortner and N. R. Kestner, in J. J. Lagowski and M. J. Sienko (Eds.), Metal–Ammonia Solutions, Butterworths, London, 1970, pp. 49–103.

23 N. R. Kestner, in J. Jortner and N. R. Kestner (Eds.), Electrons in Fluids, Springer, Berlin, 1973, pp. 1–25.

24 A. Mozumder, this volume, Chap. 5, p. 139.

25 L. Chia, L. Nemec, and P. Delahay, unpublished results.

26 J. L. Dye, in J. Jortner and N. R. Kestner (Eds.), Electrons in Fluids, Springer, Berlin, 1973, pp. 77–92.

27 J. M. Brooks and R. R. Dewald, J. Phys. Chem., **72** (1968) 2655.

28 H. Nauta and C. van Huis, J. Chem. Soc. Faraday Trans. II, **68** (1972) 647.

29 E. A. Shaede, L. M. Dorfman, G. J. Flynn and D. C. Walker, Can. J. Chem., **51** (1973) 3905.

30 N. R. Kestner, this volume Chap. 1, p. 1.

31 R. Lugo and P. Delahay, J. Chem. Phys., **57** (1972) 2122.

32 D. C. Walker, this volume Chap. 3, p. 91.

33 M. Tachiya, Y. Tabata and K. Oshima, Chem. Phys. Lett., **19** (1973) 588.

34 G. J. Hoijtink, N. H. Velthorst and P. J. Zandstra, Mol. Phys., **3** (1960) 533.

35 S. H. Glarum and J. H. Marshall, J. Chem. Phys., **52** (1970) 5555.

36 D. Huppert and K. H. Bar-Eli, J. Phys. Chem., **74** (1970) 3285.

37 T. A. Patterson, H. Hotop, A. Kasdan, D. W. Norcross and W. C. Lineberger, Phys. Rev. Lett., **32** (1974) 189.

38 W. E. Mueller and J. C. Thompson, in J. J. Lagowski and M. J. Sienko (Eds.), Metal–Ammonia Solutions, Butterworths, London, 1970, pp. 293–297.

39 J. A. Vanderhoff, E. W. LeMaster, W. H. McKnight, J. C. Thompson and P. R. Antoniewicz, Phys. Rev., A, **4** (1971) 427.

Chapter 5

DYNAMICS OF ELECTRON–SOLVENT INTERACTIONS

A. MOZUMDER

I. Introduction

In the interaction of ionizing radiations with matter, both in the gaseous and condensed phases, the electron is the most important single primary species. It appears as the primary radiation in most low-LET cases and in all cases manifests itself as an ubiquitous and unavoidable secondary radiation being produced by ionization of the medium molecules. Sometimes an electron, after being sufficiently slowed down by the medium it traverses, can attach itself to a medium molecule forming a negative ion. The negative ion so formed may have a transitory existence or in some suitable cases it can stabilize itself by further energy loss, eventually neutralizing with a positive ion, also formed in the ionization process, in the long-time limit. Anions may also be formed by dissociative electron capture either by the medium molecules themselves or by solute molecules. Alternatively, they may be formed by attachment to solute molecules followed either by ejection or stabilization. In polar media the electron and the anion undergo further stabilization by drawing and arranging the molecules of the medium around the charge distribution. Evidence is now accumulating, mainly from spectroscopic sources, that there is a certain amount of solvation for electrons in non-polar liquids and glasses also. Of course, this effect is expected to be much less pronounced in non-polar compared to polar media. Thus anions are sometimes referred to as being the negative ions which are in (local) thermal equilibrium with the medium. However, in radiation chemistry this specialization is not strictly adhered to.

Interactions of electrons and anions with the medium and as manifestations of response to external perturbations (e.g. static electric field, electromagnetic radiation, etc.) may conveniently be divided into two categories, viz. stationary and dynamical. Stationary interactions are those in which the charge as a whole is not transported away, whereas dynamical interactions deal precisely with charge transport in one way or another. Most important stationary properties of the electron (and the anion) relate to its structure and spectrum, both optical as well as ESR. Then again structure and spectra are related in quantum mechanics through (model) Hamiltonians. In fact, one of the most important reasons for the spectral study of trapped and solvated electrons in liquids and glasses is to understand their structure. This information can then be used to gain insight into the reactions of these intermediate species. Among the dynamical interactions of the electron and the anion, that leading to the mobility of the respective charge carrier is the most significant. This quantity is easily measured and, in principle, studies of mobility including the influence of temperature, composition and field are capable of revealing a harvest of information about the interaction of the charge carrier with the medium at or near thermal energies. Mobility belongs to one of the dynamical phenomena that represent *first-order responses* of the system (electron plus the medium) to an external perturbation, in this case the impressed electric field. Other first-order responses include dielectric relaxation, solvation and electron tunneling, the last mentioned item occurring only in suitable situations such as electron scavenging or neutralization in low temperature glasses. It is not clear if electron trapping should also be considered as a first-order process. If a collision analogy holds with a given trap density and a specified cross-section then perhaps it should also qualify as a first-order process.

In this chapter, which is closely related to Chaps. 1, 6 and 7, we will discuss such dynamical aspects of interaction as trapping, dielectric relaxation, mobility and (electron) tunneling. In a real sense the stationary aspects of interaction, leading to structure and spectrum as well as their interrelation, is basically quantum-mechanical in character. One must know the ground and excited state wave functions of the electron reasonably well to determine the energy and the oscillator strength of an optical transition. On the other hand, a great deal of the dynamical interaction, albeit phenomenological, can be described within a classical framework. At the same time a detailed, microscopic knowledge cannot be gained without the use of quantum mechanics. This operational dichotomy in the description of dynamical interactions will be stressed.

II. Qualitative features of electron trapping and solvation in condensed amorphous media

Trapped electrons are readily produced in amorphous, glassy media by γ- or electron-irradiation. They may also be produced by photoionization of a suitable solute such as TMPD. Their lifetimes in organic glasses at 77 K or lower temperatures easily exceed several hours so that they can be studied at leisure. Trapped electrons are also produced in inorganic glasses (e.g. borosilicate) and in many cases they are quite stable at room temperature. Although trapped electrons in inorganic glasses have been used in dosimetry they are very complicated structurally. Perhaps that is why the radiation chemistry of trapped electrons is concerned almost exclusively with organic glasses (and liquids).

Trapped electrons have been studied by radiation–chemical means over the past fifteen years or so. During this time a few good reviews have appeared on this topic. They include articles by Willard[1], Hamill[2], Ekstrom[3] and Kevan[4]. Recently, Funabashi[5] has also reviewed certain theoretical aspects of trapped electrons. The material appearing in this section is naturally related to various other topics appearing elsewhere in this book. However, the reader will be most benefited if Chaps. 1, 2 (Sect. II) and Chap. 6 (Sect. II) are read alongwith.

II.A Electron trapping as a universal phenomenon

Electron trapping and solvation in the condensed phases are well-known phenomena, having a long history behind them. Nevertheless, the advent of radiation chemistry has greatly accelerated the study of these species. Over a century ago the solvated electron was discovered in liquid ammonia when an alkali metal such as sodium or potassium was dissolved in it[6]. The solution was distinguished by its intense blue color due to absorption in the red and IR and by its conductivity which, at least in the case of dilute solutions, was correctly attributed to a stabilized electron. Although the situation at high concentrations in metal–ammonia solutions is very complicated because of the formation of a variety of complex species, the interpretation of the properties of dilute solutions in terms of isolated stabilized electrons in the liquid is well accepted. Gradually this model gained strength by the discovery of paramagnetism of the stabilized electron and by the eventual discovery of the solvated electron in other liquids such as certain amines and amides. The electron in liquid ammonia is extremely long-lived. In liquid water, which is also highly polar, the solvated or hydrated electron is short-lived because of various chemical reactions including reaction with itself, with water molecules in the surrounding and with residual impurities in the medium. Even in very pure water the lifetime is limited to ~1 ms. Therefore, it required the technique of pulse-radiolysis to observe the hydrated electron[7]. However, when observation is indeed made in the appropriate time scale (μs or less) it is seen to be produced with a good yield (~4.2–~2.7 per 100 eV dependent upon and decreasing with time) and is characterized by its absorption spectrum.

When alkali halides are irradiated with ionizing radiations (e.g. x- or γ-rays) color centers are produced. This phenomenon has also been known for a long time; in particular the F-center is believed to be an electron trapped in a negative ion vacancy. A wide variety of organic glasses—3-methyl pentane, methylcyclohexane and ethanol, to mention just a few—trap electrons at 77 K with a considerable yield[4]. These glasses are naturally divided into polar and non-polar classes. The most remarkable observation with respect to electron trapping in glasses is that even in non-polar media the saturation density is high ($\sim 10^{17} g^{-1}$); in polar media the value is still higher ($\sim 10^{19} g^{-1}$)[3]. Crystalline ice traps electrons only with a small yield and that yield decreases drastically on cooling. However, glassy alkaline ices trap electrons with a high yield which is comparable to the nanosecond yield of hydrated electrons in pure water.

In a medium which is potentially capable of trapping, trapped electrons may be generated by any suitable method of releasing electrons in the medium. A popular method is the photodetachment of an anion in solution or in the glass. It is also not unreasonable to assume that organic non-polar liquids trap electrons, although the mean life in a trap could be small. Indirect evidence of such trapping is obtained from the model of excess electron mobility in such liquids[8] where the Arrhenius-dependence of the mobility upon temperature is explainable on the basis of an electron that is partly trapped and partly free. Since some of these liquids when transformed into glass under suitable conditions will trap electrons anyhow (as evidenced by the absorption spectrum) there is also the continuity argument. Recently some direct evidence has also been obtained for short-lived electron trapping in non-polar liquids[9].

From the previous discussion it is clear that electron trapping is a universal phenomenon of the condensed media which can be brought about by a variety of means and in a variety of phases (liquids, glasses, crystals, etc.). Since crystals need some kind of imperfections (vacancies, lattice distortions, etc.) to trap electrons it may be assumed that theoretically a perfect crystal will not trap electrons. However, since perfect crystallinity is not attainable in reality, electron trapping in any crystal is ultimately a quantitative question. On the other hand, the cases of liquid methane, liquid argon, etc., seem to be semantic, i.e. either there is really no trapping or electrons are trapped in these media with too small a binding energy and/or for too little a duration before being ejected in the continuum.

At present it is customary to distinguish between the trapped and the solvated electron with the term stabilized electron being applied to either species. Trapped electrons are formed at the earliest of times or at very low temperature (4 K). They are characterized by IR or far-IR absorption. With the progress of time (or on warming) trapped electrons "relax" to form solvated electrons while their absorption spectra shift toward the visible region. In this and allied senses the trapped electron in non-polar media also eventually solvates, although the distinction between these two species in these media is not as great as that in polar media. In the opinion of Higashimura et al.[10] trapped electrons in all glassy matrices are fully solvated at 77 K in the long-time limit.

II.B Properties of trapped and solvated electrons

The most common attribute of all stabilized electrons in liquid and glassy media is their optical and ESR absorption spectra. The optical spectra are generally very broad and structureless with the solvated electron spectrum being even broader than the trapped electron spectrum in the same medium. The absorption intensity is high, a typical extinction coefficient being $\sim 10^4$. Plotted against energy or wave-number these spectra are asymmetric. Sometimes, as for the solvated electron in liquid water, the spectrum is more nearly symmetrical when plotted against the wavelength, the significance of which is not clearly understood. Nielsen et al.[11] have drawn attention to a second absorption band of the hydrated electron below ~ 200 nm. A similar second absorption is also known for irradiated alkali halide crystals and in both cases they may be attributed to the absorption by the medium modified by the presence of the electron.

In most media the wavelength for maximum optical absorption shifts toward the shorter wavelength side when the electron passes from the trapped to the solvated state; in a few media it may remain nearly unchanged[10,12]. This spectral shift is definitely related to the matrix polarity. As one examines different media from the alkanes to the alcohols in increasing order of polarity one finds increasing spectral shift (generally speaking) in that order. In addition the time required to reach the equilibrium solvated electron spectrum (actually the experimental limit) also seems to be correlated with the matrix polarity[13], i.e. decreasing with the dipole moment or the static dielectric constant. These considerations lend partial support to the involvement of C–H bond dipoles in electron solvation (or even initial trapping) in alkane glasses. Molecular reorientation is a conceivable mechanism for polar liquids and probably also for polar glasses where the constituent molecules are small (e.g. ethanol glass). The case of large polar molecules comprising media of relatively small static dielectric constant is rather special in that the spectral features of the stabilized electron in them are intermediate between those in alkanes and in highly polar media. Accordingly, the mechanism of solvation may be intermediate between the two mechanisms just referred to.

There has been a tendency in the past to associate the optical spectrum itself to matrix polarity. However, the correlation is not good and is probably not justified due to the presence of other variables. (Actually there is a better correlation of matrix polarity with spectral *shift* on solvation and the time required to reach equilibrium spectrum.) We will mention two examples to illustrate the point. First, the solvated electron in metal–ammonia solution at $-65°$ has an optical spectrum very similar to that in 3-methylpentane at 77 K. These spectra are far removed from that of the hydrated electron[14]. Onsager[15] had earlier commented that the small binding energy of the solvated electron in liquid ammonia is partly due to its low density which therefore is one of the other variables. Recently Hentz and Kenney-Wallace[16] have shown conclusively that the immediate surroundings of an electron influence the spectrum much more strongly than does the overall dielectric constant. This effect was demonstrated by the substitution of side groups in the molecules which changed the spectra considerably while affecting the dielectric constant only marginally.

In the experiment of Hase et al.[12b] the trapped electron produced and observed at 4 K in 3MP shows no clear maximum; conceivably a maximum in the spectrum could exist at a wavelength > 2000 nm. If the sample is quickly warmed to 73 K after irradiation at 4 K a spectrum with a clear maximum at ~1700 nm is obtained. In fact, this spectrum is very similar in shape to that obtained with irradiation and measurement at 73 K. By comparing the latter two spectra it is possible to conclude that no more than 10–20% of trapped electrons produced at 4 K were lost, presumably by recombination, or warming to 73 K. In almost all cases the stabilized electron at 4 K is in a trapped state but not solvated, with a possible experimental controversy in the case of deuterated ethanol[17]. On warming to 77 K they solvate. The same phenomenon is observable on time-resolved spectroscopy using pulse-radiolysis at 77 K. If the temperature is gradually increased the solvation time decreases rapidly. This solvation phenomenon is microscopic in nature and is not given by the Debye theory or, for that matter, by any other theory dealing mainly with thermal equilibrium. The usual dielectric relaxation time of most of these media at 77 K is enormously large and in fact so large compared to the duration of the experiment as to have no relevance at all. On the other hand, the microscopic solvation process is not determined by electrostatic interactions only (e.g., the charge–dipole interaction); there is some residual thermal effect too. The latter point is illustrated by the fact that the trapped electron will not solvate at 4 K even in the long-time limit. The role played by the temperature in the microscopic solvation process is as yet undetermined and remains as a challenge to the theoretician.

The ESR spectrum of the trapped electron at 4 K is, generally speaking, structureless with a line width between ~3–~7 G. Once again the line width itself is not related to the matrix polarity but the change in line width on warming to 77 K is. In the case of alcohols and alcohol–water mixtures the line width increases on warming whereas in the case of alkanes it decreases slightly. Both these effects have been attributed by Higashimura and Yoshida[12e] to solvation noting that in the OH bond the negative charge is on the oxygen atom whereas in the CH bond it is on H. This observation also then lends support to the molecular reorientation model.

Yields of stabilized electrons under γ-irradiation at 77 K are given in Table 1 for a few selected glasses. We do not have scope here for a detailed discussion of the yields for which the reader is referred to Kevan's review[4]. There are basically two methods for measuring yields. First, straightforward (double) integration of the ESR spectrum and comparison with a known standard. Second, conversion of the electrons quantitatively into anions or radicals by scavenging which are then conveniently measured. In the scavenging experiments one has to reach a plateau of the product yield curve vs. scavenger concentration to be sure that all electrons have been scavenged. In polar glasses these two methods give approximately the same yields.

TABLE 1

G-value (100 eV yield)^a of stabilized electrons under γ-irradiation at 77 K in various glasses

	Compound	Yield	Method
(a) Polar	Ethanol	2.5	Scavenging
Glasses	Ethanol	2.3	ESR
	Methanol (5% H_2O)	2.6	ESR and scavenging
	1-Propanol	2.0	Scavenging
	1-Propanol	1.5	ESR
	2-Propanol	2.0	Scavenging
	Ethylene glycol	1.5	ESR
(b) Alkane	3-Methylpentane	1.2	Scavenging
glasses	3-Methylpentane	0.6	ESR
	Methylcyclohexane	~2	Scavenging
	Methylcyclohexane	0.3	ESR
	3-Methylhexane	0.9	ESR
(c) Misc.	Cumene	0.8	Scavenging
glasses	Benzene	1.4	Scavenging
	Toluene	1.4	Scavenging
	Methylcyclohexane	1.8	Scavenging
	Various amines and ethers	~2	Scavenging

^{a}Yield measurements have uncertainties $\sim \pm 10\%$ or more.

In non-polar glasses the scavenging method consistently gives a greater yield than the ESR method, sometimes by a factor of 2 or more. This difference can be attributed to relatively rapid neutralization in non-polar glasses. In any case, ESR experiments tend to underestimate the yields because of saturation effects due either to microwave power or dose.

The G-value (100 eV yield) for trapped electrons in 3 MP by γ-radiolysis at 77 K is ~0.6 as measured by ESR. If a small amount of hole trap, such as 2-methylpentane, is added to the system, the trapped electron yield increases enormously[18]. This increase is interpreted in terms of frustration of electron–hole recombination due to hole trapping. It must be remembered, however, that theoretically it is a metastable case, therefore the measured yield may, in principle, depend upon the scavenger concentration. If this were a true escape yield scavenging of either kind of ions would have no effect on the ultimate escape probability[19]. Of course, it is understood that the true escape probability at 77 K is very nearly zero, therefore, all measured yields are in a real sense metastable yields. Understandably, addition of hole trap in a polar glass does not increase the trapped electron yield very much since there is not much electron–hole recombination to start with.

At low doses the measured yield of stabilized electrons varies linearly with the dose. However, the yield becomes sub-linear with dose at high doses starting at ~10^{20} eV g^{-1}. If the dose is further increased the yield reaches a maximum and then starts to decrease, eventually being very small at extremely high doses. Three explanations have been proposed:

(i) reaction with radiation products (Willard and co-workers[20]);

(ii) electron scavenging by radiation-produced trapped radicals through tunneling (Miller[21]) and

(iii) dielectron formation (Feng et al.[22]).

Radiolysis at low temperatures and pulse radiolysis (at or near 77 K) have helped a lot in understanding the structure and optical transition of the stabilized electron in glasses, although many details remain to be filled in. Owing to the pioneering researches by Higashimura and co-workers[10,12] it is now believed that the optical transition of the trapped electron is similar to photodetachment, i.e. the electron is ejected out. The excited state of the solvated electron, on the other hand, is believed to be discrete although it may be in the continuum and may later autoionize.

II.C Conjectures regarding the theory of trapped electrons

A sequence of steps may be envisaged in the formation of the solvated electron. First, a free or quasi-free electron is formed as a result of an ionization or a detachment process and is randomly scattered in the medium. At some suitable location the electron finds itself bound by a local potential which is either pre-existing or induced by the electron. Two conditions are necessary for successful binding. The strength of the potential must be enough to bind the electron in a stationary state and during the residence time of the electron in the locale a mechanism must exist by which the electron can get rid of the appropriate amount of excess energy, i.e. the sum of kinetic energy and the binding energy. Finally, the electron goes through the solvation process which in itself may be a sequence of quasi-stationary states providing the build-up of local polarization and deepening of the trap.

The time-dependent wave function of the excess electron may be developed in terms of the stationary wave functions localized around the nth molecule or site with coefficient $a_n(t)$ as follows[5]

$$\Psi(t) = \sum_n a_n(t) \Phi_n \tag{1}$$

The wave functions, Φ_n (Wannier), are orthonormalized in the sense that $< \Phi_n | \Phi_m > = \delta_{nm}$. The total Hamiltonian for the excess electron has diagonal and off-diagonal matrix elements represented respectively by $E_n = <\Phi_n|H|\Phi_n>$ and $V_{nm} = <\Phi_n|H|\Phi_m>$. Physical randomness of the lattice may be incorporated either in E_n, or in V_{nm} or in both. The probability of finding the electron at site n at time t is given by $a_n^2(t)$. The evolution of the probability amplitudes $a_n(t)$ is governed by the set of numbers E_n and V_{nm} with the indices n and m running throughout the medium. The problem of trapping may be stated as follows. Suppose the electron is found at the origin ($n = 0$) initially ($t = 0$). With the condition $a_0(0) = 1$ if $\lim_{t \to \infty} a_0(t)$ remains finite then the electron is considered trapped there. It has been shown that the condition is equivalent to saying that the imaginary part of the self-energy at site n vanishes[5]. Physically, the reciprocal of the imaginary part of the self-energy is proportional to the residence time. Therefore, the requirement of its vanishing means (in principle) an infinite residence time or trapping. Expression of the self-energy in terms of E and V results in an infinite series. Its summation in the case of a general random lattice is very complicated. Nevertheless Funabashi concludes that both the number and mean depth of pre-existing traps increase with the fluctuations in the matrix elements E_n and V_{nm}. From his result he conjectures that trapping will be rarer and with less depth in media consisting of spherical molecules.

An electron is capable of changing the local molecular configuration if it is slow compared to the molecular vibrations. If the interaction is strong enough the result-ant deformation may produce a sufficiently strong trap for self-trapping of the electron. Such a mechanism may take place in an otherwise perfect crystal and either acoustic or optical phonon–electron interaction may be involved[23]. However, in the opinion of this writer, a pre-existing trapping mechanism is the dominant one for electrons in organic glasses. The quasi-free electron is probably fast enough so that the local deformation is not great. On the other hand, the random arrangement of bond dipole moments in both polar and non-polar media should provide enough pre-existing trapping potential. In the rest of this section we will describe in a little detail a recent conjecture by Tachiya and Mozumder[24] regarding electron trapping in polar glasses by the pre-existing mechanism.

The model of Tachiya and Mozumder[24] considers the condensed medium as composed of a large number of cells each of which contains a fixed number of mole-cular dipoles. Two cases are explicitly considered, a tetrahedral cell and an octahedral cell. The influence of surrounding medium on a given cell is ignored and therefore in this model the binding energy is relative, no absolute value can be provided. The dipole positions are fixed with respect to the center of the cell but their directions are considered random. The motion of the electron in a cell is given by the Schrödinger equation as follows

$$[-(\hbar^2/2m)\nabla^2 + V(r)]\Psi = E\Psi \tag{2}$$

In eqn. (2), E is the total energy and the potential is given in the charge-point dipole approximation as

$$V(r) = -e\sum_{i=1}^{N}\frac{\mu_i \cdot (\mathbf{r} - \mathbf{r}_i)}{|\mathbf{r} - \mathbf{r}_i|^3}, \tag{3}$$

where N is the number of dipoles in the cell and μ_i and $\mathbf{r}_i$ are the dipole moment and position vectors of the ith dipole; $\mathbf{r}$ denotes the position of the electron. Using a variational method with normalized wave functions the energy is given by

$$E = \int \Psi^*\left(-\frac{\hbar^2}{2m}\nabla^2 + V(r)\right)\Psi d^3r \tag{4}$$

Let I_i denote the average potential energy of the electron with respect to the ith dipole. As shown in Fig. 1, let θ_i denote the angle between μ_i and $\mathbf{r}_i$; μ and a are respectively the magnitudes of μ_i and $\mathbf{r}_i$; θ and ϕ are the polar angles of $\mathbf{r}$ and θ' is the angle between μ_i and $\mathbf{r}$. Then, using a spherically symmetrical wave function and taking $\mathbf{r}_i$ as the polar axis, one gets

$$I_i = \int_0^\infty\int_0^\pi\int_0^{2\pi}|\Psi|^2\frac{(\mu r\cos\theta' + \mu a\cos\theta_i)}{(r^2 + a^2 - 2ar\cos\theta)^{3/2}}r^2\sin\theta\,dr\,d\theta\,d\Phi \tag{5}$$

Substituting the relation $\cos\theta' = -\cos\theta_i\cos\theta + \sin\theta_i\sin\theta\cos(\Phi - \Phi_i)$ into eqn. (5), we obtain

$$I_i = 2\pi\mu\cos\theta_i\int_0^\infty\int_{-1}^1|\Psi|^2\frac{(a - rx)}{(a^2 + r^2 - 2arx)^{3/2}}r^2drdx \tag{6}$$

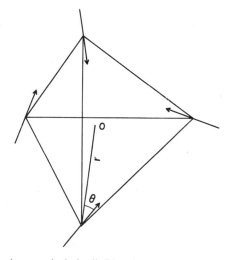

Fig. 1. Dipole disposition in a tetrahedral cell. Directions of dipoles change randomly from cell to cell.

where $x = \cos\theta$. Evaluating the x-integration in eqn. (6), we get

$$I_i = \frac{4\pi\mu\cos\theta_i}{a^2}\int_0^a |\Psi|^2 r^2 dr = \frac{\mu\cos\theta_i}{a^2}\int_0^a |\Psi|^2 d^3r \tag{7}$$

which means that the potential interaction depends only through the central component of the dipole moments which, of course, is the natural consequence of using spherically symmetrical trial wave function. From eqns. (4) and (7) we get

$$E = \int \Psi^* \left(-\frac{\hbar^2}{2m}\nabla^2\right)\Psi d^3r - \frac{e\sum_{i=1}^N \mu\cos\theta_i}{a^2}\int_0^a |\Psi|^2 d^3r \tag{8}$$

which shows that as far as the orientations of the polar molecules are concerned the total energy depends only upon the sum of the central components of the dipole moments. Taking $\Psi = (\lambda^3/\pi)^{1/2}\exp(-\lambda r)$ where λ is the variational parameter we get from eqn. (8).

$$E = \frac{\hbar^2\lambda^2}{2m} - \frac{e\mu y}{a^2}\{1 - (2\lambda^2 a^2 + 2\lambda a + 1)\exp(-2\lambda a)\} \tag{9}$$

where $y = \sum_{i=1}^N \cos\theta_i$. From eqn. (9) the only non-trivial value of λ is given by setting $\partial E/\partial\lambda = 0$ and this value of λ satisfies the following equation

$$4e\mu y(\lambda a)\exp(-2\lambda a) - \hbar^2/m = 0 \tag{10}$$

Numerical calculations performed with both the tetrahedral and octahedral cells using 1.8 Debye for μ and taking 3 Å as the distance between nearest dipoles, indicate that solution of eqn. (10) exists only for $y > 2.1$, i.e. pre-existing electron traps can occur only if the sum of the central components of the dipole moments is greater than ~3.78 Debye. To the approximation of the model the ground state is degenerate with respect to the normal components of the dipole moments, i.e. different cells, having equal values for the sum of the central components but differing in the normal component, have the same ground state energy. Figures 2 and 3 show, respectively, the dependence of λ upon y for the tetrahedral and the octahedral cells.

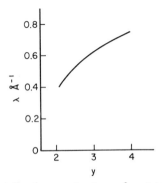

Fig. 2. Variational parameter λ vs. y for a tetrahedral cell.

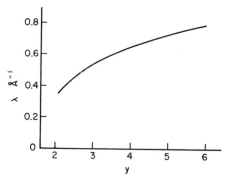

Fig. 3. Variational parameter λ vs. y for an octahedral cell.

Having established the relationship between y and λ we now turn to the problem of calculating the probability that y will be found in a certain interval. For a random distribution clearly the probability of finding $\cos\theta_i$ for a single dipole lying between y and $y + dy$ is simply given by $f_1(y)\,dy$ where

$$f_1(y) = 1/2 \text{ for } |y| < 1 \text{ and } = 0 \text{ otherwise} \tag{11}$$

If we now consider N dipoles, the probability that $\sum_{i=1}^{N}\cos\theta_i$ will lie between y and $y + dy$ is given by $f_N(y)dy$ where

$$f_N(y) = \int_{-\infty}^{\infty} f_{N-1}(y - x)f_1(x)dx = \frac{1}{2}\int_{y-1}^{y+1} f_{N-1}(x)dx \tag{12}$$

The extreme right hand side of eqn. (12), has been obtained using eqn. (11). Explicit evaluations of f_N may now be done using eqns. (11) and (12) repeatedly. The results for the tetrahedral and octahedral cells are given below

$$
\begin{aligned}
f_4(y) &= (1/96)(32 - 12y^2 + 3|y|^3) \text{ for } |y| < 2 \\
&= (1/96)(4 - |y|)^3 \text{ for } 2 < |y| < 4 \\
&= 0, \text{ otherwise}
\end{aligned}
\tag{13}
$$

and

$$
\begin{aligned}
f_6(y) &= (1/7680)(2112 - 480y^2 + 60y^4 - 10|y|^5) \text{ for } |y| < 2 \\
&= (1/7680)(1632 + 1200|y| - 1680y^2 + 600|y|^3 - 90y^4 + 5|y|^5) \\
&\qquad \text{for } 2 < |y| < 4 \\
&= (1/7680)(6 - |y|)^5 \text{ for } 4 < |y|; \\
&= 0, \text{ otherwise}
\end{aligned}
\tag{14}
$$

These probability functions peak for $y = 0$ and decrease rapidly with increase of $|y|$. As expected the probability function approaches a Gaussian for large N; it is quite evident for $N = 6$ within specified limits.

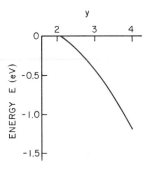

Fig. 4. Ground state energy E vs. y for a tetrahedral cell.

For a given value of y in a given type of cell the ground state energy may be calculated by first finding λ from either Fig. 2 or Fig. 3 as appropriate and then substituting this value of λ in eqn. (9). Figures 4 and 5 show the variation of E with y for the tetrahedral and octahedral cells, respectively. From these figures and either eqn. (13) or (14), as is appropriate, we may now calculate the probability $g(E)dE$ that a given cell, under the approximation of totally randomly oriented dipoles, will have a pre-existing trap for an electron with energy E. The function $g(E)$ is given by $g(E) = f(y)|dy/dE|$. The results are shown in Figs. 6 and 7, for the tetrahedral and the octahedral cells, respectively. From these figures it is clear that of all the pre-existing traps those with near zero binding energy (i.e. the shallowest traps) are the most likely to occur. They are also the cases which are likely to have the most degeneracy with respect to the normal component of the dipole moment.

In the treatment given so far there is no temperature effect as such since the dipoles are considered totally randomly oriented. Although this approximation greatly simplifies calculations it is not very realistic. On the other hand, complete dipole–

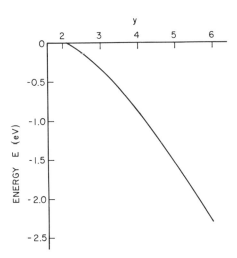

Fig. 5. Ground state energy E vs. y for an octahedral cell.

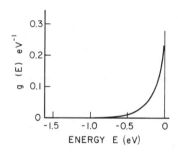

Fig. 6. Probability function per unit energy interval, $g(E)$, for an electron to have energy E in a tetrahedral cell.

dipole interaction is very difficult to handle. An intermediate approximation used in the theory of the solvated electron[25], in which only the interaction between the central components of the dipole moments are considered, has been used by Tachiya and Mozumder[24]. Using this approximation they find that at any temperature shallow traps are much more likely than deep traps. Further, the relative probability of occurrence of shallow traps (against deep traps) increases rapidly on cooling.

III. Dielectric relaxation and its possible implication in electron solvation in polar media

If the term "electron solvation" implies spectral shift toward larger wavenumbers and lowering of the ground state electron energy in a trap, then it should apply equally to polar and non-polar media. However, the effect is generally bigger for polar media and, in a restricted sense, seems to increase with polarity. That is the

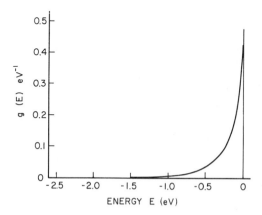

Fig. 7. Probability function per unit energy interval, $g(E)$, for an electron to have energy E in an octahedral cell.

reason for a current belief that some kind of dielectric relaxation underscores electron solvation, although precisely what this "relaxation" is has not been possible to delineate. At the outset it should be pointed out that the word "relaxation" is not used in radiation chemistry in the usual sense. In thermodynamics as well as in dielectrics[26] relaxation means the return of a system to thermal equilibrium when the external restraint is removed. Originally the system is in a stationary state, not far from equilibrium, under the influence of a small perturbation. When the perturbation is removed the system returns exponentially to equilibrium with a time constant called *the relaxation time*. However, in radiation chemistry the system is originally in thermal equilibrium and then in the presence of the perturbing influence of the electron it passes over to a quasi-stationary state which may or may not be near the equilibrium state. In mechanics such a passage from the equilibrium to a stationary state is sometimes called retardation[27] signifying that the strain developed in the system lags the external perturbation. However, such a term may not be acceptable to the practising radiation chemist on account of its connotation. It may be argued that if the system behaves linearly then the characteristic time constant should be independent of whether it passes from the stationary state to the equilibrium state or the other way (e.g. by drawing similarity to the cases of charging and discharging of a linear condenser). However, in the presence of the strong perturbation in the immediate vicinity of the electron, it is not clear that the system does behave linearly. There may be hidden a variety of non-linear effects including saturation. In any case we shall continue to use the term "relaxation" in the sense practised by the radiation chemist, i.e. the process of electron solvation.

We have remarked in Sect. II and we stress again that the relaxation, although it may be largely a local phenomenon, cannot be entirely temperature-independent. This is borne out by the fact that at 4 K an electron is always trapped and only begins to solvate on warming. In radiation chemistry then, "relaxation" is a curious combination of thermal and non-thermal effects and their role or relative importance has not been clarified. Nevertheless, we will review the theoretical picture drawing from the experimental results.

III.A Elementary descriptions

The most successful theory of dielectric relaxation (in the usual sense) is due to Debye[26]. Using the theory of Brownian motion he derived a partial differential equation for the probability density of dipole moments in a liquid, which may be written as follows

$$\zeta \frac{\partial f}{\partial t} = \frac{1}{\sin \theta} \frac{\partial}{\partial \theta} \left[\sin \theta \left\{ k_B T \frac{\partial f}{\partial \theta} + \mu f E \sin \theta \right\} \right] \tag{15}$$

In eqn. (15) E is the external field, T, the absolute temperature, k_B, Boltzmann's constant, μ, the magnitude of the dipole moment of the molecules, t, the time and f is the probability density of finding molecular dipoles per unit solid angle at θ. The quantity ζ, called the friction coefficient, is characteristic to the liquid. It is the constant of proportionality between an external torque and the uniform angular velocity of a macroscopic object set in motion in the liquid by that external agency. This angular velocity refers to the stationary state; i.e., after a sufficient lapse of time the back torque due to viscosity exactly cancels the external torque thus producing stationary angular velocity. Then $M = \zeta d\theta/dt$ where M is the external torque.

For the usual relaxation case the initial condition is that $E = E_0$ for $t < 0$ and $E = 0$ for $t > 0$. A solution of eqn. (15) may then be written as

$$f \sim 1 + \frac{\mu E_0}{k_B T}\Phi(t)\cos\theta \tag{16}$$

where $\Phi(t)$ satisfies the subsidiary equation

$$\frac{d\Phi}{dt} = -\frac{2k_B T}{\zeta}\Phi \tag{17}$$

From eqns.(16) and (17) Debye obtains

$$f \sim 1 + \frac{\mu E_0 \cos\theta}{k_B T}\exp(-2k_B Tt/\zeta) \tag{18}$$

Eqn. (18) implies that the system is reverting back to equilibrium with a time constant τ given by

$$\tau = \zeta/2k_B T \tag{19}$$

Stokes had already shown that the viscous torque on a sphere of radius a, rotating with an angular velocity $d\theta/dt$ in a liquid of viscosity η is given by $8\pi\eta a^3(d\theta/dt)$. Therefore, by definition, $\zeta = 8\pi\eta a^3$ so that eqn. (19) gives

$$\tau = 4\pi\eta a^3/k_B T \tag{20}$$

Equation (20) is the celebrated Debye formula for the dielectric relaxation time of a polar liquid. The only microscopic parameter appearing in this formula is the radius of the polar molecule, a, which actually is not well known. Therefore, eqn. (20) is sometimes written as $\tau = 3V\eta/k_B T$ where V is the volume occupied by a polar molecule. Many objections have been raised, from time to time, against the Debye theory[28] including the inapplicability of the Stokes equation for objects of molecular dimensions and complications arising from hydrogen bonding (most polar liquids are hydrogen-bonded). Nevertheless the Debye theory has a wide range of validity.

Mozumder[29] pointed out that the dielectric "relaxation" in a polar media where suddenly a charge has been created as a result of ionization should be treated somewhat differently. The reason for the modification is that the field due to the electron is not constant but decays with time. Rather the quantity that is invariant is the charge, and therefore, the electric displacement. According to the superposition principle[30], the displacement D and field E in a relaxing dielectric are related through the following equation

$$D(t) = \epsilon_\infty E(t) + \int_{-\infty}^{t} E(u)\alpha(t - u)du \tag{21}$$

In eqn. (21) the first term on the right hand side relates E and D at the same time.

The proportionality factor is then ϵ_∞, the high frequency dielectric constant. The second term represents the effect of E on D for times $u < t$. Debye essentially surmised that the decay function α must be exponential with the same usual relaxation time, i.e.

$$\alpha(t) = \frac{(\epsilon_s - \epsilon_\infty)}{\tau} \exp(-t/\tau) \tag{22}$$

where the coefficient of the exponential function has been chosen to represent the static case (d.c. field) correctly; therefore ϵ_s is the static dielectric constant. From eqns. (21) and (22) one gets the general equation connecting D and E and their time derivatives, i.e.

$$\tau(d/dt)(D - \epsilon_\infty E) + (D - \epsilon_\infty E) = (\epsilon_s - \epsilon_\infty)E \tag{23}$$

If we take the displacement D as constant, as may be appropriate for ionization in a polar medium, then $dD/dt = 0$ and $dE/dt = D(d/dt)(\epsilon^{-1})$, where $\epsilon(t)$ is the time-dependent dielectric constant. Substituting these relations in eqn. (23) one obtains

$$\frac{d}{dt}(\epsilon^{-1}) = (\epsilon_\infty \tau)^{-1}(1 - \epsilon_s/\epsilon) \tag{24}$$

On integration of eqn. (24) subject to the initial condition $t \to 0$, $\epsilon \to \epsilon_\infty$ we find

$$\epsilon^{-1} = \epsilon_s^{-1} + (\epsilon_\infty^{-1} - \epsilon_s^{-1})\exp(-t/\tau') \tag{25}$$

where

$$\tau' = (\epsilon_\infty/\epsilon_s)\tau \tag{26}$$

Equations (25) and (26) imply that in the case of constant charge or displacement the dielectric relaxes faster by the factor $\epsilon_\infty/\epsilon_s$, the ratio of the two dielectric constants. Figure 8 shows the evolution of the dielectric constant in water at room temperature for the cases of constant field and constant charge.

Using a time-dependent dielectric constant $\epsilon(t)$ one can evaluate the probability that an ionized electron in a polar liquid will solvate instead of recombining with the positive ion. The starting point is the Smoluchowski equation for the probability density P that the electron will still remain unneutralized at time t around a point distant r from the positive ion. This equation may be written as follows

$$\frac{\partial P}{\partial t} = D\left[\frac{1}{r^2}\frac{\partial}{\partial r}\left(r^2\frac{\partial P}{\partial r}\right) + \frac{e^2}{\epsilon kTr^2}\frac{\partial P}{\partial r}\right] \tag{27}$$

In eqn. (27) D is the mutual diffusion coefficient and e is the magnitude of the electronic charge. In the Gaussian approximation P is assumed to be a product of a normalized Gaussian function and a function $F(t)$ of time only which gives the probability that the electron will remain unneutralized at time t, viz.

$$P(\mathbf{r}_0; \mathbf{r}, t) = F(t)\frac{\exp\left[-(\mathbf{r} - \mathbf{r}_0)^2/4Dt\right]}{(4\pi Dt)^{3/2}} \tag{28}$$

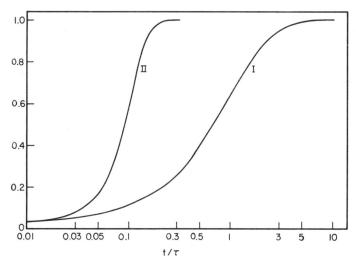

Fig. 8. Time-dependent dielectric constant for water at room temperature for constant field (I) and constant charge (II). Case (II) relaxes faster than (I). This calculation uses $\epsilon_s = 80$, $\epsilon_\infty = 2$ and $\tau = 10^{-11}$ s.

In eqn. (28) $\mathbf{r}_0$ is the initial separation of the electron. The method is illustrated here with a single initial separation, it being understood that, if necessary, an initial distribution can be incorporated by a formal integration of the final result. We are interested in evaluating the ultimate escape probability, $F(\infty)$.

Substituting eqn. (28) into eqn. (27) and carrying over the spatial integration we obtain

$$\frac{dF}{dt} = -\frac{4\pi De^2}{\epsilon k_B T} \cdot \frac{F}{(4\pi Dt)^{3/2}} \exp\left(-\frac{r_0^2}{4Dt}\right) \tag{29}$$

Now, using eqn. (25) for the time-dependent dielectric constant in eqn. (29) and carrying the integration to the ∞-time limit we get, with the initial condition $t \to 0$, $F \to 1$

$$\ln F(\infty) = -\frac{4\pi De^2}{k_B T} \int_0^\infty dt (4\pi Dt)^{-3/2} \exp(-r_0^2/4Dt) \times$$
$$\left[\epsilon_s^{-1} + (\epsilon_\infty^{-1} - \epsilon_s^{-1})\exp(-t/\tau')\right] \tag{30}$$

Now the escape probability given by eqn. (30) may be written in the standard Onsager form, viz. $F(\infty) = \exp(-e^2/\epsilon_{eff} r_0 k_B T)$, if we define the effective dielectric constant as follows

$$\epsilon_{eff}^{-1} = \epsilon_s^{-1} + (\epsilon_\infty^{-1} - \epsilon_s^{-1})\exp\left[-r_0/(D\tau')^{1/2}\right] \tag{31}$$

The results obtained so far are somewhat formal and not really applicable unless an essential modification, due to variation of D with t, is made. Time-dependence of D also originates from the relaxation process, the effective mass of the electron increasing with time due to solvation. The final result derived depends very much on the assumed dependence of D on t, since the exact relationship is not known. The simplest assumption that can be made is that $D = D_1$ for $t < \tau'$ and $D = D_2$ for $t > \tau'$ with $D_1 > D_2$. The constant D_1 is essentially the diffusion coefficient of the quasi-free electron. With this assumption and proceeding as before the escape probability $F(\infty)$ is given by[29]

$$-\ln F(\infty) = \left(\frac{e^2}{\epsilon_s r_0 kT}\right) \left[(1 + \text{erf}\xi_2 - \text{erf}\xi_1) + 1/2\{(\epsilon_s/\epsilon_\infty) - 1\}\times\right.$$
$$\{\exp(2\xi_1)\text{erfc}(\xi_1 + 1) + \exp(-2\xi_1)\text{erfc}(\xi_2 - 1)$$
$$\left.-\exp(2\xi_2)\text{erfc}(\xi_2 + 1) + \exp(-2\xi_2)\left[1 + \text{erf}(\xi_2 - 1)\right]\}\right] \qquad (32)$$

In eqn. (32) $\xi_{1,2} = r_0/(4D_{1,2}\tau')^{1/2}$, $\text{erf}(x) = 2\pi^{-1/2}\int_0^x \exp(-\xi^2)d\xi$ and $\text{erfc}(x) = 1 - \text{erf}(x)$. Actual evaluation of the escape probability then requires the knowledge of r_0 (or a distribution thereof), T, ϵ_s, ϵ_∞, D_1 and D_2 and τ, not all of which are well known. If $\epsilon_s \gg \epsilon_\infty$ and $D_1 \gg D_2$ then D_2 may not be too significant; however r_0 in this treatment must be treated as an adjustable parameter. One can only get an approximate mean value of r_0 from eqn. (32) if $F(\infty)$ is taken as the ratio of the G-value of the solvated electron in the polar medium to the G-value for ionization in the gas phase. The value of r_0 so obtained can then be used to describe the kinetics of the situation[29].

Although eqn. (32) does not allow a straightforward ab initio calculation of the escape probability in a polar medium it nevertheless shows that the static dielectric constant ϵ_s does not itself dictate the escape probability. In the past there has been a great tendency to plot solvated electron yield data as a function of the static dielectric constant for which we find the reason inadequate. For a complete description we need other data such as τ, $\epsilon_s/\epsilon_\infty$, ξ_1 and D_1/D_2. The last mentioned item (D_1/D_2) may not be too critical though, if D_1 is $\gg D_2$.

Electron solvation times are very short in polar liquids at room temperatures. Nevertheless, an upper limit has been established by Bronskill et al.[31] using a stroboscopic pulse-radiolysis technique. Later Baxendale and Wardman[32] made a clear direct observation of the electron solvation process in various liquid alcohols using nanosecond pulse-radiolysis at low temperatures (~ 180 K). Electron solvation through significant spectral changes have also been seen by Richards and Thomas[33], Kevan[34] (polar glasses) and Taub and Gillis[35]. According to Bronskill et al.[31] the electron is fully solvated in all polar media studied by them at room temperature within perhaps 10 ps, although in some cases it could take up to ~ 25 ps. The value of τ' (see eqn. (26)) for H_2O, MeOH and EtOH at this temperature is 0.2, 3.8 and 10.6 ps, respectively, showing fair agreement with theory. On the other hand, τ' for 1-propanol and 1-butanol is 49 and 72 ps, respectively showing poor agreement with experiment. Furthermore, theory predicts a large difference in the electron solvation times in 1- and 2-propanol since their dielectric constants (both ϵ_s and ϵ_∞) are nearly the same but the relaxation times vary by an order of magnitude. Such a difference has not been seen experimentally.

The low temperature experiments of Baxendale and Wardman clearly show the relaxation effect in three stages. In the first stage the absorption spectrum grows monotonically in the near-IR with no maximum within experimental limits (~1400 nm). This is characteristic of the trapped electron not yet solvated. In the next stage a peak develops in the visible with concomitant decay of the IR spectrum. This is followed by a decay of the red shoulder without any further increase at the shorter wavelength side. Finally, in orders of magnitude longer time scales the intensity of the spectrum decreases due to recombination in the spur and also due to homogeneous extra-spur recombination. However, since the spectrum remains unaltered this final stage need not concern us. Table 2 taken from Baxendale and Wardman[32] shows the measured half-life τ_s for the decay of the IR spectrum of the electron in various alcohols. These values are compared with theoretical τ' values. It is known that polar liquids behave in three different ways with respect to the distribution of relaxation times. Liquids like water have essentially one relaxation time. Those like glycerol have a continuous distribution which probably signifies that the dielectric decay function is non-exponential. Most alcohols are characterized by three relaxation times[36]. These are called τ_1, τ_2 and τ_3 (decreasing in that order) signifying, respectively, the time scales for hydrogen bond breaking in clusters, for monomeric rotation and for the rotation of the OH group. Theoretical τ' values in Table 2 are obtained using eqn. (26). From the comparison of τ_s with the various τ' Baxendale and Wardman concluded that "the solvation times are within an order of magnitude of τ_1'". Although the agreement, as seen from Table 2, is not as good as could be expected, we can nevertheless make two observations. First, the measured values are almost invariably smaller than the calculated values. Secondly, the agreement between the theory and experiment is better for lower alcohols. These are the cases of smaller molecules, shorter relaxation times and smaller viscosities. With higher alcohols we tend to lose the agreement which probably signifies the inadequacy of the macroscopic description.

Shirom and Tomkiewicz[37] studied the UV photoejection from the ferrocyanide ion in various water–alcohol mixtures using scavenging by added N_2O as a probe. In this kind of experiment one does not measure the relaxation time; instead one measures the relative yield or the escape probability. By choosing the appropriate

TABLE 2

Measured and calculated relaxation times in liquid alcohols (ns) (from Baxendale and Wardman ref. 32 Table 4)

Compound	T(K)	τ_s (measured)	τ_1'	τ_2'	τ_3'
CH_3OH	182	<1	0.3	~0.01	<0.01
C_2H_5OH	166	3	5	0.03	~0.01
$n-C_3H_7OH$	176	5	33	1.7	0.3
$n-C_3H_7OH$	152	60	750	33	2.7
$(CH_3)_2CHOH$	186	6	35	~1	~0.1
$n-C_4H_9OH$	184	4	40	0.8	~0.1

dielectric relaxation time these investigators have shown that a satisfactory agreement between experiment and theory, eqn. (32), can be obtained if one assumes a mean thermalization length of the ejected electron as 30 Å irrespective of the given water–alcohol mixture.

III.B Microscopic picture

From the discussion in the previous section a need for a microscopic description of electron solvation is obvious except perhaps in cases where the viscosity is low and the relaxation time is small. Evidence has also accumulated that the ground state energy of an electron in a hydrocarbon liquid is determined much more by the immediate surroundings than by the bulk properties of the liquid[16]. The need for a microscopic description can also be illustrated by taking the examples of very viscous liquids and glasses. The usual dielectric relaxation times in these systems may be given as an Arrhenius or a Davidson–Cole equation, i.e. of the form $\tau = A \exp(B/T)$ or $\tau = A \exp[B/(T - T_\infty)]$ where a, B and T_∞ are constants characteristic of the system and T_∞ is probably close to the glass temperature, T_G. Therefore, for very viscous liquids or glasses τ essentially is infinity or so very large (even with the modification of constant charge) as to have any relevance at all.

Unfortunately, a microscopic description of the relaxation process is hard to come by; in particular there seems to be a singular void as to its application to radiation chemistry. Mozumder[38] has attempted to construct a microscopic theory of electron solvation in a dilute solution of polar molecules in a non-polar medium. The formation process is envisaged in two stages. In the first stage, an ionized electron escaping geminate recombination is attached to a polar molecule. In the second stage the negative ion so formed draws nearest neighbor polar molecules and the solvated electron is formed by a coagulation process. Figure 9 illustrates the charge–dipole interaction and defines the polar coordinates r (separation) and θ (orientation). The electron attachment is a very fast process during which the polar molecule essentially does not move. The time scale of this process is $t_f = \epsilon r_0^4/8\mu u_e$ where ϵ is the dielectric constant (essentially that of the pure non-polar medium), μ, is the dipole moment of the polar molecule, u_e is the electron mobility and r_0 is the initial separation of the electron and the polar molecule[39]. The attachment time when averaged over a random distribution of initial angles and separations subject to a given dipole concentration n may be written as $t_1 = 4.395 \times 10^{-2}(\epsilon n^{-4/3}/\mu u_e)$. Computed values of t_1 as a function of the molarity of the solution is given in Table 3 for $\epsilon = 2$, $\mu = 2$ Debye and $u_e = 0.35 \text{ cm}^2\,\text{V}^{-1}\,\text{s}^{-1}$.

Fig. 9. Charge–dipole interaction. Radius vector **r** is considered positive when directed away from the charge. The charge **e** may be considered either as an electron or a negative ion.

TABLE 3

Concentration dependence of formation times of solvated electrons in dilute dipolar solutions ($\epsilon = 2$, $\mu = 2$ Debye, $u_e = 0.35$ cm^2 V s^{-1} and $\eta = 0.01$ poise)

Concentration (M)	Attachment time (ps) t_1	Coagulation time (ns) t_2
0.03	8.9	4.65
0.02	15.2	7.84
0.01	38.4	19.14
0.007	61.8	30.30
0.005	96.8	46.74
0.002	328	152.15
0.001	827	371.58

The equations of the coagulation stage may be given as follows

$$\frac{d\theta}{dt} = -B_\theta \left(\frac{e\mu}{\epsilon r^2}\right) \sin \theta \tag{33}$$

and

$$\frac{dr}{dt} = -B_r \left(\frac{2e\mu}{\epsilon r^3}\right) \cos \theta \tag{34}$$

In eqns. (33) and (34) r is the separation between the negative ion and the nearest polar molecule and θ is the orientation of the dipole moment relative to the direction of r. The rotational and linear mobilities (relative) are indicated by B_θ and B_r, respectively. Using the Stokes–Debye equation they are given respectively by $(8\pi\eta a^3)^{-1}$ and $(3\pi\eta a)^{-1}$ where, a is the size of the polar molecule and η is the medium viscosity. In this treatment only drift velocities (linear and rotational) are considered, all fluctuations being ignored. Noting the r-dependence in eqns. (33) and (34), it is clear that the angular alignment occurs first, followed by drift. Computer calculations have been performed starting with an initial angle and separation[38]. Coagulation time is defined when $r = 2a$. Finally, the so calculated coagulation time is averaged over a random distribution of initial angles and separations consistent with a given molarity M of the solution. Results of this final calculation are shown in Table 3. Figure 10 shows the variation of angle-averaged coagulation time with initial separation, r_0. When this coagulation time is further averaged over distribution of r_0 for a given concentration then the result can be empirically fitted with the equation t_2 (ns) $= 0.0508$ M$^{-1.288}$. Since t_1 is $\ll t_2$, t_2 may be called the formation time of the solvated electron. If we take $M = 55$ molar for neat water (for which the model admittedly breaks down) then this equation gives $t_2 \sim 0.3$ ps which is consistent experimentally, although this does not justify the procedure.

The few experiments that are available for the formation of solvated electrons in alcohol–alkane mixtures[40] do not support the theory positively, although some experiments are consistent with the theory[40b,d]. The difficulty seems to be that for

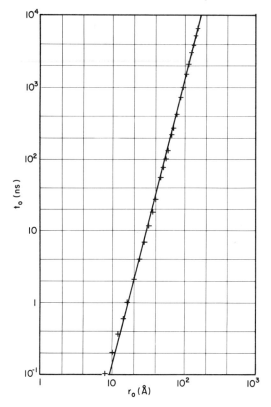

Fig. 10. Log–log plot of the angle-averaged coagulation time (ns) vs. initial separation (Å) with $\epsilon = 2$, $\mu = 2$ Debye, $a = 2$ Å, $\eta = 1$ cP. The data are well-fitted by the equation $t_2(\text{ns}) = 1.96 \times 10^{-5} r_0^{-3.8646}$ (Å).

the simple theory of monomeric coagulation one has to go to very dilute solution with which experiments cannot be done because of vanishing yields. On the other hand, at reasonable alcohol concentrations polymer configurations should exist already which are potentially capable of solvating electrons without the necessity of further coagulation. However, a certain amount of coagulation is always expected on general grounds, but the theoretical calculation including polymers becomes very complicated.

Another microscopic picture of dielectric relaxation is being developed by this writer and Tachiya[41] for use in neat polar liquids. Here we start at $t = 0$ with a trapped electron which is supposed to be the precursor of the solvated electron. The electron is initially considered trapped with a small binding energy in a pre-existing trap as described in ref. 24. The classical torque acting on the polar molecules of the cell is calculated and the change in their orientation over an infinitesimal time period is computed. This gives at the next instant of time the sum y of the central components of the dipole moments. Remembering that the ground state electron energy depends only on this sum, we then evaluate this energy at that next instant of time. This energy is somewhat lower than initial because of the increase of y due to the dipolar orientation. We continue these calculations through a "march of steps" method until no significant change is obtained in the ground state energy, i.e., the electron is fully solvated. From the curve of binding energy against time one can then compute the relaxation time through a suitable definition. We feel that this calculation may be adequate for a polar liquid, but not for a polar glass. The time-determining step for the polar glass is probably not dipolar orientation, but H-bond breaking for which no adequate theoretical description in the presence of an electron seems to be available.

IV. Considerations related to the application of mobility theories in radiation chemistry

In this section we shall not concern ourselves with the experimental situation regarding mobility measurement or the specific model(s) related to experiments. For these the reader is referred to Chap. 7. However, we will make a few comments regarding the application of mobility theories. Some of these relate to radiation chemistry, i.e. the transient electron mobility induced by ionizing radiation in hydrocarbon liquids.

It is customary to compare measured mobilities with the pre-exponential factor $e\lambda^2/h$ in an Eyring-type expression such as

$$u = (e\lambda^2/h)(F_{\ddagger}/F) \exp(-\epsilon_0/k_B T) \tag{35}$$

In eqn. (35), u is the electron mobility, λ is the mean distance between equilibrium positions, ϵ_0 is the potential barrier against flow and $F_{\ddagger}$ and F are the partition functions in the activated and equilibrium states, respectively. If λ is taken as a typical intermolecular separation (~ 4 Å) then this factor is ~ 1 cm^2 V^{-1} s^{-1}. Mobilities are termed high or low according as their values are $\gg$ or $\ll 1$ cm^2 V^{-1} s^{-1}, respectively. At either limit a good theory exists—a gas-like approach for the high mobility case and an activation model for the low mobility case. The intermediate case (~ 1 cm^2 V^{-1} s^{-1}), to which many radiation–chemical experiments belong, is the most difficult to analyze theoretically. Although some exact relationships such as Kubo's equations exist[42], considerable uncertainties still remain in relating the model to meaningful calculation for this case.

In the radiation chemistry of hydrocarbon liquids, ionization is produced by the absorption of energy from a penetrating charged particle. In the longest time limit some of the ionized electrons escape recombination with positive charged ions to become free to conduct electricity. The yield of free ions, G_{fi}, is defined as the number of free carriers produced per 100 eV of absorbed energy. Experimentally, a correlation has been found between G_{fi} and u, the mobility of the same free carriers. G_{fi} increases with mobility[43], being 4.4 for liquid Ar ($u = 475$ cm^2 V^{-1}s^{-1}), 1.0 for neopentane ($u = 50$ cm^2 V^{-1}s^{-1}) and 0.1 for n-hexane ($u = 0.1$ cm^2 V^{-1}s^{-1}). This correlation is in apparent disagreement with Onsager's theory[44] in terms of which the free-ion yields are usually discussed. Once the initial distribution is given Onsager's theory predicts that the free-ion yield should be independent of the mobility and that this fact is unaltered in the presence of external fields or small amounts of impurities (electron scavengers)[45]. Thus the only possible rationalization within the framework of the Onsager theory lies in the supposition that a certain crucial factor controls both mobility and free-ion yield, the latter only through the establishment of the initial (or thermalization) distribution. Most workers in this field seem to have missed this important point; however, Hentz[46] has noted this in terms of energy loss of epithermal electrons which is identified to be the common factor. A preliminary analysis[47] reproduced the desired correlation qualitatively, but failed to give quantitative agreement with experiments.

In the radiation chemistry of liquid hydrocarbons the transient electron mobility can be expressed by an Arrhenius equation where the measured activation energy decreases with the mobility[48]. Most investigators describe such mobilities in terms of inhomogeneous or two-carrier transport models. Kestner and Jortner[49] have proposed easy and difficult regions of conduction in the liquid with the fraction of volume in the easy regions showing an Arrhenius temperature-dependence. Hentz[46] and Minday et al.[48b] propose equilibrium between the trapped and quasi-free states of the electron in the liquid although in detail there are great differences between these models. Both models are based on the Frommhold[50] equation relating the observed mobility u with the quasi-free mobility u_{qf} as follows

$$u = u_{qf}/(1 + v\tau_t) \approx u_{qf}/v\tau_t \qquad (36)$$

where v is the trapping frequency and τ_t is the average residence time in traps. The value of u_{qf} is nearly the same in different media in the model of Minday et al. but not necessarily so in Hentz's model. Likewise, the critical field for non-linearity (the external field beyond which the mobility becomes field-dependent) is nearly constant in the model of Minday et al. but decreases with mobility, as is experimentally seen, in Hentz's model. The quasi-free mobility is described by Hentz in terms of a simple random-walk model. Minday et al.'s description of the quasi-free mobility is based on the theory of Cohen and Lekner[51], extended to include polyatomic molecules[52]. In terms of this theory the quasi-free mobility may be expressed as

$$u_{qf} = \frac{2}{3}\left(\frac{2}{\pi m k_B T}\right)^{1/2}\frac{e}{[n4\pi a^2 S(0)]} \qquad (37)$$

where n is the number density of scatterers, a is the scattering length and the structure factor at zero momentum transfer, $S(0) = nk_B T\chi_T$, the isothermal compressibility being denoted by χ_T. In general the structure factor is related to the pair-correlation function $g(R)$ as follows

$$S(K) = 1 + n\int d^3\mathbf{R} \exp(i\mathbf{k}\cdot\mathbf{R})[g(R) - 1] \qquad (38)$$

Bakale and Schmidt[53] found that, by and large, non-linearity in mobility is signalled by the drift velocity rather than by the field. The mobility starts to become non-linear when the drift velocity exceeds 3–5 times the sound velocity in the liquid. A clear relevance of the electron–phonon interaction in the liquid is indicated and this aspect must be incorporated into theory. In this writer's opinion it should be possible to construct a homogeneous, one-carrier model for electron mobility in liquid hydrocarbons. However, it may be necessary to construct two kinds of models; one gas-like for high mobility liquids and another solid-like for low mobility liquids. If we assume the individual steps of transfer as a photon-assisted tunneling from a trap to an adjacent trap and further assume that the successive steps are not strongly correlated then an Arrhenius form for mobility can be expected. Electron tunneling is well known in scavenging and recombination processes. It has been proposed as an alternative mechanism for transport[54], but the proposal has not been pursued further. It should be pointed out in this context that the quasi-free electron state in hydrocarbon liquids has no direct experimental verification.

V. Electron tunneling in glasses

V.A Evidence of electron tunneling

Tunneling refers to transfer of particles through a classically forbidden region where the potential energy exceeds the total energy. The usual implication is that the motion is classically allowed in both the initial and final states but not in the intervening region. Qualitatively, tunneling is a consequence of the uncertainty principle. If the energy in a stationary state can be given extremely precisely, the location of the particle will be correspondingly indeterminate and therefore, the probability of existence of the particle even in classically forbidden regions cannot be denied. If the energy has some uncertainty the location will have a corresponding uncertainty as given by the approximate wave function describing the pseudo-stationary state. On the other hand, it is customary to consider tunneling as a *rate process*. The latter occurs because usually there is a (more or less) fast relaxation following particle transfer. In the absence of the fast relaxation process the particle will tunnel back and forth with equal rapidity (principle of detailed balancing) thus establishing a stationary state with conserved energy. Examples of such cases are found in the tunneling of the N-atom in NH_3, of the OH bond in the CH_3OH molecule, etc.

It is clear that tunneling is a quantum-mechanical phenomenon and a barrier, i.e. a region where classical motion is forbidden, is required for its description. It should be noted that whereas, at or above the top of the barrier, the classical transmission probability is unity the corresponding quantum-mechanical probability is less than unity, i.e. there is a finite probability of reflection even at the top of (or above) the barrier in the same way as there is a finite probability of penetrating the barrier below the top. Importance of proton tunneling in chemistry was pointed out by Bell[55]. Proton tunneling over the H-bond in ice with a small barrier was used by Eigen[56] to explain the high proton mobility in ice. Recently a detailed calculation has been available for the neutralization of H_3O^+ and OH^- in ice based on the tunneling description[57].

Recently in radiation chemistry, considerable interest has been developed in the tunneling model for some of the reactions of the trapped and solvated electrons in frozen glasses. Many of these reactions are with scavengers[58–61]; but there are recombination reactions too[62,63]. As pointed out by Buxton et al.[64] a 10 M NaOH glass remains rigid at 135 K and even at 178 K diffusion does not become important until $\sim 10^{-4}$ s after the electron is solvated[59]. A similar situation undoubtedly exists in organic glasses and it can be said generally that the temperature has to be considerably above the glass temperature for diffusion to be the dominant mechanism of motion. The phenomenon is easily understood if diffusion is considered to be an activated process. Thus, at low temperatures where glasses behave essentially as rigid, some other mechanism is required for faster reactions and in many cases tunneling seems to be the answer.

Experimental evidence of electron tunneling, in both aqueous and organic glasses, is based on the following observations[58–62]:

(i) (relative) lack of temperature-dependence of the rate of electron decay over a wide temperature range in both scavenging and recombination experiments;

(ii) exponential-dependence of the trapped (or solvated) electron yield with scavenger concentration at a given time;

(iii) at a given concentration, dependence of the electron yield with the logarithm of time (actually with its cube, vide infra) and

(iv) the extreme sensitivity of the lifetime of trapped (or solvated) electrons on scavenger concentration but not on the nature of the scavenger.

Of these items the first implies that the decay curves can be superposed if normalized at one time. The actual yields are temperature-dependent, being higher at higher temperatures which could be attributed to a variety of reasons including reaction before the initial trapping. The second item is in disagreement with the diffusion picture according to which the yield should decrease proportionately to the square root of scavenger concentration for small concentrations. It should be noted that in many scavenging experiments in rigid glasses the electron does not decay appreciably in a certain time scale, although in the same time scale it readily decays in the presence of a scavenger of as small a concentration as 1 mM. The third item follows logically from a simple tunneling model (vide infra). It is hard to understand the fourth item in terms of classical diffusion although Miller[58,59] has justified the same on a simple tunneling picture. On the other hand, his stated relationship of reactivity with the electron affinity of the scavenger could equally well apply to diffusion-controlled reactions.

Some authors[65] prefer to use the term "electron transfer reaction" rather than electron tunneling. This is especially significant since sometimes tunneling is not a good description although it may be a quantum-mechanical transfer process and in any case tunneling alone does not determine the reaction rate. The Franck–Condon factor and relaxation processes contribute very significantly to the velocity of the reaction. However, in this writer's opinion, electron tunneling, used in a broader sense, is a convenient term to use in as much as it clearly distinguishes itself from classical diffusion.

V.B Application of the tunneling model in radiation chemistry

In Chap. 6, Steen will briefly review the experimental situation with respect to the applicability of the tunneling model for electron reactions in radiation chemistry. Here we will complement Steen's review and also examine some theoretical aspects. The reader is encouraged to read the following discussion together with Sect. III.B of Chap. 6.

Recently Miller, in a series of articles[58,59], has attempted to provide overwhelming support in favor of the tunneling model for electron reactions in glasses, both organic and aqueous. His argument rests mainly on two points. First, the logarithm of the trapped electron yield is linear with scavenger concentration (at a fixed time). Apparently the reciprocal of the yield shows considerable curvature at small scavenger concentrations which is taken to deny diffusion-controlled scavenging of mobile (or quasi-free) electrons. Secondly, the time-dependence of the trapped electron yield at relatively small scavenger concentrations (~ 10 mM) presumably cannot be fitted to a $\sim t^{-1/2}$ dependence which is characteristic of diffusion. Instead it depends on $\log t$ which can be justified on the tunneling model. In addition, Miller[58] points out that in some cases the presence of a second (not-so-good) scavenger actually enhances the yield of the negative ion formed out of a good scavenger and in fact this enhancement increases with the concentration of the second scavenger when that of the first is kept constant. Envisaging trapped electrons in a glass being scavenged by several scavengers $i(i = 1, \cdots n)$ at concentration c_i with cross-section σ_i, the probability of scavenging by the ith scavenger will be proportional to $c_i\sigma_i$ if competitive (diffusion-controlled or not) reactions are considered. If the yield of ith scavenging reaction be denoted by G_i and G_i^0 in the presence and in the absence of other scavengers, respectively, then it is easily shown that[58]

$$G_i^{-1} = (G_i^0)^{-1}\left[1 + (c_i\sigma_i)^{-1}\sum_{j\neq i} c_j\sigma_j\right] \tag{39}$$

Equation (39) shows that in the case of two scavengers the reciprocal yield of one reaction at a fixed concentration should vary linearly with the concentration of the other scavenger. More importantly, its yield is reduced by competition. However, Miller[58] has shown that in γ-irradiated 3MP glass the yield of 9-methylanthracene anions actually increases with the concentration of added benzene in the range 0–20 mM. This is readily explained in the tunneling model in terms of additional channels opened up for electron transfer. The explanation of the phenomenon is that tunneling occurs simultaneously to all available scavengers followed by a secondary transfer, presumably by tunneling also, to the more efficient scavenger, viz. 9 MA. Although electron transfer from one scavenger to another is not unknown, it is hard to visualize the increase of 9 MA$^-$ with benzene concentration on a competitive scavenging model.

Miller considers a barrier of fixed height V_0 and electron energy E_0, giving a binding energy $\epsilon = V_0 - E_0$. Using a well width of 6 Å, $V_0 = 1$ eV and the usual electron mass he then obtains approximately

$$\log \omega = 15 - 0.443 \, a \, \epsilon^{1/2} \tag{40}$$

where ω is the tunneling rate (s^{-1}), a is the separation between the trapped electron and the scavenger (Å) and ϵ is in eV. He further simplifies the problem by assuming that in time t all electrons within a distance $a(t)$ from a scavenger will be scavenged and the others outside this distance will be intact in their traps. This value of a is a function of time and in this description $\log t = -\log \omega$ giving

$$a(t) = [15 + \log t]/[0.443 \, \epsilon^{1/2}] \tag{41}$$

With typical scavenger concentrations ~ 10 nM in 3 MP, t is ~ 5 mins. and ϵ is taken as 0.53 eV, equal to the energy at photobleaching threshold. From eqn. (41), one then gets a value of ~ 54 Å, the typical tunneling distance in this case subject to the uncertainties of the model. Later Miller[58] corrected eqn. (41) by adding the size of the electron distribution on its right hand side; however, since it is small (~ 5 Å) the correction is not significant unless the tunneling distance is small. The relationship between the probability W of not being scavenged at a fixed time (say, ~ 5 mins.) and scavenger concentration c is obtained on the basis of random disposition of scavengers as

$$\ln W = -(4\pi/3)cr^3 \tag{42}$$

or

$$\log W = -1.09 \times 10^{-3} \, Mr^3 \tag{43}$$

where M is the molarity of the scavenger and r is the capture (or tunneling) radius also referred to the same time. If experiments refer to the same time and concentrations, the values of a and r in eqns. (41)–(43) should nearly be the same. In practice, however, a, determined by eqn. (41) from the decay of trapped electron in the presence of a scavenger at a given concentration and r given by eqns. (42) and (43) from the variation of yield with scavenger concentration (in the same range as before) at a given time sometimes agree, but not always, although they are of the same order of magnitude. Values of a and r range in most cases between 30 and 60 Å (see ref. 58, Table 1). In short, a simple tunneling model works qualitatively, but not quantitatively.

Electron tunneling has also been invoked for 6 M NaOH glasses at 77 K[58]. Using nanosecond pulse radiolysis it has been shown that there is a continuous electron transfer reaction for times ~ 10 ns to mins. Reactions at still shorter times presumably occur but are not observable in ns-pulse-radiolysis. These very fast reactions probably also occur in non-polar glasses meaning that much of the decay actually happens during steady-state (or slower pulse-) radiolysis. In aqueous alkaline glasses it has been found that scavenging by CrO_4^{2-} is in good agreement with the tunneling model, but not the scavenging by NO_3^- which is an inefficient scavenger. The inefficiency of scavenging at low temperatures has been attributed by Miller[59] to restrictions due to the Franck–Condon effect and symmetry. Relaxation or the gradual degradation of energy in the negative ion state is probably a contributing factor also. In its absence the electron will have a good chance of tunneling back to the trap. In any case the most important observation in aqueous glasses is the insensitivity of the decay rate to temperature. The temperature can be raised up to a factor ~ 2 from 77 K to ~ 150 K without essentially altering the decay rate. One has to go considerably ($\sim 25°$) above the glass temperature to observe diffusive reaction.

Kroh and Stradowski[60] measured the decay of solvated electrons in 8 M NaOH glassy ice at $\lambda_{max} = 585$ nm. The stated interpretation of their experiments is complex depending partly on the competition for mobile electrons by traps and scavengers and partly on the tunneling of trapped or solvated electrons to the scavenger molecules. However, they take the tunneling rate to be proportional to $c^{1/3}$ (or $c^{1/2}$ for organic scavengers) where c is the scavenger concentration. In the opinion of this writer this dependence cannot be justified theoretically. The tunneling rate, apart for being time-dependent, should be proportional to the product of c and the yield of trapped electron at that time (vide infra). In a later paper Kroh and Stradowski[60] used the simple tunneling model similar to Miller's treatment. The tunneling rate constant is taken as the product of the oscillation frequency of the trapped (or solvated) electron and the barrier penetration probability. The oscillation frequency is taken as the frequency of optical absorption which seems reasonable to a first approximation. They have also attempted to explain the wavelength-dependence of the decay of the trapped electron.

Zamaraev et al.[61] appear to be the first to invoke tunneling for the decay of the trapped (in fact, solvated) electron in alkaline glassy ice. In the beginning they observed the simultaneous decay of the trapped electron and of the radical anion O^- in a process which resembles recombination in this system. Later they extended their considerations to reactions with added scavengers. These investigators drew attention to the important phenomenological features of the tunneling model as applied to their experiments. In particular they pointed out the independence of reaction rates on temperature between 120 K and 77 K and possibly to lower temperatures. This particular feature was independent of dose, quality of radiation (β- or γ-radiolysis) or the species studied (e_t or O^-). Experimentally Zamaraev et al. obtained a logarithmic-dependence of the yield of trapped electrons with time as follows

$$[e_t]/[e_t]_0 = 1 - B \ln(vt) \qquad (44)$$

In eqn. (44) $[e_t]_0$ refers to the trapped electron yield at the initial time v^{-1}, not at $t = 0$; B and v are constants. Out of various possibilities for explaining the logarithmic-dependence eqn. (44), Zamaraev et al. chose the dependence of the tunneling rate k on distance r given as

$$k = v \exp(-r/a) \qquad (45)$$

where a is the relaxation length associated with the decay of the trapped electron wave function. It was already known from saturation experiments that the average separation between e_t and O^- is ~ 35 Å. Zamaraev et al. made the (rather implausible) assumption that the distribution of separation distance is uniform, bounded between the limits r_1 and r_2 with the calculated result that

$$[e_t]/[e_t]_0 = \frac{a}{(r_2 - r_1)} \left(\frac{r_2}{a} - \ln vt \right) \qquad (46)$$

Notice that the theoretical formula (eqn. (46)) is not exactly the same as the experimental equation (eqn. (44)); however, if $r_2 \gg r_1$ then the results are nearly identical. Zamaraev et al. assume $a \sim 1$ Å (typical relaxation length for atomic wave functions) and $v \sim 6 \times 10^{14}$ s^{-1} (corresponding to optical absorption energy of 2.2 eV). With $(r_1 + r_2)/2 = 35$ Å they then estimate $r_1 = 20$ Å and $r_2 = 50$ Å by comparing eqn. (46) with experiments.

Kieffer et al.[62] obtained evidence of tunneling in the recombination of trapped electrons in methylcyclohexane containing $\sim$10 mM biphenyl. The isothermal luminescence decay curves, after normalization at 4 min, were superimposable at 4.2, 66 and 77 K, thus lending support to the tunneling mechanism. Possibility of thermal detrapping was considered at a higher temperature. However, there seems to be some objection to the methylcyclohexane work based on spectroscopic evidence[66]. On the other hand, in 3 MP containing biphenyl the ITL following recombination can be attributed to tunneling only at 4 K but not at 77 K[62]. In any case, the tunneling distances deduced by Kieffer et al. from their experiments lie in the range 60–100 Å, which appears to be very large.

Brocklehurst[63], in a recent review, has correctly pointed out the unsatisfactory state of the theory of electron transfer rates in condensed (non-polar) media. Following Marcus[65], he points out that the tunneling model only calculates one component of the total transfer rate, leaving out such important details as the Franck–Condon factor, electronic structural consideration and vibrational relaxation in the acceptor molecule. Brocklehurst's discussion of these topics is illuminating; however, his expression of the tunneling rate is still (somewhat of necessity) subject to the errors and uncertainties of the WKB approximation. He applies the tunneling model to recombinations of aromatic cations with solvated electrons or anions in alkanes. He argues that a large transfer distance ($\sim$50 Å) is possible in rigid media. In liquids the final step in the recombination process is tunneling and the critical distance may be estimated by equating the tunneling rate to the diffusion rate. This critical distance increases (proportionately) with viscosity. Therefore, in highly viscous liquids again large tunneling distances are to be expected.

Recently Tachiya and Mozumder[67] have examined somewhat rigorously the theory of simultaneous tunneling of a trapped electron to all neighboring scavengers. If the tunneling rate to a scavenger i placed at a distance r_i be denoted by $k(r_i)$ then the probability that the electron will remain unscavenged at time t is given by

$$P(t) = \int_v \cdots \int_v \exp\left[-t \sum_{i=1}^{N} k(r_i) \right] \phi(\mathbf{r}_1 \cdots \mathbf{r}_i \cdots \mathbf{r}_N) \times$$
$$dv_1 \cdots dv_i \cdots dv_N \qquad (47)$$

where N is the total number of scavengers, dv_i is the volume element corresponding to scavenger i and ϕ denotes the probability density of finding scavenger 1 at $\mathbf{r}_1, \cdots i$ at $\mathbf{r}_i$, etc. Considering a random distribution $\phi = v^{-N}$, where v is the total sample volume one obtains from eqn. (47).

$$P(t) = \left[1 - v^{-1} \int_0^R \{1 - \exp[-tk(r)]\} 4\pi r^2 dr\right]^{cv} \tag{48}$$

In eqn. (48) v is considered to be a sphere of radius R and c is the (average) scavenger concentration. Taking the limit $R \to \infty$ at fixed c and remembering that $k(r)$ falls rapidly to zero as $r \to \infty$ (cf. eqn. (45)) one gets from eqn. (48)

$$P(t) = \exp[-cf(t)] \tag{49}$$

where

$$f(t) = \int_0^\infty \{1 - \exp[-tk(r)]\} 4\pi r^2 dr \tag{50}$$

Equation (49) gives the exponential-dependence of the trapped electron yield with scavenger concentration at a given time which is observed in many experiments. However, it is to be stressed that this dependence arises out of the random disposition of the scavengers, rather than from any specific tunneling property. The only way essential quantum mechanics enters into this result is that simultaneous tunneling to all scavengers is allowed which will not be so classically. In this respect it resembles simultaneous decay of an excited nucleus via multiple channels.

The exact significance of the parameters v and a in eqn. (45) is not known but the Franck–Condon factor should be incorporated in v. Substituting eqn. (45) into eqn. (50), eqn. (49) may be written as

$$P(t) = \exp\left[-\frac{c}{c_0} g(vt)\right] \tag{51}$$

where $c_0 \equiv 3/(4\pi a^3)$. The functional form of g is complex and related to successive derivatives of the gamma function. However, for all significant values of t, $vt \gg 1$, in which limit it is given quite accurately by[67]

$$g(z) = (\ln z)^3 + 1.73(\ln z)^2 + 5.93(\ln z) + 5.44 \tag{52}$$

with $z = vt$. Miller[58] considers tunneling only to the nearest scavengers for which the random distribution is given by[68] $\phi = c \exp(-4\pi/3 \, cr^3)$. The survival probability is then given from eqn. (47) (without the summation) as follows

$$P(t) = \int_0^\infty \exp[-k(r)t] \exp\left(-\frac{4\pi}{3} cr^3\right) 4\pi cr^2 dr \tag{53}$$

Miller further simplifies the problem by assuming that $\exp[-k(r)t] = 1$ for $t < k(r)^{-1}$ and $= 0$ for $t > k(r)^{-1}$. He then obtains from eqn. (53)

$$P(t) = \exp\left[-\frac{c}{c_0}(\ln vt)^3\right] \tag{54}$$

Miller found that the kinetics of trapped electron decay is well described by eqn. (54)

over the time scale of ns–min. if v is taken as $\sim 10^{15}$ s^{-1}. If indeed v is as large as 10^{15} s^{-1} then $g(z)$ is well represented by the right hand side of eqn. (52), i.e., $g(z) = (\ln z)^3$. Substitution into eqn. (51) then reproduces eqn. (54) so that in this limit Tachiya and Mozumder's expression reduces to that of Miller although the former is not limited to tunneling to nearest neighbors only. In this sense the entire time scale now available for experimentation can be taken as the long time limit except perhaps for poor scavengers.

In conclusion, it can be said that electron tunneling to scavengers is now an established concept in low temperature glasses. In polar glasses tunneling probably first occurs with the trapped electron and then with the solvated electron. If we take the binding energy of a trapped electron as ~ 1 eV then at a temperature of 77 K ($k_B T \approx 0.006$ eV) the classical transition probability per second into the mobile state is given in the order of magnitude by $f_v \exp(-1 \cdot /0.006) \approx f_v 10^{-72}$ where f_v is the vibrational frequency of the trapped electron. Even if we take f_v as $\sim 10^{15}$ s^{-1} the resultant probability of promotion to the mobile state is too small to be significant during the time scale of experiments. It should be noted that the tunneling per second should be relatively independent of the nature of the scavenger. However, the Franck–Condon factor and the relaxation processes are rather specific to the acceptor molecule; thus the net electron transfer rate becomes rather sensitive to the nature of the scavenger molecule. At present it is not clear why recombination by tunneling is so rare. It is possible that in many cases the energy level of the trapped electron is not close enough to a stationary energy level in the neutral molecule which results on neutralization. Electron tunneling in the liquid state is not ruled out completely, but it is probably overshadowed by diffusion, which is much faster in the liquid state. However, the final step in the nuetralization process, even in liquids, is probably electron tunneling[63].

VI. A list of problems of current interest

In the opinion of this writer the following constitutes a list of problems, covering the material presented in this chapter, which are most challenging from the theoretical point of view.

VI.A Electron trapping in non-polar media

The physical mechanism has not yet been delineated clearly. Although the involvement of the bond–dipole moment (such as that of CH) has been envisaged from time to time it has not been clearly established. Similar questions can be raised with respect to spectral relaxation in non-polar media. An alternative to such a description is matrix trapping due to irregularities. Sufficient work is not available at present to distinguish between these mechanisms.

VI.B Microscopic dielectric relaxation

Development of microscopic models is a must for understanding the formation of solvated electrons in glasses and highly viscous liquids. There are two principal difficulties to overcome: (i) inclusion of dipole–dipole interaction and (ii) treatment of H-bond breakage as induced by the field of an electron.

VI.C Yields and relaxation times

Obviously the yield of solvated electrons (*G*-values) and the relaxation times should be related in some manner. Without the knowledge of the distribution of initial distances this relationship is not reliably available at present.

VI.D Electron mobility in dielectric liquids

Description of electron mobility in terms of the trapped and the quasi-free electron is widely used; however, the quasi-free state is hard to establish in dielectric liquids. The Arrhenius-dependence of mobility on temperature, which is the strongest argument for such a model, may also be explained by other models such as trap-to-trap tunneling assisted by phonon interaction. If indeed the electron exists in two states (trapped and quasi-free) then most of the technique of inverse Laplace transformation relating scavenging experiments with lifetime distribution may be rendered meaningless except at low concentrations.

VI.E Electron scavenging by tunneling in glasses

The fundamentals of this situation now seem to be well-settled, that is there is no doubt that a large part of electron scavenging in glasses occurs through the tunneling mechanism. On the other hand, the true transfer rate theory is far from satisfactory. To obtain that, it is not sufficient to treat the barrier problem alone. In addition, (i) the Franck–Condon effect must be treated adequately and, (ii) one must also take into account the deactivation process in the negative ion (relaxation) which either prevents or considerably slows down the back tunneling.

References

1 J. E. Willard, in P. Ausloos (Ed.), Fundamental Processes in Radiation Chemistry, Interscience, New York, 1968, pp. 599–649.

2 W. H. Hamill, in E. T. Kaiser and L. Kevan (Eds.), Radical Ions, Interscience, New York, 1968, pp. 321–416.

3 A. Ekstrom, Radiat. Res. Rev., **2** (1970) 381.

4 L. Kevan, in M. Burton and J. L. Magee (Eds.), Advances in Radiation Chemistry, Vol. 4, Wiley, New York, 1974, pp. 181–305.

5 K. Funabashi, in M. Burton and J. L. Magee (Eds.), Advances in Radiation Chemistry, Vol. 4, Wiley, New York, 1974, pp. 103–180

6 W. Weyl, Ann. Phys. (Leipzig), **197** (1863) 601; Chem. Zentralbl., **35** (1864) 601.

7 (i) E. J. Hart and J. W. Boag, J. Amer. Chem. Soc., **84** (1962) 4080; (ii) J. W. Boag and E. J. Hart, Nature (London), **197** (1963) 45; (iii) J. P. Keene, Nature (London), **197** (1963) 47.

8 (a) R. R. Hentz, Proceedings of the International Meeting on Primary Radiation Effects in Chemistry and Biology, Comision Nacional de Energia Atomica, Buenos Aires, Argentina, 1970, pp. 147–165; (b) H. T. Davis, L. D. Schmidt and R. M. Minday, Chem. Phys. Lett., **13** (1972) 413.

9 (a) I. A. Taub and H. A. Gillis, J. Amer. Chem. Soc., **91** (1969) 6507; (b) J. H. Baxendale and E. J. Rasburn, J. Chem. Soc. Faraday I, **70** (1974) 705.

10 T. Higashimura, M. Noda, T. Warashina and H. Yoshida, J. Chem. Phys., **53** (1970) 1152.

11 S. O. Nielsen, P. Pagsberg, E. J. Hart, H. Christensen and G. Nilson, J. Phys. Chem., **73** (1969) 3171.

12 (a) H. Hase, M. Noda and T. Higashimura, J. Chem. Phys., **54** (1971) 2975; (b) H. Hase, T. Higashimura and M. Ogasawara, Chem. Phys. Lett., **16** (1972) 214; (c) H. Hase, T. Warashina, M. Noda, A. Namiki and T. Higashimura, J. Chem. Phys., **57** (1972) 1039; (d) T. Ito, K. Fueki, A. Namiki and H. Hase, J. Phys. Chem., **77** (1973) 1803; (e) H. Yoshida and T. Higashimura, Can. J. Chem., **48** (1970) 504.

13 L. Kevan, Advan. Radiat. Chem., **4** (1975) 205.

14 A. Ekstrom, Radiat. Res. Rev., (1970) 384 (Fig. 1).

15 L. Onsager, Radiat. Res. Suppl., **4** (1964) 13.

16 R. R. Hentz and G. A. Kenney-Wallace, J. Phys. Chem., **76** (1972) 2931; **78** (1974) 514.

17 (a) H. Hase and T. Warashina, J. Chem. Phys., **59** (1973) 2152; (b) L. M. Perkey, Farhataziz and R. R. Hentz, J. Chem. Phys., **61** (1974) 2979.

18 (a) J. B. Gallivan and W. H. Hamill, J. Chem. Phys., **44** (1966) 2378; (b) M. A. Bonin, J. Lin, K. Tsuji and F. Williams, Adv. Chem. Ser., ACS Publication No., **82** (1968) 269.

19 A. Mozumder and M. Tachiya, J. Chem. Phys., **62** (1975) 979.

20 A. Eckstrom, R. Suneram and J. E. Willard, J. Phys. Chem., **74** (1970) 1888.

21 J. R. Miller, J. Chem. Phys., **56** (1972) 5173.

22 D. F. Feng, K. Fueki and L. Kevan, J. Chem. Phys., **58** (1973) 3281.

23 (a) Y. Toyozawa, Prog. Theor. Phys. **26** (1961) 29; (b) T. Holstein, Ann. Phys., **8** (1959) 343.

24 M. Tachiya and A. Mozumder, J. Chem. Phys., **66** (1974) 3037; **61** (1974) 3890.

25 D. A. Copeland, N. R. Kestner and J. Jortner, J. Chem. Phys., **53** (1970) 1189.

26 P. Debye, Polar Molecules, The Chemical Catalog Company, New York, 1929, p. 83.

27 See, for example, (a) A. M. Freudenthal and H. Geiringer, in S. Flugge (Ed.), Handbuch d. Physik VI, Springer-Verlag, Berlin, 1958, p. 269 and (b) M. Reiner, in S. Flugge (Ed.), Handbuch d. Physik VI, Springer-Verlag, Berlin, 1958, p. 463 and 481.

28 See, for example, W. Kauzmann, Rev. Mod. Phys., **14** (1942) 12.

29 A. Mozumder, J. Chem. Phys., **50** (1969) 3153, 3162.

30 See, for example, H. Fröhlich, Theory of Dielectrics, University Press, Oxford, 1958, 2nd edn.

31 M. J. Bronskill, R. K. Wolff and J. W. Hunt, J. Chem. Phys., **53** (1970) 4201.

32 J. H. Baxendale and P. Wardman, Nature (London), **230** (1971) 449; J. Chem. Soc. Faraday I, 3 (1973) 584.

33 J. T. Richards and J. K. Thomas, J. Chem. Phys., **53** (1970) 218.

34 L. Kevan, J. Chem. Phys., **56** (1972) 838.

35 I. A. Taub and H. A. Gillis, J. Amer. Chem. Soc., **91** (1969) 6507.

36 See, for example, S. K. Garg and C. P. Smyth, J. Phys. Chem., **69** (1965) 1294.

37 M. Shirom and M. Tomkiewicz, J. Chem. Phys., **56** (1972) 2731.

38 A. Mozumder, J. Phys. Chem., **76** (1972) 3824.

39 For mathematical details, which are omitted here, please see ref. 38.

40 (a) T. J. Kemp, G. A. Salmon and P. Wardman, in M. Ebert, J. P. Keene, A. J. Swallow and J. H. Baxendale (Eds.), Pulse Radiolysis, Academic Press, London, 1965, pp. 247–257; (b) L. B. Magnuson, J. T. Richards and J. K. Thomas, Int. J. Radiat. Phys. Chem., **3** (1971) 295; (c) B. J. Brown, N. T. Barker and D. F. Sangster, J. Phys. Chem. **75** (1971) 3639; (d) see ref. 16; (e) see ref. 9 (b).

41 A. Mozumder and M. Tachiya, J. Chem. Phys., forthcoming.

42 R. Kubo, J. Phys. Soc. Jap. **12** (1957) 570; R. Kubo, in W. E. Brittin and L. S. Dunham (Eds.) Lectures in Theoretical Physics, Vol. I, p. 120, Interscience, New York, 1959; R. Kubo, Rep. Progr. Phys., **29** (1966) 255.

43 An apparent exception to this rule is tetramethylsilane which could be due to impurities.

44 L. Onsager, Phys. Rev., **54** (1938) 554.

45 A. Mozumder and M. Tachiya, J. Chem. Phys., **62** (1975) 979, Appendix.

46 (a) R. R. Hentz, Proceedings of the International Meeting on Primary Radiation Effects in Chemistry and Biology, Buenos Aires, 1970, p. 147–165; (b) R. R. Hentz (private communication).

47 A. Mozumder, unpublished.

48 (a) W. F. Schmidt and A. O. Allen, J. Chem. Phys., **52** (1970) 4788, (b) R. M. Minday, L. D. Schmidt and H. T. Davis, J. Chem. Phys., **54** (1971) 3112.

49 N. R. Kestner and J. Jortner, J. Chem. Phys., **59** (1973) 26.

50 L. Frommhold, Fortschr. Phys., **12** (1964) 597; Phys. Rev., **172** (1968) 118.

51 J. Lekner, Phys. Rev., **158** (1967) 130; M. H. Cohen and J. Lekner, Phys. Rev., **158** (1967) 305.

52 H. T. Davis, L. D. Schmidt and R. M. Minday, Chem. Phys. Lett., **13** (1972) 413.

53 G. Bakale and W. F. Schmidt, Chem. Phys. Lett. **22** (1973) 164.

54 R. M. Minday, L. D. Schmidt, and H. T. Davis, J. Phys. Chem., **76** (1972) 442.

55 R. P. Bell, Acid–Base Catalysis, Oxford, London, 1941.

56 M. Eigen, Z. Elecktrochem., **64** (1960) 115; M. Eigen and L. deMaeyer, Proc. Roy. Soc., Ser. A, **247** (1958) 505.

57 M. C. Flanigan and J. R. de la Vega, J. Chem. Phys., **61** (1974) 1882.

58 J. R. Miller, J. Chem. Phys., **56** (1972) 5173; J. R. Miller, Chem. Phys. Lett., **22** (1973) 180.

59 J. R. Miller, J. Phys. Chem., **79** (1975) 1070.

60 J. Kroh and Cz. Stradowski, Radiochem. Radioanal. Lett., **9** (1972) 169; J. Kroh and Cz. Stradowski, Int. J. Radiat. Phys. Chem., **5** (1973) 243.

61 K. I. Zamaraev, R. F. Khairutdinov, A. I. Mikhailov and V. I. Gol'danskii, Dokl. Akad. Nauk S.S.S.R. **199** (1971) 640; K. I. Zamaraev and R. F. Khairutdinov, Chem. Phys., **4** (1974) 181.

62 F. Kieffer, C. Meyer and J. Rigaut, Chem. Phys. Lett. **11** (1971) 359; F. Kieffer, C. Lapersonne-Meyer and J. Rigaut, Int. J. Radiat. Phys. Chem., **6** (1974) 79.

63 B. Brocklehurst, Chem. Phys., **2** (1973) 16.

64 G. V. Buxton, F. C. R. Catell and F. S. Dainton, Trans. Faraday Soc., **67** (1971) 687.

65 See, for example, R. A. Marcus, Ann. Rev. Phys. Chem., **15** (1964) 155.

66 J. Moan, quoted by H. B. Steen in Chapter 6.

67 M. Tachiya and A. Mozumder, Chem. Phys. Lett., **28** (1974) 87.

68 P. Hertz, Math. Ann., **67** (1909) 387.

FORMATION MECHANISMS AND PRIMARY REACTIONS OF EXCESS ELECTRONS IN CONDENSED POLAR MEDIA

H. B. STEEN

I. Introduction

Interest has recently been renewed in the earliest events in the radiolysis of condensed media, and especially the reactions of the electrons. Current models of radiolysis are based largely on reactivities and yields measured in dilute solutions on the microsecond time scale or by steady-state radiolysis. A substantial body of experimental data has now become available indicating that, with respect to the earliest events in radiolysis, these models may need significant revision. Various studies of the radiolysis of highly concentrated solutions have shown that the reactivities of electrons on the picosecond time scale may be significantly different from those extrapolated from measurements in dilute solutions[1–5]. A theoretical remedy for these discrepancies was attempted by Hamill[6] who proposed a new model for the radiolysis of water in which he assumed that the electron may react prior to solvation. He termed this unsolvated electron the "dry electron" (e^-) rather than the "mobile electron" (e_m^-), a term that had already been used for glasses. The distinction between e^- and e_m^- appears to lack a physical basis but descriptively these terms refer to the precursors of solvated electrons (e_s^-) in liquids and stabley trapped electrons (e_t^-) in glasses, respectively. This distinction seems to have taken root and will be retained here.

Hamill's model appeared to be corroborated by studies of e_t^- in glasses. The yield of e_t^- ($G(e_t^-)$) as well as the yield of other species for which the electron was believed to be the precursor, were found to be reduced by the presence of electron scavengers. This quenching of $G(e_t^-)$ was attributed to the capture of e_m^-[7].

Subsequent studies by Hunt and co-workers[8–13], using a pulse-radiolysis technique with a time resolution of the order of 10 ps, seemed to support the idea of presolvation electron reactions. They found that electron scavengers in the concentration range above 0.1 M may significantly reduce the initial yields of e_s^- ($G_i(e_s^-)$), indicating that the electrons may react as e^-. However, several workers have proposed alternative interpretations of these picosecond pulse-radiolysis data by which the introduction of a new chemical species, i.e., e^-, is avoided. It has been suggested that the effect of time-dependent reaction rates may modify the reactivities of e_s^- at picosecond times significantly and thereby explain these data[14]. The data have also been attributed to electron tunneling from e_s^- at very short times[15] and to the fact that at high solute concentrations a significant fraction of e_s^- may be formed so close to the scavenger molecules that the reaction may occur without any preceding diffusion[16].

Correspondingly, the importance of c_m^- as a reactive precursor of e_t^- has been called into question primarily on the basis of observations that a considerable fraction of e_t^- may react with the scavenger at short times[15,17-20], presumably by electron tunneling. In the present chapter some results are described that may elucidate the possible role of e^- and e_m^- in radiolysis, and various interpretations of these data are discussed.

The subject of the primary electron reactions in the radiolysis of condensed media is intimately connected with the mechanisms by which the electrons are produced. In trying to fit the experimental information into a consistent picture, it has become increasingly clear to this author that one of the basic assumptions underlying all current radiolysis models may be erroneous. Thus, it is generally taken for granted that the electrons formed in the radiolysis of condensed media are initially in a quasi-free, mobile state, so that solvation or trapping occurs only after the electrons have migrated some distance away from their origin. There is hardly any sound theoretical basis or any firm experimental evidence to support this assumption for the case of polar media. On the contrary, several experiments seem to be at variance with it. It has been reported that in some systems $G(e_t^-)$ and $G_i(e_s^-)$ are constant for large variations in the temperature[21,22]. These observations are interpreted here to indicate that ion recombination prior to trapping or solvation is negligible. Furthermore, studies of recombination luminescence[23] indicate that the distribution of cation-electron distances is independent of whether the ionization is brought about by high energy radiations or by photoexcitation of levels close to the gas phase ionization potential (I_g).

On the other hand, experiments with photoionization in liquid and glassy solutions demonstrate that substantial yields of e_s^- and e_t^- may be formed from excited solute states well below I_g of the solute[24-26]. Thus, it is tentatively suggested here that the majority of e_s^- and e_t^- produced by radiolysis are formed directly from similar quasi-bound states of the medium and not via an e^- or e_m^- state. It appears that this hypothesis is qualitatively in accordance with the experimental data.

This chapter is not meant to be a complete review of the subject. The experimental results that are described have been chosen to show that none of the current models can reasonably account for the data on the primary processes of radiolysis, and hopefully, to spur consideration of alternative models.

II. Optical properties of solvated and trapped electrons

II.A Optical absorption of e_s^-

The solvated electron in polar media exhibits a pronounced optical absorption which covers the entire visible region. Its absorption spectrum is broad, asymmetric and structureless (a few reports about fine structure have not been confirmed) and usually peaks somewhere between 500 and 700 nm in alcohol and aqueous solvents with ϵ_{max} being of the order of $1-2 \cdot 10^4$ M^{-1} cm^{-1}. Studies with picosecond radiolysis[10] have shown that in water and various alcohols at room temperature this spectrum is present in its final form within 10 ps after absorption of high energy radiation. Studying e_s^-, formed by photoionization of the ferrocyanide ion in aqueous solution by means of a picosecond flash-photolysis technique, Rentzepis et al.[27] found that an IR absorption was present at 2 ps after the absorption of light and that the normal spectrum of e_s^- developed in about 4 ps. It is not known, however, if this is really the solvation time of a free electron. Thus, as discussed below (Sect. VII), it is possible that the growing in of the spectrum of e_s^- observed in this experiment reflects the electron transfer from the excited solute to the solvent rather than the solvation of a free charge.

The initial absorption in the IR and its gradual shift towards the visible have been observed on a longer time scale and in more detail by pulse-radiolysis studies of various alcohols for which the viscosity was greatly increased by cooling to near the glass transition temperature[28,29]. Similar results have been reported from pulse-radiolysis studies of various glasses at 77 K[17,18]. For electrons produced by photoionization of N, N'-tetramethylphenylenediamine (TMPD) in ethanol and in 3-methylhexane glasses at 77 K, Richards and Thomas[30] found the same initial IR absorption and the same subsequent transition to the spectrum of the stable e_t^- as originally reported for electrons produced by radiolysis. Hence, the trapping process appears to be independent of how the electrons are formed.

There appears to be unanimous agreement that the initial absorption in the IR is caused by electrons in shallow traps and that the shift of this absorption towards the visible reflects the dipolar relaxation of the trap which is induced by the presence of the electron. It is of particular interest that in 6-10 M NaOH glass at 77 K, which has a viscosity several orders of magnitude larger than the above mentioned alcohols, no transient IR absorption has been observed. Thus, the final e_t^- spectrum appears to be fully developed within 10^{-7} s[17,19].

II.B Optical absorption of e_t^-

The absorption spectrum of e_t^- in glasses is generally similar to that of e_s^- in the liquid phase. This observation is taken to indicate that the structure of the polarized shell of molecules surrounding the electron is essentially the same in both phases, the main difference between e_s^- and e_t^- being the larger stability of e_t^- caused by the high viscosity of the glass that prevents diffusion. This similarity is the justification for the extrapolation of data and conclusions from one phase to the other.

In many glasses the spectrum of e_t^- is somewhat broader than that of e_s^- in the corresponding liquid, especially on the IR side. This broadening appears to reflect a variation in trap-depth. Thus, by optical bleaching with monochromatic light at the IR end of the spectrum the broadening in this part of the spectrum can be removed, apparently by a transfer of electrons to deeper traps[31,32], as shown for equivolume ethylene glycol/water glass (EG/H$_2$O) in Fig. 1.

In some cases the spectrum of e_t^- can be separated into several components as shown in Fig. 1, indicating that the broadening of the spectrum is due to the existence

178

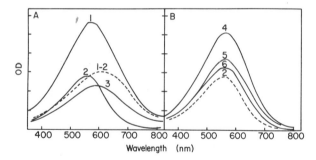

Fig. 1. Spectra of e_t^- in equivolume EG/H$_2$O irradiated and measured at 77 K (A) and at 130 K (B). 1 and 4, no photobleaching; 2 and 6, after bleaching at 700 nm; 3 and 5, after bleaching at 366 nm.

of distinctly different types of traps rather than reflecting a continuous distribution of trap-depths. Similar results have been reported for 8–10 M NaOH glass[33,34]. The existence of two distinct components in the absorption spectrum is indicated also by the decay kinetics of e_t^- during optical bleaching at intermediate wavelengths.[35] As shown in Fig. 2, this decay can be decomposed into two exponential components. (It should be noted that in order to obtain the true decay kinetics of e_t^-, or of any other absorbing species, the optical density (OD) of the sample at the bleaching wavelength must be small, i.e. OD $<$ 0.2, otherwise the decay kinetics will depend markedly on OD and consequently on the bleaching time[35].) Upon increasing the temperature, the red component in the absorption spectrum of e_t^- disappears and only one exponential component remains in the bleaching kinetics (Fig. 2) while the total number of e_t^- is unaltered, presumably because the lower viscosity allows complete relaxation of the shallow type of traps. It should be noted that, according to the spectra, the shallow traps that give rise to this red component are much deeper

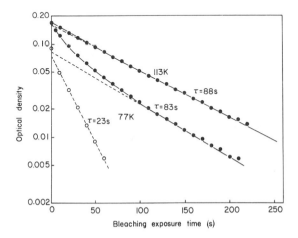

Fig. 2. The decay of X-ray-induced e_t^- in equivolume EG/H$_2$O as measured by OD (585 nm) during bleaching at 546 nm. Data are from ref. 35.

than those associated with the IR absorption observed by pulse-radiolysis (Sect. II.A).

At the higher temperature the spectrum of e_t^- cannot be changed by bleaching, indicating a uniform trap-depth (see Fig. 1), neither is the shape of this spectrum affected by the presence of electron scavengers[36]. Hence, the spectrum observed at this temperature, as well as that of e_s^-, is probably due to electrons in a single type of trap. This conclusion is supported by the failure of an attempt to "burn" narrow gaps in the spectrum of e_t^- by the use of intense monochromatic light[37].

At very low temperatures, e.g. 4 K, various irradiated glasses exhibit a permanent absorption in the IR, which is somewhat similar to that observed on the microsecond time scale close to the glass transition temperature[21,38–43]. This absorption is apparently due to electrons in shallow traps which are kept unrelaxed by the high viscosity of the medium that prevents molecular rotation at this temperature. It should be emphasized, however, that even at 4 K a significant fraction of e_t^- may be in deep traps. In ethanol, for example, ESR measurements show that ~25% of e_t^- at 4 K are in deep traps[21].

It has also been found[21] for ethanol, that $G(e_t^-)$ for radiolysis at 4 K is equal to $G(e_t^-)$ at 77 K where all e_t^- are in deep traps. This observation was taken to indicate that the shallow traps exist prior to the irradiation. Apparently it was argued that if the traps were formed only by the presence of the electron and in competition with charge recombination, a significant temperature-dependence of $G(e_t^-)$ would be expected since both dielectric relaxation, i.e. trapping, and recombination are temperature-dependent processes. However, it is conceivable that the initial trapping is induced by the presence of the electron by electronic polarization. Since this is an intramolecular process, which does not involve molecular rotation, it is essentially independent of temperature. Hence, the temperature-independence of $G(e_t^-)$ does not necessarily imply pre-existing traps. The initial trapping may be followed by rotation of polar molecular groups, notably OH, which probably occurs even at 4 K, and subsequently by molecular rotation if the viscosity is low enough. Pre-existing traps are likely to result from the fortuitous arrangement of the dipolar molecules of the medium if these molecules are randomly distributed[44]. It is not obvious, however, that the latter assumption is correct, especially in dealing with polar media.

II.C Excited states of e_t

In Fig. 3 is depicted the quantum yield of optical bleaching ϕ_B of e_t^- in EG/H_2O glass as a function of the wavelength of the bleaching light. It can be seen that below a certain wavelength, λ_i, there is a pronounced increase in ϕ_B. The photon energy corresponding to λ_i is usually taken as the trap-depth of e_t^-, i.e. the energy needed to bring the trapped electron into the conduction band. This interpretation is confirmed by the observation that photobleaching of e_t^- with wavelengths below λ_i generates significant electric conductivity, as also shown in Fig. 3.

It can also be seen that the absorption spectrum of e_t^- extends to wavelengths well above λ_i. Hence it appears that e_t^- in EG/H_2O has one or more bound excited

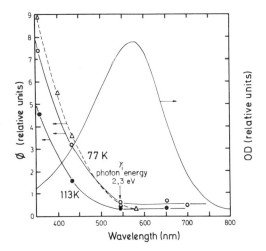

Fig. 3. The quantum yield of photobleaching of e_t^- in equivolume EG/H_2O, as measured by the rate of decay of e_t^- during bleaching at 77 and 113 K, as functions of the wavelength of the bleaching light (full curves), and the photocurrent induced by this bleaching at 77 K (broken curve). Also shown is the absorption spectrum of e_t^- at 113 K. Bleaching data from refs. 35 and 36. The photocurrent data are from ref. 45.

states (e_t^{-*}). Similar data have been reported for other alcohol glasses[46], as well as for single crystal ice[47], whereas it has been claimed that e_t^- in NaOH glass has no bound excited states[48,49]. The observation that ϕ_B does not drop to zero for wavelengths above λ_i, but remains at a small constant value (Fig. 3) indicates that e_t^{-*} is reactive. Studying the optical absorption of e_s^- in water using the extremely high light intensity of a 694 nm ruby laser beam, Kenney-Wallace and Walker[50] found no light saturation effects and concluded that the lifetime of e_s^{-*} is less than $6 \cdot 10^{-12}$ s.

As discussed in detail elsewhere[35], the wavelength-dependence of ϕ_B (Fig. 3) and the effect of scavengers on ϕ_B strongly indicate that, at least in EG/H_2O glass, the transfer of electrons from e_t^- to the conduction band, induced by photons with energies larger than the trap-depth of e_t^-, occurs via superexcited states rather than by a direct transition. The transition from these superexcited states to the conduction band competes with efficient internal conversion back to the ground state of e_t^-.

III. Reactions of electrons in glasses

III.A The mobile electron hypothesis

The primary experimental basis for the assumption that electrons in condensed polar media may react prior to trapping, that is, while they are still in a mobile state, is the observation that electron scavengers reduce $G(e_t^-)$. $G(e_t^-)$ is usually measured by optical absorption or by ESR spectroscopy at some time after irradiation. Thus, the inference of a reactive mobile electron is based on two tacitly accepted assumptions, namely (1) that e_t^- is stable so that the reduction of $G(e_t^-)$ can be attributed to the reaction of the precursor of e_t^- rather than to the reaction of e_t^- itself, and (2) that this reactive precursor of e_t^- is really a mobile electron. As we shall see below, the evidence for these assumptions is not convincing. Furthermore, no direct observation of the reaction of e_m^- in radiolysis has been reported. Thus, the role of e_m^- has been inferred only from measurements on e_t^- and other species for which e_m^- is believed to be the precursor.

The reactivity of e_m^- with an electron scavenger S as inferred from experiments with steady-state irradiation is often given in terms of the concentration of S ($[S]_{1/2}$) needed to halve $G(e_t^-)$, or more properly, the inverse of this, i.e. $[S]_{1/2}^{-1}$, which is called here the scavenging efficiency. Some values of $[S]_{1/2}$, as measured by the rate of increase of e_t^- during X-irradiation of EG/H$_2$O at 77 K (see Fig. 4), are given in Table 1. It can be seen that for the most efficient scavenger yet encountered, that is for CrO_4^{2-}, $[S]_{1/2} = 5 \cdot 10^{-3}$ M. The average distance between the S molecules at the concentration $[S]_{1/2}$ is commonly taken as a rough measure of the mean distance traveled by e_m^- in the direction away from its origin before trapping when no S is present, it being tacitly assumed that this distance is much larger than the encounter radius of the $e_m^- - S$ pair, and furthermore, that the reaction occurs on first encounter. On these assumptions the value of $[S]_{1/2}$ for CrO_4^{2-} indicates that e_m^- moves in the order of 65 Å away from the cation before it is trapped by the medium.

The kinetics of the reduction of $G(e_t^-)$ by scavengers as observed for various S in EG/H$_2$O at 77 K, are shown in Fig. 5. It can be seen that the reaction deviates noticeably from so-called homogeneous kinetics (Fig. 5A), the deviation apparently becoming larger for more efficient S, so that for the most efficient scavengers approximately exponential kinetics is observed (Fig. 5B). A similar variation of the scavenging kinetics has been reported for NaOH glass[52-55]. Other workers, however,

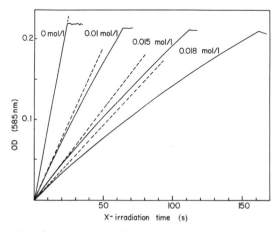

Fig. 4. The concentration of e_t^- as measured by OD (585 nm) in equivolume EG/H$_2$O glass at 77 K containing various concentrations of FeCl$_3$ as a function of the irradiation time. $G(e_t^-)$ is proportional to the slope of the curves.

TABLE 1

Scavenger concentrations ($[S]_{1/2}$) needed
to halve $G(e_t^-)$ in equivolume EG/H$_2$O
at 77 K. The observation of e_t^- was made on
the time scale of a few seconds (see Fig. 4).
(Data from refs. 36 and 51)

Solute	$[S]_{1/2}$(M)
K$_2$CrO$_4$	0.005
FeCl$_3$	0.010
CuCl$_2$	0.014
CCl$_3$COOH	0.022
CCl$_3$COONa	0.027
NaNO$_3$	0.033
H$_2$O$_2$	0.047
ClCH$_2$COOH	0.26
HCl	0.41

have found homogeneous kinetics for NaOH glass[56]. Homogeneous kinetics has also been reported for various alcohols and alcohol–water mixtures at 77 K[57].

The scavenging kinetics observed by optical absorption measurements in EG/H$_2$O glass at 77 K (Fig. 5) does not seem to be significantly affected by the variation in trap-depth that is present at this temperature. Similar kinetics has been observed at 113 K[36] where the trap-depth appears to be approximately uniform (Sect. II.B).

Theoretically, homogeneous scavenging kinetics is predicted under the assumption that the electron is a well localized entity, i.e. with a diameter of a few angstroms, which is moving by random diffusion like a molecule in a liquid solution. If there is a simple competition between trapping of this electron by pre-existing traps (T)

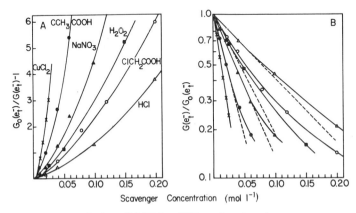

Fig. 5. The yield of e_t^- in equivolume EG/H$_2$O at 77 K as a function of scavenger concentration plotted according to "homogeneous" (A) and exponential (B) scavenging kinetics. $G(e_t^-)$ was measured by the initial rate of increase of OD (585 nm) during X-irradiation (see Fig. 4). Data from refs. 36 and 51.

on the one hand and capture by S on the other, the expression for homogeneous kinetics

$$G_0(e_t^-)/G(e_t^-) = 1 + k_S[S]/k_T[T] \tag{1}$$

is obtained. Here, k_S and k_T are the rate constants for capture by S and trapping in T, respectively.

The deviation from homogeneous kinetics has been explained by assuming that a certain reaction volume V_S is associated with each S molecule within which there can be no trapping in T. Hence, as S increases the effective number of T decreases and a deviation from eqn. (1) in the observed direction will result[52,55]. However, in order to account for the deviation from homogeneous kinetics that is observed for the most efficient scavengers, V_S must be of the order of 10^5 Å^3. This volume corresponds to an encounter radius $r_e \approx 30$ Å which is certainly too large to be compatible with the assumption underlying homogeneous kinetics, namely that the reactive entity is well localized. In fact, the assumption that e_m^- is well localized implies that it has a fairly large energy. Thus, for a free electron to have a de Broglie wavelength of 5 Å it must have an energy of ~6 eV. On the other hand, a free thermalized electron at 77 K has a de Broglie wavelength of ~120 Å. Although the effective mass of an electron in a conduction band may be larger than that of a free electron[58] by a factor of 2–4 because of its dielectric interaction with the medium, its de Broglie wavelength will still be several tens of angstroms. This means that if e_m^- is assumed to be a thermal electron, no motion of this electron has actually to be assumed in order to account for its reactivity.

On the assumption that e_m^- has thermal energy and is correspondingly delocalized, exponential scavenging kinetics is expected. To see this, assume that inside a certain radius r_e, which is of the order of magnitude of half the de Broglie wavelength of e_m^-, the S molecule has a probability p of capturing e_m^- in competition with trapping, whereas outside r_e, $p = 0$. If S is randomly distributed, then the probability $P(s)$, that an electron having s S molecules within r_e, will not react with an S prior to trapping by the matrix is given by the Poisson formula

$$P(s) = (1 - p)^s \, e^{-\bar{s}} \cdot \bar{s}^{-s}/s \tag{2}$$

where the mean value of s is $\bar{s} = (\tfrac{4}{3})\pi r_e^3 \cdot [S]$.

Taking the sum over the entire distribution of s one obtains

$$G(e_t^-)/G_0(e_t^-) = \sum_{s=0}^{s=\infty} P(s) = e^{-\bar{s}} \cdot e^{(1-p)\bar{s}} = e^{-p\bar{s}} = e^{-4/3\pi r_e^3 \cdot p[S]} \tag{3}$$

That is, exponential scavenging kinetics is expected for all scavenging efficiencies. Using the observed value of $[S]_{1/2}$ for CrO_4^{2-} (Table 1) and assuming $p = 1$ for this solute, one obtains $r_e \approx 40$ Å which is of the same order of magnitude as half the de Broglie wavelength of a thermal electron. However, in this simple form this model cannot account for the deviation from exponential kinetics observed for less efficient scavengers.

As noted above, the inference that electrons in polar glasses may react prior to trapping, i.e. in the mobile state, is based partly on the assumption that there is no decay of e_t^- between its formation and detection, so that the observed number of e_t^- is really the number of electrons trapped initially. This assumption does not hold true. Thus, a noticeable decay of e_t^- has been observed subsequent to steady state irradiation in virtually all glasses that have been studied in this respect. This decay is significantly enhanced by electron scavengers in the concentration range above 10^{-2} M[53,54,59-63].

Obviously, the decay of e_t^- must go on during the steady-state irradiation as well. An example of this is depicted in Fig. 4, which shows an apparent decrease of $G(e_t^-)$ with increasing irradiation time. Similar results were obtained by Kroh and Plonka[64] for NaOH glass. They found that $[S]_{1/2}^{-1}$ as obtained by measuring $G(e_t^-)$ subsequent to ^{60}Co γ-irradiation appeared to increase with the time of irradiation, and they attributed this increase to the reaction of e_t^- with S.

When experiments like these were carried out with better time resolution, that is, by pulse-radiolysis rather than steady state irradiation, it was found that there is a significant decay of e_t^- even on the microsecond time scale. Richards and Thomas[17] measured $[S]_{1/2}$ of CCl_4 for e_t^- in 3-methylhexane at 77 K by the optical absorption of e_t^- about $1\,\mu s$ after a 12 ns electron pulse. They obtained a value of $[S]_{1/2}$ which is about five times larger than that measured some 5 mins. after ^{60}Co γ-irradiation[15]. Thus, it appears that more than 80% of the reduction of $G(e_t^-)$, as observed after steady-state irradiation, is due to reactions taking place subsequent to trapping. Hence, at least in this glass, only a minor fraction, i.e. less than 20% of the reduction of $G(e_t^-)$ by CCl_4, can be attributed to capture of e_m^-. Recently, Miller[19] corroborated this conclusion when he was able to monitor the concentration of e_t^- in 10 M NaOH glass at 77 K from below 10^{-7} s to several minutes. He found that in the presence of scavengers there was a significant decay of e_t^- all through this time period. Hence, a considerable part of the reduction of $G(e_t^-)$, observed on the time scale of seconds or minutes in this glass, must be due to reaction of e_t^- rather than e_m^-. Previously, a similar decay of e_t^- produced by flash-photolysis of naphthalene anions in 2-methyltetrahydrofuran glass at 87 K had been followed upwards from the millisecond range[20].

On the other hand, it has recently been found that electron scavengers significantly reduce the photocurrent produced by optical bleaching of e_t^- in a γ-irradiated 5 M K_2CO_3 glass at 77 K. Noda and Kevan[65] found that $2.5 \cdot 10^{-2}$ M KNO_3 reduces the photocurrent by $\sim 75\%$, indicating that when e_m^- are formed by photobleaching they are indeed efficiently captured by NO_3^-. Similar results have been obtained for acrylamide in EG/H_2O glass and for NO_3^- in 6 M NaOH at 77 K[36]. In the latter glass 0.02 M of NO_3^- reduces the photoconductivity by $\sim 90\%$, whereas Miller[19] reported that under identical conditions this concentration of NO_3^- reduces $G(e_t^-)$ by only 20%. $G(e_t^-)$, in this case, was measured at $\sim 1\,\mu s$ after the irradiation at which time the vast majority of e_m^- should be trapped. It therefore appears that a concentration of NO_3^- that captures $\sim 90\%$ of the e_m^- reduces $G(e_t^-)$ only by 20%. Hence, if the photocurrent observed by optical bleaching of e_t^- is really due to e_m^-, e_m^- cannot be the precursor of the main fraction of e_t^- produced by the radiolysis of NaOH glass.

Furthermore, if the electrons are trapped in pre-existing traps, whether they are shallow or deep, and if the main scavenging occurs prior to trapping, one would expect the scavenging efficiency to be largely independent of temperature. This expectation is not borne out by experiments. Higashimura et al.[66] found that the scavenging efficiency of benzyl chloride in ethanol at 4 K is approximately four times larger than at 77 K. A decrease of the scavenging efficiency with increasing temperature has also been observed for various scavengers in EG/H_2O glass[67]. These results are further discussed in Sect. III.B.

It must be concluded that, at least in some glasses, reactions of e_m^- with solutes are much less important than previously anticipated from steady-state radiolysis studies. As discussed below, a major fraction of the quenching of $G(e_t^-)$ observed at long times appears to be due to reactions taking place subsequent to trapping, viz. by electron tunneling from both shallow and deep traps.

III.B The electron tunneling hypothesis

The decay of e_t^- as observed both with pulse-radiolysis[19] and subsequent to steady-state irradiation[53] exhibits an approximately logarithmic dependence on time. For the entire time period from 1 μs to several hours the decay kinetics can be reasonably well described by eqn. (4)[53].

$$[e_t^-]/[e_t^-]_0 = 1 - A\ln(t/t_0) \tag{4}$$

where t_0 is the time at the end of irradiation and A is a constant. Thus, the decay rate decreases rapidly with increasing time. This somewhat unusual decay kinetics is probably the main reason why the decay of e_t^- was not noticed in the earlier studies with steady-state radiolysis, since it implies that most of the decay occurs during the irradiation.

Several authors[15,52,53,59,61] have suggested that the decay of e_t^- is due to a tunneling of the electron through the energy barrier separating e_t^- from S, and it has been shown[15,53] that the observed decay kinetics (eqn. (4)) is reasonably accounted for by this hypothesis.

The tunneling model predicts that the rate of tunneling should be independent of temperature provided the trap-depth is much larger than thermal energies. This prediction has been confirmed by Mikhailov et al.[53] who found that the decay of e_t^- in 10 M NaOH, presumably caused by the reaction $e_t^- + O^-$, was independent of temperature from 77 to 120 K. The results of these workers indicated that the decay rate was the same even at 4.2 K. Similar results were obtained by Kroh and Plonka[64] who observed the decay of e_t^- to be independent of temperature between 113 and 173 K.

Decay of e_t^- may be observed also by the luminescence resulting from the geminate recombination of e_t^- and the parent cation (Sect. V.B). This type of luminescence exhibits roughly the same type of decay kinetics as observed for the decay of e_t^- by reaction with scavengers[68–70]. For methylcyclohexane (MCH) as well as for NaOH glass it has been reported that this decay is independent of temperature between 4 and 77 K, indicating that the recombination occurs by electron tunneling. As noted by Moan[71], however, the case of MCH glass is somewhat surprising in this respect since the absorption spectrum of e_t^- is quite different at 4 K and 77 K, showing that whereas at 4 K the majority of e_t^- are in shallow traps, all e_t^- are in deep traps at 77 K[42]. Hence, disregarding the possibility of experimental artifacts, it appears that the rate of decay is independent of trap-depth which should not be the case with electron tunneling. The relevance of recombination luminescence decay data with regard to electron tunneling in this glass may thus be questioned.

From a simple tunneling model, where e_t^- is represented by a two-dimensional square well separated from S by a square barrier, it can be derived that the distance r_e from S within which an e_t^- will have been captured by tunneling after a time t, is given approximately by eqn. (5), using the units Å, s and eV.

$$r_e(t) = r_0 + (15 + \log t)/0.443 \sqrt{I_e} \tag{5}$$

where I_e is the trap depth of e_t^- and r_0 is the sum of the radii of e_t^- and S, so that the width of the barrier is $r_e(t) - r_0$. Although $r_e(t)$ is a function of the diameter of the potential well representing e_t^-, it is only slightly sensitive to this parameter. Thus, the only alteration in eqn. (5) obtained by changing this diameter from 6 Å to 3 Å is that the number 15 is replaced by 16.

Assuming that e_t^- and S are randomly distributed, one finds, by the same reasoning as for eqn. (3) that

$$\log P = \log ([e_t^-]/[e_t^-]_0) = -1.09 \cdot 10^{-3} r_e(t)^3 [S] \tag{6}$$

where P is the probability that e_t^- is not captured by tunneling and $[e_t^-]$ and $[e_t^-]_0$ are the concentrations of e_t^- at the time t in the presence and absence of S, respectively, $r_e(t)$ is in Å and $[S]$ is in M. Equation (5) was originally derived by Mikhailov[72] and eqns. (5) and (6) were derived independently somewhat later by Miller[15]. From eqns. (5) and (6) it can be seen that whereas $r_e(t)$, and hence the apparent scavenging efficiency, increases with t, the scavenging kinetics will remain the same, so that exponential kinetics is to be expected in pulse-radiolysis studies as well as in experiments with steady-state irradiation. For a variety of organic glasses studied with steady-state radiolysis Miller[15] found that eqn. (6) holds true, whereas Richards and Thomas[17] observed such kinetics on the microsecond time scale, thus supporting the tunneling hypothesis. However, as noted in Sect. III.A, exponential kinetics is not observed for all S in EG/H$_2$O and NaOH studied by steady-state radiolysis.

The most obvious weakness of this simple tunneling model is that it predicts the same scavenging efficiency for all scavengers, whereas in actual fact values of $[S]_{1/2}$ for different S may vary by more than one order of magnitude (Table 1). This shortcoming results from the assumption made in this model that there is perfect resonance between e_t^- and S, in the sense that there is a continuum of electron-accepting orbitals in S and no symmetry restrictions on the electron tunneling. Clearly, these conditions are usually not fulfilled in the real case. The rate of tunneling, and consequently the value of r_e, will therefore generally be smaller than anticipated from eqn. (5). Hence, the value of $[S]_{1/2}$ that can be calculated from eqn. (6) should be regarded as a lower limit, that is, it will apply only to the most efficient scavengers. This is verified by Miller's pulse-radiolysis study of NaOH glass[19] which showed that with CrO_4^{2-} as a scavenger the decay of e_t^- followed that calculated from eqns. (5) and (6) fairly accurately, whereas for the less efficient scavenger NO_3^- the decay was much slower than predicted from these equations. Thus, variations in the values of $[S]_{1/2}$ between different scavengers are to be expected as a result of their different distribution of electron-accepting orbitals. However, it seems reasonable to assume that generally, the density of such orbitals will tend to become larger and more uniform with increasing energy difference between the electron affinity of the scavenger and the e_t^- trap-depth. Hence, the deviations from eqns. (5) and (6) are likely to decrease with the trap-depth of e_t^-. This expectation also seems to imply that variations in the reactivity, i.e. the tunneling rate, between different scavengers should tend to be smaller for electrons in shallow traps than for electrons in deep ones. The tunneling model is discussed in more detail in Chap. 5.

The scavenger concentration needed to capture 50% of e_t^- by tunneling, as calculated from eqns. (5) and (6), is plotted in Fig. 6 as a function of the time at which e_t^- is observed. It can be seen that when observed with steady-state radiolysis, i.e. by the type of experiment used to obtain the data in Table 1, the lower limit of $[S]_{1/2}$ should be $1.5 \cdot 10^{-2}$ M for the case of EG/H_2O where the electron trap-depth is 2.3 eV (Sect. II.C). According to Table 1, however, the observed values go as low as $5 \cdot 10^{-3}$ M. Hence, it appears that electron tunneling from these traps cannot account for more than about 30% of the reduction of $G(e_t^-)$ observed at long times. This conclusion is supported also by the observation that the scavenging efficiencies obtained for different scavengers with steady-state radiolysis, both in NaOH glass and in methanol/water glass, are generally not proportional to the reactivities of these scavengers measured by their effect on the rate of decay of e_t^- under conditions where this decay appears to be due to tunneling[56].

The tunneling model also predicts that $[S]_{1/2}$ should be independent of temperature. As noted in Sect. III.A, this is not confirmed experimentally[66,67]. In Fig. 7

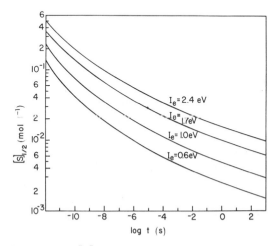

Fig. 6. The scavenger concentration $[S]_{1/2}$ needed to capture 50% of e_t^- by electron tunneling within a time t, as calculated from eqns. (5) and (6) for various values of the trap depth of e^-, I_e, assuming $r_0 = 4$ Å. This estimate should be regarded as a lower limit for $[S]_{1/2}$.

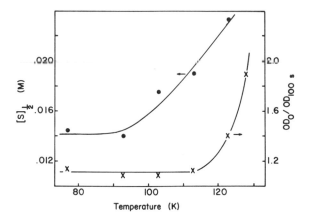

Fig. 7. The concentration of CuSo$_4$ needed to halve $G(e_t^-)$ in equivolume EG/H$_2$O and the rate of decay of e_t^- measured subsequent to the irradiation as functions of the temperature. Data from ref. 67.

is shown the temperature-dependence of $[CuSO_4]_{1/2}$ in EG/H$_2$O. It appears that above 90 K there is a significant increase of $[CuSO_4]_{1/2}$. Similar results were obtained for other S[67]. On the other hand, it can also be seen from Fig. 7 that the decay of e_t^-, as observed subsequent to steady-state irradiation in the same glass and under identical conditions, was independent of the temperature up to about 113 K. The latter result indicates that below 113 K the decay of e_t^- is due to tunneling, as is suggested also by the decay kinetics. Hence, it seems clear that at least in this particular case the reduction of $G(e_t^-)$ by S is not due exclusively to tunneling from the same type of e_t^- that is observed to decay subsequent to the irradiation.

This temperature-dependence of $[S]_{1/2}$, as well as the other discrepancies between the tunneling model and the experimental data that are noted above, can be at least qualitatively accounted for by the assumption that electron tunneling takes place not only from the deep, stable traps that can be observed subsequent to steady-state irradiation, but also from the shallow, transient traps in which the electrons appear to be trapped initially. Although the rate of tunneling from the shallow traps is in itself independent of temperature, this tunneling competes with the dielectric relaxation of the medium by which the shallow traps are transformed into deep ones. And since the rate of this relaxation is likely to increase with the temperature, the fraction of e_t^- that has time to react by tunneling while in shallow traps, will decrease correspondingly and cause an increase of $[S]_{1/2}$.

As noted above, it is to be expected that variations in the reactivity of different S with electrons in shallow traps should be smaller than for electrons in deep traps. There is not sufficient experimental data to check if this prediction holds true. If it is correct, however, it also follows that for less efficient scavengers, a larger fraction of the electrons that react with the scavenger should do so from shallow traps. Hence, the temperature-dependence of the scavenging efficiency is likely to be larger for less efficient scavengers. This expectation is indeed borne out by the data shown in

TABLE 2

Scavenger concentrations ($[S]_{1/2}$) needed to halve the yield
of X-ray induced e_t^- in equivolume EG/H$_2$O (Data are from
ref. 67)

Scavenger	$[S]_{1/2}(M)$		$[S]_{1/2}^{113K}/[S]_{1/2}^{77K}$
	77 K	113 K	
NaNO$_3$	0.033	0.052	1.58 $\pm$ 0.11
CCl$_3$COONa	0.027	0.040	1.48 $\pm$ 0.13
CuSO$_4$	0.014	0.019	1.36 $\pm$ 0.15
FeCl$_3$	0.010	0.012	1.20 $\pm$ 0.13

Table 2 from which it appears that the temperature-dependence of the scavenging
efficiency increases with decreasing scavenger efficiency.

The hypothesis that electrons react from transient, shallow traps is depicted
schematically in Fig. 8. It can be seen that if one compares two electron scavengers
with different scavenger efficiency and with their concentrations adjusted so that they
produce the same final reduction in $G(e_t^-)$, the decay of e_t^- as observed subsequent
to dielectric trap relaxation should be larger for the more efficient scavenger. This
prediction is borne out by Miller's observations[19] of the decay of e_t^- in NaOH glass
from 10^{-7}s onwards. He reports that with concentrations of CrO$_4^{2-}$ and NO$_3^-$ that

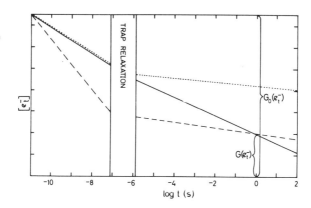

Fig. 8. A schematic depiction of how electron scavengers (S) may reduce $G(e_t^-)$ as observed at long
times, i.e. at log t (s) ≥ 0 by electron tunneling. The left and right portions of the diagram represent
the periods before and after dielectric relaxation from shallow, transient traps to deep, stable traps. The
full curve and the upper broken curve represent the decay of electrons caused by tunneling to S in the
presence of equal concentrations of an efficient and a less efficient S, respectively. The lower, broken
curve represents the decay with a larger concentration of the less efficient S. It can be seen that when
the rate of transition from shallow to deep traps increases, i.e. when the period of trap relaxation moves
towards the left, the resulting variation in $G(e_t^-)$ is larger with the less efficient scavenger, thus accounting
for the temperature dependence of the scavenging efficiency and why this effect is larger for less efficient
scavengers. (See Table 2.)

give approximately the same final reduction of $G(e_t^-)$, the decay is much more pronounced for CrO_4^{2-}, the scavenging efficiency of which is about seven times that of NO_3^-.

However, even when the tunneling model is extended to include tunneling from shallow traps it is not, in its present form, able to account for the non-exponential scavenging kinetics observed in some glasses. Also in the case where tunneling takes place first from shallow traps in the prerelaxation period and then from deep traps at longer times, exponential scavenging kinetics should be observed provided the tunneling from both types of traps obeys this type of kinetics. To see this, let us assume that during the prerelaxation period $[e_t^-]$ drops from the initial value $[e_t^-]_0$ to $[e_t^-]_1$ by tunneling from shallow traps and that subsequent to the trap relaxation $[e_t^-]$ drops further to $[e_t^-]_2$. Now, if in accordance with eqn. (6) $[e_t^-]_1 = [e_t^-]_0\, e^{-k_1}[S]$ and $[e_t^-]_2 = [e_t^-]_1\, e^{-k_2}[S]$, obviously

$$[e_t^-]_2 = [e_t^-]_0\, e^{-(k_1 + k_2)[S]}$$

So, it is difficult to account for the non-exponential scavenging kinetics on the basis of the simple electron tunneling model only, even if tunneling from shallow traps is taken into account.

In conclusion it appears that the reduction of $G(e_t^-)$ by scavengers, as observed with steady-state radiolysis, is primarily due to electron tunneling from the shallow traps prior to dielectric relaxation and subsequently from the deep, stable traps, rather than to capture of e_m^-. However, some quenching of the precursor of e_t^- may be involved, especially with less efficient scavengers. The latter process may possibly explain the deviation from exponential scavenging kinetics observed for these scavengers.

IV. Photoionization

It is well known that when a solute molecule is excited to an energy above or close to its gas phase ionization potential (I_g), it may lose an electron to the solvent. In both rigid and liquid polar media photoionization may occur for excitation energies several electron volts below I_g. This phenomenon has been observed for a large number of solutes.

In contrast to high energy radiolysis, photoionization can be carried out with a discrete photon energy which can be varied at will. Photoionization, therefore, is an efficient method for studying the ionization process in more detail. Parameters such as the probability of ionization and the distribution of cation–electron distances can be studied as functions of the energy involved in the ionization process. It appears that photoionization data may be a valuable supplement to the information obtained from radiolysis studies.

IV.A Photoionization in glasses

Photoionization in glasses is most easily obtained by biphotonic excitation via the triplet state (T_1) of aromatic solutes, as shown in Fig. 9, or by excitation of

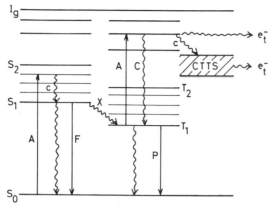

Fig. 9. Energy level diagram for photoionization of aromatic solutes (S) by biphotonic excitation via the triplet state. The letters A, C, F, P and X denote light absorption, internal conversion, fluorescence, phosphorescence and intersystem crossing, respectively. Two different modes of formation of e_t^- are shown in order to indicate that it is not known whether e_t^- is formed directly from the Franck–Condon state or from an intermediate (CTTS) state.

various metal complexes like the ferrocyanide ion[73]. Concentrations of e_t^- that are readily detectable with optical absorption or ESR spectroscopy can be produced by the use of continuous light sources like high pressure mercury and argon arcs fitted with suitable filters or monochromators. As discussed in Sect. V.B the photoinduced electrons can also be detected by recombination luminescence. It should be noted, however, that studies of photoionization in glasses are highly susceptible to artifacts resulting from optical bleaching of the products and especially bleaching of e_t^-[74,75]. Thus, the light used to induce photoionization will also cause some e_t^- to disappear. Only with relatively high light intensities and short exposure times (with biphotonic ionization the exposure time should preferably be less than the decay time of the phosphorescent triplet state) can this effect be reasonably neglected.

The photoionization efficiency in glasses has been measured as a function of the excitation energy for a number of systems[23,25,76,77]. Typical results obtained by biphotonic excitation of tryptophan in EG/H$_2$O at 77 K are shown in Fig. 10. It can be seen that above the ionization threshold (I_c) the photoionization efficiency increases rapidly with excitation energy. For excitation energies approximately equal to I_g the probability that the "second photon" will produce an electron approaches unity. The values of I_c obtained from such experiments seem to depend primarily on the solvent. Thus, I_c increases with decreasing solvent polarity[23,25].

For tryptophan in EG/H$_2$O I_c is approximately 2.4 eV smaller than I_g, which is ~8 eV[78] (see Fig. 10). Mobile, thermal electrons in condensed media may have a certain excess energy due to the electron affinity of the medium, which effectively means that the conduction band may extend below I_g. The magnitude of this excess energy in polar media is not well known, but is probably not larger than ~0.5 eV[79]. For H$_2$O, Henglein[80] recently estimated a value of 0.2 eV. Hence, the observed value of $I_c = 2.4$ eV means that an e_t^- may be formed from an excited state 2–2.5 eV

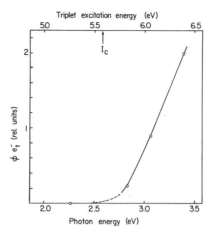

Fig. 10. The quantum yield of e_t^- produced by biphotonic excitation via the triplet state of tryptophan in equivolume EG/H_2O at 77 K as a function of the photon energy. The data are from ref. 26.

below the conduction band. This energy difference is comparable to the trapping energy I_e of e_t^-, i.e. ~2.3 eV in EG/H_2O (Fig. 3), indicating that $I_c \approx I_g - I_e$. It must be concluded from this that e_t^- may be formed without any intermediate passage through a conduction band state.

A second point which is of particular interest in the present context is that the formation of e_t^- occurs from a highly excited (triplet) state (see Fig. 9). Thus, the formation of e_t^- from this state, or at least the first step in this process, competes efficiently with internal conversion to the lowest excited triplet state (Fig. 9). This implies that the transfer of the electron to the solvent, or the formation of an intermediate state from which this transfer can occur, takes place in a time which is of the order of 10^{-11} s. On the other hand, the above mentioned pulse-radiolysis studies[18] indicated that trap relaxation in this glass takes place on the microsecond time scale. Hence, it must be concluded that either (1) a sufficient concentration of deep traps must pre-exist in the glass to accommodate direct electron transfer from the highly excited triplet state, or that (2) the formation of e_t^- from the excited solute occurs via an intermediate state (Fig. 9), i.e. a CTTS (charge transfer to solvent) state, having sufficient lifetime to allow formation of deep traps, i.e. a lifetime of the order of microseconds. There is no evidence to support the existence of a CTTS state with such a long lifetime. As to the first alternative it seems conceivable that a direct electron transfer from the excited state to the pre-existing trap may occur by tunneling. Assuming that this tunneling occurs within 10^{-11} s^{-1} (to compete with internal conversion) and that the probability of formation of e_t^- from an excited state 2 eV below the conduction band is 0.01, in accordance with the observations, it can be calculated from eqns. (5) and (6) (assuming $r_0 = 4$ Å) that the concentration of pre-existing traps 2 eV deep must be of the order of $4 \cdot 10^{-3}$ M. As judged from the yield of e_t^- in deep traps in ethanol at 4 K observed by Hase et al.[21], such a concentration of deep, pre-existing traps is not completely unreasonable. It should be noted that the hypothesis that e_t^- is formed by direct transition to pre-existing traps does not exclude that this transition occurs via a CTTS state. Clearly, the mechanism by which the electron is transferred from the excited solute molecule to the trapped state in the solvent glass is not understood.

IV.B Photoionization in solutions

The photoionization of molecules in solutions has been studied mostly by flash-photolysis, and by photoelectron emission spectroscopy (see Chap. 4). The photoionization efficiency has been measured as a function of the excitation energy only in a few cases. Recent results[81] obtained for an aqueous solution of tryptophan are shown in Fig. 11. The low light intensity used in these experiments excludes bi-photonic processes, so that ionization is induced by a single photon. It can be seen from Fig. 11 that for wavelengths above 250 nm the ionization quantum yield ϕ_{e^-} is constant, indicating that in this excitation region ionization takes place only after the molecule has reached the lowest vibrational level of the S_1 state. The excitation energy of S_1 in tryptophan is 4 eV, which is in turn ~ 4 eV less than I_g[78]. To account for this energy difference it seems reasonable to assume that some kind of inter-mediate state or reaction must be involved in the ionization process. For the case of aqueous solutions of phenols Grossweiner and Joschek[24] suggested that a proton transfer precedes the formation of e_s^-.

It can be seen from Fig. 11 that below 250 nm, ϕ_{e^-} increases rapidly with the photon energy, demonstrating that in this wavelength region the ionization occurs from higher excited singlet states, that is, it competes with internal conversion to S_1. The e_{aq}^- produced below 250 nm was found to be homogeneously distributed in the

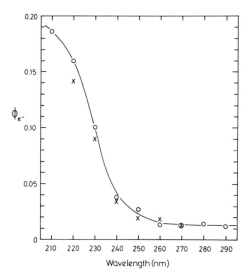

Fig. 11.   The quantum yield of e_{aq}^- produced by single photon photoionization of tryptophan in an acidic, aerated solution of $FeSO_4$ in H_2O at 300 K. O——O——: pH $= 1$, x——x—— pH $= 3$. The data are from ref. 81.

sense that the yield of e_{aq}^- picked up by H^+ was independent of $[H^+]$ from pH = 3 to pH = 1, as shown in Fig. 11.

The wavelength at which the increase of ϕ_{e^-} begins, i.e. ~ 250 nm, corresponds to an excitation energy which is approximately 2.4 eV below I_g. At 210 nm, where $\phi_{e^-} \approx 0.2$ (Fig. 11), the excitation energy is still 2 eV below I_g. Hence, it can be concluded that, as with the photoionization of glasses, the electrons are formed directly in the solvated state and not via the conduction band, and furthermore, that this ionization, or at least its first step, e.g. the formation of a CTTS state, is so rapid that it can compete favourably with internal conversion. The main difference between photoionization in liquid solutions and in glasses is that the probability of ionization from a certain excitation level is larger in the liquid phase.

As to the mechanism of the ionization of the highly excited molecule, an electron transfer by some intermediate chemical reaction, as suggested for the ionization from the S_1 state, seems unlikely in view of the very rapid rate of the process. Assuming that e_{aq}^- is formed by a direct tunneling to pre-existing traps, taking place in about 10^{-11}s, it can be calculated from eqns. (5) and (6) that the concentration of such traps with a trap-depth of 2 eV (assuming $r_0 = 4$ Å and $P = 0.2$ according to Fig. 11) must be of the order of 0.1 M, which may seem an unreasonably large value.

Hence, it may appear that the photoionization in this case occurs largely via a CTTS state, as suggested by several authors[24,82–84]. It is conceivable that a CTTS state with a high dipole moment and/or an extended electron orbital may polarize its surroundings, and thereby facilitate a transition of an electron to the solvent, in a time which may be of the order of 10^{-11}–10^{-10} s. This hypothesis is in accordance with the observation by Struve et al.[85] that the solvation time of an excited dipolar molecule (p-dimethylaminobenzonitrite) in ethanol is about $4 \cdot 10^{-11}$ s.

V. Recombination luminescence

When glasses containing a chromophore, such as an aromatic solute, are exposed to ionizing radiations, emission of light can be observed both during the irradiation and subsequent to it. In several polar glasses, like alcohols containing an aromatic solute in concentrations larger than say 10^{-3} M, the emission stems almost exclusively from the solute. This luminescence has been attributed primarily to ion recombination. Hence, it may give information about various aspects of ionization such as the spin correlation of the recombining ions, the distribution of their mutual distances and the probability of recombination.

The recombination luminescence that will be discussed here can be divided into three categories, although in some cases the distinction between these categories may be somewhat diffuse.

(i) The spontaneous luminescence that can be observed during irradiation.

(ii) The delayed, isothermal luminescence (afterglow) that may last for minutes and even hours after the irradiation.

(iii) The thermoluminescence TL that is emitted upon heating of the irradiated medium.

Afterglow and TL can also be observed when the ionization has been brought about by photoionization of the solute, e.g. by biphotonic triplet excitation as described in Sect. IV.A.

We shall mention here only some features of recombination luminescence that seem to be of particular significance in the present context. The discussion will be limited primarily to one system that has been rather extensively studied in this respect, namely EG/H_2O with an indole solute. However, similar results are obtained for other solutes as well as in other alcohol glasses[74,86].

V.A *Spontaneous irradiation-induced luminescence*

As shown in Fig. 12, the spectrum of the X-ray-induced luminescence of EG/H_2O glass containing indole exhibits a fluorescence and a phosphorescence band which are identical to those observed upon exposure with low intensity UV light, the wavelength of which is absorbed only by the solute and which produces virtually no ionization. Hence, the emission is exclusively from the radiative S_1 and T_1 levels of the neutral solute molecule. The decay time of the X-ray-induced fluorescence has been found to be less than $5 \cdot 10^{-5}$ s[87], whereas that of the X-ray-induced phosphorescence is identical to that observed for the phosphorescence induced by UV light[88]. It appears that energy transfer from the solvent glass to the solute does not contribute significantly to the radiation-induced solute excitation[89,90]. The combined yield of X-ray-induced S_1 and T_1 excitation in the solute is $G \approx 4$ excitations/100 eV absorbed in the solute[87].

As seen from Fig. 12, the phosphorescence to fluorescence ratio (P/F) observed

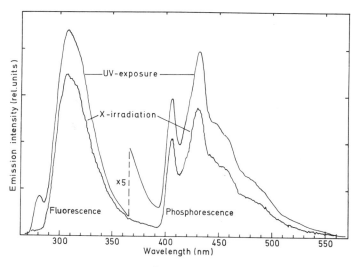

Fig. 12.   Uncorrected emission spectra of 10^{-2} M indole in equivolume EG/H_2O at 77 K recorded during irradiation with X-rays and with 280 nm UV light. The small peak at 280 nm is scattered exciting light. Note that for the UV-induced phosphorescence the scale of intensity has been divided by 5.

during X-irradiation is considerably enhanced as compared to that observed with UV exposure. This enhancement has been taken to indicate that a major fraction of the radiation-induced solute excitation results from electron–cation recombination. Studies of the effect of electron scavengers on the yields of the X-ray-induced excitation of indoles in EG/H_2O at 77 K[87] and of the temperature-dependence of these yields[91], independently confirm this conclusion. From the results of these studies it was calculated that at 77 K approximately 90 % of the triplet excitation and 70 % of the singlet excitation of indole resulted from geminate ion recombination, the remainder being attributed to excitation directly from the ground state. The observation that the decay time of the X-ray-induced fluorescence of these solutes is below $5 \cdot 10^{-5}$ s shows that the ion recombination takes place within this time. Intersystem crossing between higher excited levels and direct excitation of the triplet manifold were found to be of minor importance in this system. The conclusion that the yield of direct triplet excitation (presumably caused by slow electrons) is small is in accordance with a theoretical estimate by Platzman[92]. Unfortunately, the details of his calculations do not seem to have been published.

From the study of the effect of scavengers on the excitation yields it was found also that the ratio of triplets to singlets formed by the recombination is $T/S = 1.2$ for indole in EG/H_2O. (This value applies to the situation prior to S_1–T_1 intersystem crossing.) Similar values of T/S have been obtained for other systems as well[74,86]. Since the ion pairs are presumably singlets initially and since the recombination is geminate, such a large value of T/S indicates that a considerable degree of spin relaxation of the ion occurs before recombination. An alternative explanation has been put forward by Magee[93,94], who suggested that significant yields of triplets may result from the recombination of ions originating from different ion pairs, that may occur in regions of high ionization density, i.e. in spurs. Thus, since the different ion pairs are likely to be uncorrelated with regard to spin, recombination between them should give rise to triplets and singlets in the statistical ratio 3:1. However, at least for the case of indole in EG/H_2O glass, T/S was found to be completely independent of the radiation energy and hence of the ionization density[95]. Hence, the substantial triplet yield observed in this system cannot be attributed to the spur effect.

It therefore appears that substantial spin relaxation occurs prior to recombination. This means that the lifetime of the recombining ion pair τ_R must be of the same order of magnitude or larger than the spin relaxation time. According to the ESR spectrum of e_t^- in EG/H_2O the spin–spin relaxation time T_2 in this glass is $>2 \cdot 10^{-8}$ s. Hence, the recombining electrons cannot be free or in a conduction band, since in that case recombination should take place within 10^{-11} s. It is conceivable, therefore, that the electrons recombine from a trapped state. On the other hand, this trapped state cannot be that of the stable e_t^-. Recombination from this state, as observed by afterglow and TL (Sect. V.B), gives rise only to excitation of the T_1 level of the solute, apparently because the energy released by the recombination is too small to excite the S_1 level. It therefore appears that if the electrons recombine from a trapped state, these traps must be shallow ones. This idea is compatible with the observation that electrons in shallow traps in this glass seem to have a lifetime of the order of 10^{-7} s[18], which should be sufficient to allow spin–spin relaxation. The spin-lattice relaxation time T_1, on the other hand, is about $5 \cdot 10^{-4}$ s[96] which is considerably larger than τ_R, according to the lifetime of the X-ray-induced fluorescence. For this situation, i.e. when $T_2 \ll \tau_R \ll T_1$, Brocklehurst[97] predicted $T/S = 1$, which is in good agreement with the observed value for indole in EG/H_2O, i.e. $T/S = 1.2$.

If the spontaneous solute excitation is due to recombination of electrons from shallow traps, having lifetimes in the region 10^{-8}–10^{-6} s, it follows from the tunneling model (see Fig. 6) that $[S]_{1/2}$ for these electrons should be in the region 1–$3 \cdot 10^{-2}$ M (assuming 0.6 eV $< I_e < 1.0$ eV). Experimental values of $[S]_{1/2}$ are given in Table 3. It can be seen that these values are roughly one order of magnitude larger than the theoretical estimate. Although the theoretical value of $[S]_{1/2}$ is a lower limit, this difference between theory and experiment is so large that it may indicate that the reactive species involved here is not a trapped electron (see also Sect. VII.A). This conclusion is corroborated by the observation that $[S]_{1/2}$ seems to be somewhat dependent on the solute[87]. However, the data are too scarce to completely exclude the possibility that the spontaneous luminescence is due to recombination from shallow traps.

An alternative interpretation of the data is that the reactive precursor of the radiation-induced solute excitation is an excited state of the solute molecule, i.e. what might be called an ion pair state or a CTTS state. This hypothesis is supported by the fact that the values of $[S]_{1/2}$, observed for the radiation-induced solute excitation, are of the same order of magnitude as those found for the quenching of the UV-induced fluorescence under the same conditions, that is, in the absence of diffusion (see Table 3). This quenching of the fluorescent S_1 level seems to occur

TABLE 3

Scavenger concentrations ($[S]_{1/2}$) needed to halve the quantum yield of the UV-induced fluorescence of indole in equivolume EG/H_2O at 77 K and to halve the yield of X-ray-induced T_1 excitation of indole under identical conditions (Data are from refs. 36, 87 and 115)

Solute	$[S]_{1/2}$(M)	
	UV excited fluorescence	X-ray-induced triplet yield
CCl_3COONa	0.26	0.15
Acrylamide	0.4	0.2
$NaNO_3$	–	0.2
H_2O_2	1.1	0.7
$ClCH_2COOH$	5.2^a	0.8
HCl	15^a	$> 7^a$

[a]Values determined by extrapolation from lower concentrations.

by an electron transfer to the scavenger[98]. Furthermore, this hypothesis is in accordance with the observation that the quenching of the radiation-induced excitation seems to obey exponential scavenging kinetics[87]. This type of kinetics generally applies to the quenching of excited states, i.e. the quenching of fluorescence and phosphorescence, of molecules in rigid glasses[99], that is, under conditions where there is no diffusion.

Hence, it is possible that the recombining ions are so closely correlated that the ion pair is essentially similar to the CTTS state that appears to be involved in the photoionization of this solute (Sect. IV.A). What has been called recombination may thus be described as internal conversion from this CTTS state to the radiative levels. This hypothesis implies that a major fraction of the ionization, induced by high energy radiation, occurs via such CTTS state(s).

If it is assumed that the rate of intersystem crossing in the CTTS state is similar to that of the S_1-T_1 transition in this solute, i.e. $\sim 5 \cdot 10^7 \, s^{-1}$, the value of $T/S = 1.2$ implies that the lifetime of the CTTS state is about $2 \cdot 10^{-8} \, s$. It is known[99], however, that intersystem crossing between higher excited levels may well be two orders of magnitude larger than for the S_1-T_1 transition. Hence, the lifetime of the CTTS state may be as low as $10^{-10} \, s$.

Another notable feature of the spontaneous radiation-induced solute luminescence, as seen from Table 3, is that it is not affected by the presence of H^+. Thus, even 2 M HCl does not noticeably reduce the yield of this luminescence. It is interesting that the UV excited S_1 level of indole in low temperature glasses is also unaffected by H^+ (Table 3), whereas it is significantly quenched by other electron scavengers, supposedly by an electron transfer reaction. In liquid solution, on the other hand, that is, when molecular diffusion is allowed, H^+ is the most efficient of all scavengers of the S_1 level. Hence, it may appear that for H^+ to act as an electron scavenger it must be able to diffuse. It should be noted that this difference in the reactivity of H^+ in the liquid and the solid phase is much too large to be accounted for only by the "superficially" high diffusibility of $H^{+\bullet}$ in aqueous liquids[100].

V.B Afterglow and thermoluminescence

There is ample evidence that the afterglow and TL of EG/H_2O and various alcohol glasses containing suitable solutes, are due to geminate recombination between solute cations and e_t^-[101,102]. In passing, it may be interesting to note that ion-pair concentrations as low as 10^{-13} M may easily be studied by such measurements. The TL glow curve has typically two peaks as shown in Fig. 13. The peak at the lower and the higher temperature we denote TL_1 and TL_2, respectively. The latter peak is approximately at the glass transition temperature. (This temperature does not seem to be associated with a true phase transition. It is rather the somewhat ill-defined point where the viscosity, which decreases gradually with increasing temperature, becomes low enough to allow molecular diffusion.) As noted in Sect. III.B the recombination that gives rise to the afterglow probably occurs by electron tunneling, whereas at least for TL_2, diffusion of e_t^- seems to be involved as well[102].

For EG/H_2O glass containing indoles or phenols, the spectrum of the afterglow

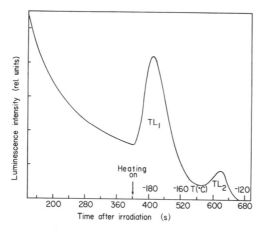

Fig. 13. The afterglow and thermoluminescence of equivolume EG/H$_2$O containing 10^{-2} M indole as observed after X-irradiation at 77 K.

and TL is identical to that of the solute phosphorescence[102]. Hence, only the T_1 level of the solute is excited by the recombination in this matrix. The reason for this seems to be that the energy of the recombining ion pair is not sufficient to excite the S_1 level. Thus, in less polar glasses, where the trap-depths are smaller and the recombination energy correspondingly larger, the afterglow and TL spectra include the fluorescence band as well[23].

The afterglow and the TL peaks appear to represent different parts of the distribution of cation–electron distances. Thus, the afterglow and the TL$_1$ peak seem to be due to recombination of electrons which are located considerably closer to the parent cation than the electrons that give rise to the TL$_2$ peak. This is indicated by the fact that the TL$_2$ peak shows a larger thermal activation energy, and, more significantly, by the finding that the TL$_2$ peak is more susceptible to quenching by electron scavengers than the afterglow and the TL$_1$ peak. The scavenger concentration needed to halve the intensity of TL$_1$ is roughly twice as large as that which halves the TL$_2$ peak[102,103]. This is accounted for by the assumption that the electrons giving rise to the TL$_2$ peak may also react while they are diffusing back to the cation.

The combined yield of afterglow and TL in EG/H$_2$O glass is of the order of 1% of that of the spontaneous radiation-induced luminescence[102]. Hence, only a very small fraction of e_t^- seems to recombine even when the viscosity becomes low enough to allow diffusion. Furthermore, it appears that the afterglow and TL$_1$ peak are associated with a small fraction of e_t^-, i.e. less than 5%, which have a fairly high probability of recombination, whereas the TL$_2$ peak is associated with the bulk of e_t^-, for which the probability of recombination (leading to light emission) is of the order of 0.1%. This is inferred from the observation[102] that upon heating to above the TL$_1$ peak, the total number of e_t^- decreases by less than 5%, whereas all e_t^- disappear when the temperature reaches that of the TL$_2$ peak. On the other hand, a measureable decrease of the optical absorption at the red side of the spectrum of

e_t^- occurs at about the temperature TL_1, indicating: (1) that TL_1 is indeed due to recombination of e_t^-, and (2) that these electrons are in somewhat more shallow traps than the bulk of e_t^-, possibly because they are located so close to the cations that the traps are noticeably modified by the Coulomb field.

In view of the conclusion that the afterglow decay kinetics and the TL glow curve seem to reflect the distribution of cation–electron distances, it is interesting to compare the results obtained with different types of radiation. Moan[23] has reported that the afterglow as well as the TL glow curve of tryptophan in EG/H_2O induced by X-irradiation closely resemble those observed when the ionization is produced by UV induced triplet excitation of the solute under otherwise identical conditions. The photon energy used in the latter experiment, i.e. $\lambda = 254$ nm, brings this solute close to I_g. Hence, it appears that the distribution of cation–electron distances is independent of the radiation energy at least down to about I_g. That is, the distance travelled by the electron seems to be independent of the energy used in the ionization process.

This result is hardly compatible with the commonly accepted idea that the electrons produced by the absorption of high energy radiation in condensed media are ejected as free charges with an initial average energy of the order of 10–20 eV. On the other hand, it is in accordance with the assumption that for all types of radiation the ionization of this solute (i.e. indole or its amino acid form tryptophan) occurs via an excited state below I_g from which the electron is formed directly in a trapped state. Further implications of this hypothesis are discussed in Sect. VII.

VI. Electron reactions in liquids

VI.A Picosecond observations of e_s^-: the dry electron hypothesis

Recent developments of the method of pulse-radiolysis have brought the time resolution of the observation of radiation-induced e_s^- into the picosecond region. By an ingenious stroboscopic technique, which utilizes the fine structure of the electron pulse from an electron linear accelerator, Hunt and co-workers were able to observe e_s^- from ~ 20 ps[8,9]. Presently, other workers have obtained a comparable time resolution by related techniques[104].

The most significant observations reported by Hunt and co-workers[10–13] are the following:

(1) The absorption spectrum of e_{eq}^- is fully developed within 10 ps. The same result is obtained for e_s^- in various alcohols (see Sect. II.A).

(2) When present in concentrations above 0.1 M most electron scavengers noticeably reduced $G_i(e_{aq}^-)$, where $G_i(e_{aq}^-)$ is the value found by extrapolation to zero time on the assumption that the decay of e_s^- below 20 ps is similar to that observed above this time.

(3) For a variety of scavengers, covering a fairly wide range of scavenging efficiencies, the scavenging kinetics is exponential.

(4) The efficiencies of various scavengers in reducing $G(e_{aq}^-)$ are not proportional

to the corresponding rate constants $k(e_{aq}^- + S)$ as observed by the decay of e_s^- above 20 ps. However, there is a definite trend that the efficiency for decreasing $G_i(e_{aq}^-)$ increases with $k(e_{aq}^- + S)$.

(5) Among several exceptions to this trend the most notable one is the hydronium ion. Thus, $HClO_4$ did not affect $G_i(e_{aq}^-)$ even in concentrations of 2 M. This result has been confirmed by others[104].

(6) The value of $k(e_{aq}^- + H_3O_{aq}^+)$ is independent of the concentration of acid between 1 and 5 M.

(7) In the absence of scavengers there is no decay of e_{aq}^- or e_s^- in ethanol between 20 and 350 ps within the experimental accuracy of 5%.

Hunt and co-workers[11] interpreted these results as being due to a reaction of the precursor of e_{aq}^-. They assumed, in accordance with Hamill's new model for radiolysis[6], that this precursor of e_s^- is the quasi-free, mobile electron which they termed, again following Hamill[6], the dry electron (e^-). In fact, they implied that e^- reacts prior to thermalization. However, in order to account for the exponential scavenging kinetics they had to make the rather improbable assumption that in the absence of scavenger all e^- travel essentially the same distance and experience the same number of collisions before being trapped by solvation.

Several other cases have been reported where the electron reaction rate constant previously obtained for dilute solution does not seem to be applicable when the solute concentration exceeds about 0.1 M[1-5]. These observations have been taken to indicate that electrons may react prior to solvation, that is, in the e^- state.

VI.B Alternative interpretations of the picosecond data

Several authors, being reluctant to accept a new chemical species, i.e. e^-, have tried to interpret the data of Hunt and co-workers on the assumption that the reactive species is actually e_s^- after all. Schwarz[14] suggested that the data could be explained in terms of the time-dependence of the reaction rate constants of e_s^-, which is to be expected theoretically for diffusion controlled reactions involving highly concentrated solutes. However trying to fit his hypothesis to the experimental data, Schwarz was not able to obtain convincing agreement. On the other hand, the theory of time-dependent reactions is not so well worked out that this lack of agreement is sufficient to exclude Schwarz's hypothesis. It should be noted, however, that the whole concept of time-dependent electron reactions is so far based primarily on theoretical considerations and has little support from experimental data.

Czapski and Peled[16] have suggested another explanation of the picosecond data which is closely related to that of Schwarz. They pointed out that with solutes in the concentration range above 0.1 M a significant fraction of e_s^- (now assumed to be formed with no reactive precursor) will be so close to a scavenger molecule that no diffusion is needed for the reaction to occur. That is, some e_s^- is formed within the reaction radius of the scavenger, which is to say that the Smoluchowski equation breaks down at high scavenger concentrations. In such encounter pairs reaction is likely to occur well within the resolution time of picosecond pulse-radiolysis and may therefore be observed as a reduction of $G_i(e_t^-)$. The fraction of e_s^- not formed in such encounter pairs can be shown to be proportional to $e^{-V_s} \cdot [S]$, where V_s is the molar volume of the encounter pair. This follows simply from the assumptions that the reactants are randomly distributed and that V_s has a sharp boundary (see eqns. (2) and (3)). Thus, the encounter pair hypothesis is consistent with the exponential scavenging kinetics observed by Hunt and co-workers.

Czapski and Peled found that although their hypothesis alone cannot explain the picosecond pulse-radiolysis data, a fairly consistent picture is obtained, at least for most of the solutes, if the effect of time-dependent reaction rates is taken into account as well. However, for some other solutes like acetic acid, for which the discrepancies between the reactivities measured at picosecond and microsecond times are particularly large, their hypothesis is hardly satisfactory. The encounter pair hypothesis is in turn closely related to another interpretation of the data of Hunt and co-workers namely that they may be explained on the basis of electron tunneling from e_s^- to the scavenger[15]. With scavenger concentrations above 0.1M substantial tunneling may occur even within the time resolution of picosecond pulse-radiolysis. Assuming a trap-depth of $e_{aq}^- I_e = 1.7$ eV, it can be seen from Fig. 6 that approximately 0.2 M of an efficient scavenger would be needed to halve $G(e_{aq}^-)$ as observed at 10 ps. This estimate is compatible with the experimental values reported by Hunt and co-workers[12] which are $[S]_{1/2} = 0.27$ M for the most efficient scavengers. The decay of e_s^- caused by such tunneling between 20 and 350 ps, that is, during the period that the concentration of e_s^- was actually monitored in the picosecond pulse-radiolysis experiments, would be of the order of 10% of that resulting from the diffusion controlled reaction and could therefore easily have escaped notice. Furthermore, the tunneling model accounts for the exponential scavenging kinetics observed by Hunt and co-workers (see eqn. (6)). With this model the differences between the e^- reactivities determined by Hunt and co-workers and the reactivities of e_s^- can be attributed to differences in the degree of resonance between e_s^- and S, as discussed in Sect. III.B. It should be emphasized that the effects of time-dependent reaction rates, encounter pair reactions and electron tunneling should be regarded as complementary rather than alternative interpretations of the picosecond data.

VI.C Shortcomings of the current model of condensed phase ionization

All current models for the radiolysis of condensed media are actually based on the assumption, which is generally and usually tacitly accepted, that the major fraction of the ionization occurs by the formation of quasi-free, mobile electrons with initial energies somewhere in the range 10–30 eV, on average. It is further assumed that these electrons are gradually slowed down to thermal energy and that they move in the medium with this energy until they become trapped or solvated, a process which is believed typically to occur in about 10^{-11}s in polar media. According to this model the initial yield of e_t^- or e_s^- is given by

$$G_i(e_s^-) = P \cdot G(e^-) \tag{7}$$

where $G(e^-)$ is the yield of ionization and P is the probability that the electron will not recombine prior to trapping or solvation. Usually it is assumed that

$$G(e^-) \approx 100/W \tag{8}$$

where W is the average energy per ionization in the gaseous phase of the medium. Thus, the mechanism of ionization in condensed media is assumed to be essentially similar to the ionization of gases.

In Table 4 are given values for W and $G_i(e_s^-)$ for various alcohols and for H_2O together with values of the liquid phase static dielectric constant ϵ_s. The values of $G_i(e_s^-)$ were obtained with a radiolysis technique[104] that has a time resolution of about 30 ps. According to Wolff et al.[13], there is no detectable decay of e_s^- (i.e. less than 5%) between 20 ps and 350 ps in the absence of electron scavengers. Hence, it appears that these values of $G_i(e_s^-)$ are really the initial yields.

It can be seen from Table 4 that for H_2O $G_i(e_s^-)$ is about 25 % larger than the value expected from eqn. (8), whereas for the alcohols the values of $G_i(e_s^-)$ are only about half the gas phase yields. According to the currently accepted mechanism of ionization in condensed media, this difference between H_2O and alcohols, as well as the differences in $G_i(e_s^-)$ between the different alcohols, are primarily due to differences in the probability of escaping geminate ion recombination.

According to Mozumder[105] this probability is given by eqn. (9) (see Chap. 5 Sect. IIIA, eqn. (31)).

$$P = \exp\left(-e^2 / k_B \cdot T \cdot r_0 \cdot \epsilon_{eff}\right) \tag{9}$$

where e is the electron charge, k_B Boltzmann's constant, T absolute temperature, r_0 the distance from the parent cation at which the electron becomes thermalized and ϵ_{eff} the effective dielectric constant. ϵ_{eff} is the time-integrated value of the instantaneous dielectric constant which varies with time from the optical value ϵ_{opt} to the static value ϵ_s. ϵ_{eff} therefore depends on the dielectric relaxation time τ_d of the medium and on the diffusion constant D of the electron. It is usually assumed[105], however, that $\epsilon_{eff} \approx \epsilon_s$. On this assumption it follows from eqn. (9) that P, and hence

TABLE 4

Liquid phase static dielectric constant (ϵ_s) and yields of ionization in the liquid and gaseous phase of water and various alcohols at room temperature

Medium	ϵ_s	$G_i(e_s^-)^a$ (rel. units)	$G_i(e_s^-)$ ($e_s^-/100$ eV)	$100/W^d$ (ion pairs/100 eV)
Water	80	1.0	4.1^b 4.0^c	3.3
Methanol	34	0.43	1.8^b 1.6^c	
Ethanol	24	–	1.6^c	
1-propanol	21	0.59	2.4^b 1.8^c	4.2 ± 0.2^e
2-propanol	19	0.58	2.4^b 1.5^c	
1, 2-ethanediol	39	0.51	2.1^b	

[a] Values measured at about 30 ps, ref. 104.
[b] The values of ref. 104 assuming $G_i(e_{aq}^-) = 4.1$ according to ref. 107.
[c] Values from ref. 13.
[d] Values from ref. 106.
[e] Average of values from seven alcohols, ref. 106.

$G_i(e_s^-)$, should increase with ϵ_s, at least within a certain group of substances. As shown by the data in Table 4, however, this expectation is not borne out by the experiments. On the contrary, these data show no correlation between $G_i(e_s^-)$ and ϵ_s. The same conclusion was reached by Pikaev and Brodskii[108] for another series of substances. The lack of correlation between $G_i(e_s^-)$ and ϵ_s does not necessarily imply that the current theory for ionization in condensed media is wrong. Thus, it may be explained if the approximation $\epsilon_{eff} \approx \epsilon_s$ is not valid.

According to eqn. (9), P, and hence $G_i(e_s^-)$, should be sensitive also to variations in T. Hence, it is interesting to note that the limited amount of data that exist indicate that $G_i(e_s^-)$ is essentially independent of T. Literature values of $G_i(e_s^-)$ for various alcohols at different temperatures, are given in Table 5 together with values of ϵ_s. It can be seen that by a variation of T by about a factor of two, $G_i(e_s^-)$ is constant to within 10%. Admittedly, ϵ_s and T vary in opposite directions. But, as seen from Table 5, there is still a considerable variation in the product of these parameters with temperature, which should have given rise to a similar variation in $G_i(e_s^-)$. It should be noted, however, that a considerable variation in $G(e_t^-)$ at temperatures approaching the glass transition point, has been reported[109].

Even more conspicuous is the finding by Hase et al.[21] that $G(e_t^-)$ in ethanol is independent of temperature from 4 K to 77 K. Indeed, $G(e_t^-)$ at 4 K appears to be within 20% of $G_i(e_s^-)$ at 343 K. Hase et al.[110] also measured $G(e_t^-)$ in three different alkane glasses, and found that in all cases it was constant to within experimental limits ($\pm 20\%$) between 4 K and 71 K. For the alkanes in particular this temperature-independence of $G(e_t^-)$ cannot be explained away by our lack of exact knowledge about the value of ϵ_{eff}, since for these non-polar substances $\epsilon_s \approx \epsilon_{opt}$, which means that ϵ_{eff} should be practically independent of T. Hence, one should expect from eqn. (9) a very significant variation in $G(e_t^-)$ between 4 K and 71 K.

The fact that such a temperature-dependence is not observed, strongly indicates that recombination of electrons prior to trapping or solvation is insignificant in these media. Hence, the large differences between the ionization yield in the condensed phase and the gas phase, which for non-polar substances may be about one order of magnitude, must be attributed to some other mechanism. On the other hand, it seems clear from the theoretical point of view that if the electron passes through the mobile

TABLE 5

Liquid phase static dielectric constants (ϵ_s) and ionization yields ($G_i(e_s^-)$) for alcohols at different temperatures (T) (The data are from ref. 22)

Medium	ϵ_s	T(K)	$\epsilon_s \cdot T(10^2)$	$\epsilon_1 \cdot T_1/\epsilon_2 \cdot T_2$	$G_i(e_s^-)$	
					pH = 11	neutral
Ethanol	57	172	98	1.59	2.0	1.8
Ethanol	18	343	62		2.2	1.6
Methanol	69	183	126	1.42	2.2	1.9
Methanol	23	385	88		2.3	1.9

state at all, substantial recombination has to occur, especially in media with low polarity. As discussed below, a possible solution to this apparent self-contradiction is that the major fraction of e_s^- and e_t^- is not formed via the quasi-free, mobile state. Finally, it should be noted that eqn. (9) predicts that when the electron is formed with thermal energy, i.e. $r_0 = 0$, its probability of becoming trapped or solvated should be zero. The photoionization experiments, however, demonstrate that this probability may approach unity.

VII. Alternative model for the condensed phase ionization mechanism

The data on $G_i(e_s^-)$ discussed in Sect. VI.C, as well as data indicating that the distribution of cation–electron distances in glasses is independent of the energy of the radiation (Sect. III.A and V.B), cannot be reasonably accounted for on the basis of the generally accepted assumption that ionization occurs by the initial formation of quasi-free unsolvated electrons in essentially the same way as ionization occurs in the gas phase. Although the data obtained from one type of experiment on one system may perhaps be fitted into the current theory by adjusting unknown parameters appropriately, or explained away as experimental artifacts, it appears that, when seen collectively, these data clearly call for a reconsideration of the earliest processes of the radiolysis of condensed, polar media.

In the remainder of this chapter we shall briefly discuss an alternative mechanism of ionization. We shall assume that instead of being formed via a quasi-free, unsolvated state, the major fraction, i.e. more than 90%, of e_s^- and e_t^- is formed directly from a quasi-bound excited state M*, or a band of such states, having an energy smaller than $I_g - E_s$ where E_s is the free electron affinity of the medium. By definition M* is a CTTS state. The possibility that the precursor of e_s^- is an excited molecular or collective state rather than a free electron, has previously been noted in passing[11,108], but does not seem to have been seriously considered.

The present model is depicted schematically in Fig. 14. The initial energy absorption is of the order of 20 eV on the average, and a molecule of the medium is thereby brought to the continuum of superexcited states M** above I_g. M** will very rapidly lose energy to surrounding molecules by internal conversion and thereby give rise to the formation of M*. The internal conversion to M* may compete with internal conversion to other excited states and to the ground state and possibly with various modes of dissociation as well. The formation of e_s^- or e_t^- from M* may compete with internal conversion of M* to lower excited levels or to the ground state. In contrast to the case of gases, this internal conversion from the superexcited states is much faster than ionization so that formation of quasi-free electrons is small. This difference between the condensed and the gaseous phase can be accounted for by the continuous interaction with neighboring molecules in the condensed phase. In fact, because of the strong intermolecular coupling in condensed, polar media, it seems possible that the initial excitation has a collective character so that at no stage is an energy larger than I_g localized on any particular molecule. This possibility was first noted by Fano[111], and recently it has been supported by a study of the optical properties of liquid H_2O in the vacuum UV region by Heller et al.[112] who

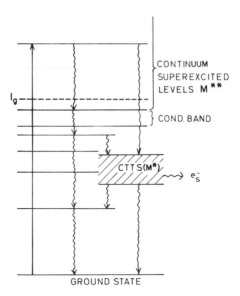

Fig. 14. An energy level diagram of an alternative model for the formation of the major fraction of the trapped and solvated electrons that are produced by the radiolysis of condensed media. The model assumes the following: the main fraction of the radiation energy is deposited by low energy inelastic scattering of secondary electrons which initially gives rise to excited states M**. M**, which may possibly be of collective nature, is on the average 10–20 eV above the gas phase ionization potential I_g. In contrast to the case of the gas phase, ionization of M** in the condensed phase is negligible because of very rapid internal conversion (wavy arrows) to a quasi-bound excited state M* as well as to other excited states and the ground state. M*, which is located below the conduction band, is a charge transfer to solvent (CTTS) state and may transfer an electron directly to a trapped or solvated state in the medium, with no intermediate passage through the conduction band. In general, the trapped or solvated electrons are initially produced in a shallow trap which subsequently undergoes dipolar reorientation to form a fully relaxed solvation shell. The rate of the latter process depends primarily on the viscosity of the medium and determines the extent to which the electron may react while it is in the shallow, transient trap. Thus, reaction from shallow traps may be of particular importance in low temperature glasses. When formed within a certain encounter radius of a reactant or when brought to within this radius by diffusion, the trapped or solvated electron may react by quantum-mechanical tunneling. In polar media typical encounter radii may be 9 Å, 18 Å and 28 Å for reactions taking place at the time scale of picoseconds, microseconds and seconds, respectively.

The conduction band in H_2O and organic media is probably not comparable to that found in ordinary semiconductors. The term is used here to designate the energy region of thermal, quasi-free electrons, viz. those which are believed to give rise to the photocurrent that can be observed upon optical bleaching of trapped electrons.

interpret the prominent peak in the absorption spectrum of H_2O at about 21 eV as being due to collective excitations and conclude that "this loss represents an extremely important energy absorption mechanism in water".

As noted in Sect. II.A it appears that e_s^- and e_t^- are initially formed mainly in shallow traps which subsequently undergo dipolar relaxation to form the fully solvated species. It should be noted that the relaxation time of e_s^- in liquids may be faster than that anticipated by Rentzepis et al.[27] from their studies of the formation of the absorption spectrum of e_s^- produced by photoionization of ferrocyanide anion. According to the present model, the growing in of the spectrum observed by these workers may represent the transition from the CTTS state to the solvated state, rather than the solvation of a free charge. This possibility was noted in their paper. In the liquid phase, therefore, the deepening of the electron trap may be so fast that the electrons do not necessarily participate in chemical reactions while they are in the shallow traps, even at the highest solute concentrations. In glasses, on the other hand, and in highly viscous liquids, where dielectric relaxation is much slower, the electron may be captured by a scavenger molecule, presumably by electron tunneling, in competition with relaxation of the shallow trap. As noted in Sect. III.B, such tunneling from shallow traps, together with the tunneling from deep traps at a later stage, may account for most of the observations previously attributed to reactions of e_m^-. Hence, the difference in electron reactivities between liquids and glasses may be attributed to the different rate of dielectric relaxation in the two phases.

The mechanism by which the electron is transferred from M* to the solvated or trapped state in the medium is not known. However, as noted in Sect. IV, experiments with photoionization of solutes demonstrate that such an electron transfer from excited levels located 1–3 eV below I_g may occur with appreciable probability in both liquids and low temperature glasses, and, furthermore, that this formation of e_s^- and e_t^- is fast enough to compete efficiently with internal conversion of a highly excited state like M*.

But, the observation of photoionization from quasi-bound excited states is not limited to aromatic solutes. Photoionization of liquid H_2O has been observed with photon energies as low as about 6.5 eV[113,114], which is some 6 eV less than I_g, although at this energy the quantum yield of e_{aq}^- is only of the order of 1%. However, the quantum yield increases considerably with the excitation energy so that at 10 eV, which is still about 2.6 eV below I_g, it is of the order of 10%[113].

Since the properties of M* are independent of its mode of formation, i.e. on whether the molecule is excited by secondary electrons produced by high energy radiation or by photons with energies in the neighborhood of I_g, the distribution of cation–electron distances should be independent of the radiation energy provided this energy is sufficient to produce M* at all. This expectation is borne out by the observations of afterglow and TL (Sect. V.B) as well as by the fact that the effect of scavengers on $G(e_t^-)$ is independent of whether e_t^- is induced by X-rays or by photoionization[75].

In contrast to the currently accepted mechanism of ionization in condensed media, the present model is not at variance with the temperature-independence of $G_i(e_s^-)$ and $G(e_t^-)$ in various media. Thus, according to this model $G_i(e_s^-)$ and $G(e_t^-)$ depend primarily on two factors which may both be independent of temperature, namely the yield of M* and the probability of formation of e_s^- or e_t^- from M*. As seen from Fig. 14, the yield of M* is determined by the relative rates of internal conversion from M** to M* and to other states, respectively. Since in the condensed phase internal conversion is likely to be largely independent of temperature, the yield of M* should be so too. If the formation of e_s^- or e_t^- occurs initially by electron tunneling from M* to pre-existing traps and/or by electronic (i.e. intramolecular) polarization of the medium, both of which are essentially temperature-independent processes, the electron yield should be independent of temperature.

The present model is also qualitatively in accordance with the picosecond observations by Hunt and co-workers. In terms of this model, the reactive species implied by these authors may be either e_s^- reacting at short times according to the tunneling or encounter pair hypothesis mentioned in Sect. VI.B, or M* reacting with the scavenger, or, perhaps more probably, both types of reaction occur. Hence, what has been interpreted as a capture of e^- by the scavenger, may actually be a quenching of M* in competition with formation of e_s^- or e_t^-. There is virtually no diffusion taking place at picosecond times, and as noted in Sect. V.A, the quenching of solute fluorescence in glasses indicates that in the absence of diffusion the quenching kinetics of excited states is exponential. Hence, the exponential scavenging kinetics observed by Hunt and co-workers is to be expected also if M* is the reactive precursor of e_s^-. The observation that the reactivities of the precursor of e_s^- are not proportional to those of e_s^- would be a natural consequence of the hypothesis that this precursor and e_s^- are not only different states of the electron, but two distinctly different species.

In conclusion it appears that the present model is in accordance with most of the data on the early events of radiolysis that have been discussed in the previous sections. Admittedly, this agreement may be attributed partly to the crude and qualitative character of the model, and, to the scarcity of precise experimental information on several of the processes that are assumed to be involved. It is felt, however, that the obvious shortcomings of the currently accepted model for radiation-induced formation of electrons in condensed media call for a consideration of alternative mechanisms.

Acknowledgements

I am indebted to Dr Magne Kongshaug for many stimulating discussions and to Mrs. Gerd Hammerstad for invaluable assistance in typing the manuscript. Work was partially supported by NATO Research Grant No. 683.

References

1 T. Sawai and W. H. Hamill, J. Phys. Chem., **74** (1970) 3914.
2 S. Khorana and W. H. Hamill, J. Phys. Chem., **75** (1971) 3081.
3 E. Peled and G. Czapski, J. Phys. Chem., **75** (1971) 3626.
4 H. Ogura and W. H. Hamill, J. Phys. Chem., **78** (1974) 504.
5 B. Cercek and O. I. Micic, Nature Phys. Sci., **238** (1972) 74.
6 W. H. Hamill, J. Phys. Chem., **73** (1969) 1341.
7 See for example: L. Kevan, in G. O. Phillips (Ed.), Energetics and Mechanisms in Radiation Biology, Academic Press, London, 1968, pp. 21–33.
8 M. J. Bronskill and J. W. Hunt, J. Phys. Chem., **72** (1968) 3762.
9 M. J. Bronskill, W. B. Taylor, R. K. Wolff and J. W. Hunt, Rev. Sci., Instrum., **41** (1970) 333.
10 M. J. Bronskill, R. K. Wolff and J. W. Hunt, J. Chem. Phys., **53** (1970) 4201.
11 R. K. Wolff, M. J. Bronskill and J. W. Hunt, J. Chem. Phys., **53** (1970) 4211.
12 J. E. Aldrich, M. J. Bronskill, R. K. Wolff and J. W. Hunt, J. Chem. Phys., **55** (1971) 530.

13 R. K. Wolff, M. J. Bronskill, J. E. Aldrich and J. W. Hunt, J. Phys. Chem., 77 (1973) 1350.

14 H. A. Schwarz, J. Chem. Phys., 55 (1971) 3647.

15 J. R. Miller, J. Chem. Phys., 56 (1972) 5173.

16 G. Czapski and E. Peled, J. Phys. Chem., 77 (1973) 893.

17 J. T. Richards and J. K. Thomas, J. Chem. Phys., 53 (1970) 218.

18 L. Kevan, J. Chem. Phys., 56 (1972) 838.

19 J. R. Miller, Chem. Phys. Lett., 22 (1973) 180.

20 J. R. Miller and J. E. Willard, J. Phys. Chem., 76 (1972) 2641.

21 H. Hase, T. Warashina, M. Noda, A. Namiki and T. Higashimura, J. Chem. Phys., 57 (1972) 1039.

22 K. N. Jha, G. L. Bolton and G. R. Freeman, J. Phys. Chem., 76 (1972) 3876.

23 J. Moan, J. Chem. Phys., 60 (1974) 3859.

24 L. I. Grossweiner and H.-I. Joschek, Advan. Chem. Ser., 50 (1965) 279.

25 A. Bernas, M. Gauthier and D. Grand, J. Phys. Chem., 76 (1972) 2236; Chem. Phys. Lett., 17 (1972) 439.

26 J. Moan, Chem. Phys. Lett., 18 (1973) 446.

27 P. M. Rentzepis, R. P. Jones and J. Jortner, J. Chem. Phys., 59 (1973) 766.

28 J. H. Baxendale and P. Wardman, Nature (London), 30 (1971) 449.

29 J. H. Baxendale and P. Wardman, J. Chem. Soc. Faraday Trans. I, 69 (1973) 584.

30 J. T. Richards and J. K. Thomas, Chem. Phys. Lett., 8 (1971) 13.

31 F. S. Dainton, G. A. Salmon and U. F. Zucker, Chem. Commun., 8 (1968) 1172.

32 B. G. Ershov and A. K. Pikaev, Radiat. Res. Rev., 2 (1969) 1.

33 G. V. Buxton, F. S. Dainton, T. Lantz and P. Sargent. Trans. Faraday Soc., 66 (1970) 2962.

34 H. Hase and L. Kevan, J. Chem. Phys., 54 (1971) 908.

35 H. B. Steen and J. Moan, J. Phys. Chem., 76 (1972) 3366.

36 H. B. Steen, unpublished data.

37 K. K. Ho and L. Kevan, Int. J. Radiat. Phys. Chem., 3 (1971) 193.

38 D. R. Smith and J. J. Pieroni, Can. J. Chem., 45 (1967) 2723.

39 T. Highashimura, M. Noda, T. Warashina and H. Yoshida, J. Chem. Phys., 53 (1970) 1152.

40 H. Hase, M. Noda and T. Higashimura, J. Chem. Phys., 54 (1971) 2975; J. Chem. Phys., 55 (1971) 5411.

41 H. Hase and T. Warashina, J. Chem. Phys., 59 (1973) 2152.

42 H. Hase, T. Higashimura and M. Ogasawara, Chem. Phys. Lett., 16 (1972) 214.

43 K. Kawabata, H. Horii and S. Okabe, Chem. Phys. Lett., 14 (1972) 223.

44 M. Tachiya and A. Mozumder, J. Chem. Phys., 60 (1974) 3037.

45 J. Moan, Chem. Phys. Lett., 17 (1972) 565.

46 A. Bernas, D. Grand and C. Chachaty, Chem. Commun., (1970) 1667.

47 L. Kevan, J. Phys, Chem., 76 (1972) 3830.

48 I. Eisele and L. Kevan, J. Chem. Phys., 53 (1970) 1867.

49 P. Hamlet and L. Kevan, J. Amer. Chem. Soc., 93 (1971) 1102.

50 G. Kenney-Wallace and D. C. Walker, Ber. Bunsenges. Phys. Chem., 75 (1971) 634.

51 H. B. Steen, O. Kaalhus and M. Kongshaug, J. Phys. Chem., 75 (1971) 1941.

52 R. F. Khairoutoinov, A. I. Mikhailov and K. I. Zamaraev, Dokl. Akad. Nauk S.S.S.R., 200 (1971) 905.

53 A. I. Mikhailov, R. F. Khairoutoinov, K. I. Zamaraev and V. I. Goldansky, in J. Dobo and P. Hedvig (Eds.), Proc. Third Tihany Symp. on Radiation Chem., Vol. 2, Akademai Kiado, Budapest, 1972, pp. 1201–1211.

54 E. L. Girina and B. G. Ershov, Izv. Akad. Nauk S.S.S.R., (1972) 278.

55 H. Hase and L. Kevan, J. Phys. Chem., 74 (1970) 3358.

56 J. Kroh and Cz. Stradowski, Radiochem. Radioanal. Lett., 9 (1972) 169.

57 T. Sawai, Y. Shinozaki and G. Meshitsuka, Bull. Chem. Soc. Jap. 45 (1972) 984.

58 B. G. Ershov and A. K. Pikaev, Advan. Chem. Ser., 81 (1968) 1.

59 B. G. Ershov and E. L. Tseitlin, High Energy Chem. (U.S.S.R.), 4 (1970) 165.

60 H. B. Steen and M. Kongshaug, J. Phys. Chem., 76 (1972) 2217.

61 J. Kroh and Cz. Stradowski, Int. J. Radiat. Phys. Chem., 5 (1973) 243.

62 K. I. Zamaraev, R. F. Khairoutoinov, A. I. Mikhailov and V. I. Goldanski, Dokl. Akad. Nauk S.S.S.R. **199** (1971) 640.
63 O. Kaalhus, Kjeller Report, **146** (1972) 17.
64 J. Kroh and A. Plonka, Int. J. Radiat. Phys. Chem., **6** (1974) 2454.
65 S. Noda and L. Kevan, J. Phys. Chem., **78** (1974) 2454.
66 T. Higashimura, A. Namiki, M. Noda and H. Hase, J. Phys. Chem., **76** (1972) 3744.
67 H. B. Steen, J. Chem. Phys. (1 Oct. 1974).
68 J. Moan, J. Lum., **6** (1973) 256.
69 F. Kieffer, C. Lapersonne-Meyyer and J. Rigaut, Int. J. Radiat. Phys. Chem., **6** (1974) 79.
70 B. G. Ershov and F. Kieffer, Chem. Phys. Lett., **25** (1974) 576.
71 J. Moan, in H. I. Adler, O. F. Nygaard and W. K. Sinclair (Eds.), Proc. 5th Int. Conf. of Radiation Research, Academic Press, New York, 1975.
72 A. I. Mikhailov, Dokl. Phys. Chem., **197** (1971) 223.
73 M. Shirom and M. Tomkiewicz, J. Chem. Phys., **56** (1972) 2731.
74 H. B. Steen, O. I. Sørensen and J. A. Holteng, Int. J. Radiat. Phys. Chem., **4** (1972) 75.
75 J. Moan and O. Kaalhus, Int. J. Radiat. Phys. Chem., **5** (1973) 441.
76 G. Bomchil, P. Delahay and I. Levin, J. Chem. Phys., **56** (1972) 5194.
77 M. Shirom and Y. Siderer, J. Chem. Phys., **57** (1972) 1013.
78 By a somewhat indirect experimental method M. A. Slifkin and A. C. Allison, Nature (London), **215** (1967) 950 estimated I_g = 8.45 eV for tryptophan, whereas I. Fischer-Hjalmars and M. Sundbom, Acta. Chem. Scand., **22** (1968) 607 calculated theoretically I_g = 7.86 eV for indole which should be similar to tryptophan in this respect.
79 J. Jortner, Actions Chim. Biol. Radiat., **14** (1970) 7.
80 A. Henglein, to be published.
81 H. B. Steen and M. Kongshaug, Proc. Int. Conf. on the Excited States of Biological Molecules, John Wiley, Lisbon, 1974, in press.
82 G. Stein, Advan. Chem. Ser., **50** (1965) 230.
83 M. Ottolenghi, Chem. Phys. Lett., **12** (1971) 339.
84 M. Shirom and G. Stein, J. Chem. Phys., **59** (1973) 1753.
85 W. S. Struve, P. M. Rentzepis and J. Jortner, J. Chem. Phys., **59** (1973) 5014.
86 P. I. Hønnas and H. B. Steen, Photochem. Photobiol., **11** (1970) 67.
87 H. B. Steen and O. Strand, J. Chem. Phys., **60** (1974) 5043.
88 H. B. Steen, Photochem. Photobiol., **6** (1967) 805.
89 H. B. Steen, Photochem. Photobiol., **8** (1968) 47.
90 H. B. Steen and O. Strand, Int. J. Radiat. Phys. Chem., in press.
91 H. B. Steen, Radiat. Res., **41** (1970) 268.
92 R. L. Platzman, in G. Silini (Ed.), Radiation Research, North-Holland Publ. Co., Amsterdam, 1967, pp. 20–42.
93 J. L. Magee, in M. Burton, J. S. Kirby-Smith and J. L. Magee (Eds.), Comparative Effects of Radiation, John Wiley, New York, 1960, pp. 130–150.
94 J. L. Magee and J. J. Huang, J. Phys. Chem., **76** (1972) 3801.
95 H. B. Steen, manuscript in preparation.
96 M. Bowman, private communication.
97 B. Brocklehurst, Nature (London), **221** (1969) 921.
98 R. F. Steiner and E. P. Kirby, J. Phys. Chem., **73** (1969) 4130.
99 J. B. Birks, Photophysics of Aromatic Molecules, Wiley-Interscience, London, 1969.
100 R. A. Robinson and R. H. Stokes, Electrolyte Solutions, Butterworths, London, 1959, p. 463.
101 B. Brocklehurst, Radiat. Res. Rev., **2** (1970) 149.
102 J. Moan and H. B. Steen, J. Phys. Chem., **75** (1971) 2887.
103 J. Moan, Act. Chem. Scand., **26** (1972) 399.
104 S. C. Wallace and D. C. Walker, J. Phys. Chem., **76** (1972) 3780.
105 A. Mozumder, J. Chem. Phys., **50** (1969) 3153; A. Mozumder, in M. Burton and J. L. Magee (Eds.), Advances in Radiation Chemistry, John Wiley, New York, 1969, pp. 1–102.
106 G. G. Meisels and D. R. Ethridge, J. Phys. Chem., **76** (1972) 3842.

107 J. W. Hunt, R. K. Wolff, M. J. Bronskill, C. D. Jonah, E. J. Hart and M. S. Matheson, J. Phys. Chem., **77** (1973) 425.

108 A. K. Pikaev and A. M. Brodskii, High Energy Chem., **6** (1972) 20.

109 G. V. Buxton, F. C. R. Cattell, and F. S. Dainton, Trans. Faraday Soc., **67** (1971) 687.

110 H. Hase, T. Warashina, M. Ogasawara, and T. Higashimura, J. Chem. Phys., **61** (1974) 843.

111 U. Fano, in M. Burton, J. S. Kirby-Smith and J. L. Magee (Eds.), Comparative Effects of Radiation, John Wiley, New York, 1960, pp. 14–21.

112 J. M. Heller, Jr., R. N. Hamm, R. D. Birkhoff and L. R. Painter, J. Chem. Phys., **60** (1974) 3483.

113 N. Getoff and G. O. Schenck, Photochem. Photobiol., **8** (1968) 167.

114 J. W. Boyle, J. A. Gormley, C. Hochanadel and J. F. Riley, J. Phys. Chem., **73** (1969) 2886.

115 H. B. Steen, J. Phys. Chem., **74** (1970) 4059.

Chapter 7

ELECTRON MIGRATION IN LIQUIDS AND GLASSES

W. F. SCHMIDT

I. Introduction

The electronic properties of disordered materials have received increasing interest during recent years and the discovery of electrical switching phenomena in chalcogenide glasses (Ovshinsky effect) has led to many experimental and theoretical investigations on the behavior of electrons in solid amorphous materials[1]. Electronic processes in liquids and glasses have been studied for over a decade from different points of view. Radiation chemists in their research on the interaction of radiation with condensed matter became interested in the generation and migration of excess electrons while investigating the primary processes which follow the absorption of energy. Later, the study of chemical reactions induced by electrons led to measurements of electron transport properties as mobility, reaction rates, etc. in liquids and glasses.

Excess electrons in liquified rare gases aroused interest because in liquid argon ionization pulses due to single α-particles and γ-quanta could be observed indicating the possibility of using liquid filled ionization chambers for particle detection and dosimetry[2]. Quite recently multi-wire proportional chambers filled with liquid xenon have been applied for track measurements of high energy elementary particles[3,4]. Excess electrons in liquid helium have received much interest because they can serve as a probe for liquid structure. The study of electronic conduction in dielectric liquids and solid organic insulators like polyethylene is of special relevance for the application of these materials for high voltage insulation. Investigations of electronic and ionic transport properties in high electric fields are necessary prerequisites. Excess electrons in liquid ammonia have been known for over a century. Their transport properties in the various concentration ranges of metal–ammonia solutions are still the object of very intense research efforts[5]. Excess electrons in other polar solvents such as water, alcohol, etc. became amenable to experimental observation with the introduction of pulse-radiolysis[6,7].

In non-polar liquids, which are perfect insulators in the pure state, drift mobility experiments can be carried out with electron concentrations as low as 10^5 electrons cm^{-3}. The electric field strength can usually be increased up to several hundred kilovolts per centimeter to allow the investigation of non-thermal electrons. Temperature studies give additional information on the transport mechanism. Polar liquids usually exhibit a self-conductivity due to dissociation and greater concentrations of excess charge carriers are necessary in order to detect conductance changes, from which mobility data are mainly obtained. Electron migration in glasses is observed when a γ-irradiated sample is illuminated with light or warmed up. Since the glass exhibits a low natural conductance, drift experiments with small electron concentrations are possible. Hall effect measurements usually require a greater concentration of charge carriers. With a special, very sensitive modulation technique, however, the Hall mobility of electrons in glassy ice has been measured.

In the present chapter the topics mentioned in this introduction are treated with different depth. Excess electrons in liquefied rare gases have been studied for over a decade and excellent reviews have appeared from time to time[6,8–11]. The same is true for excess electrons in liquid ammonia where the proceedings of the Colloquia Weyl give a comprehensive survey of the current status[12,14]. These two topics are treated in a cursory manner only. The main emphasis is being laid on the experimental techniques and the results of electron mobility in liquid hydrocarbons and non-polar and polar glasses (Sect. II). The results obtained with these methods are summarized in Sects. III and VI and models for electron transport in non-polar liquids and glasses are discussed in Sect. VII. Results on electron mobility in polar liquids and models are reviewed in Sect. IV. In Sect. V anions obtained by electron scavenging are discussed. It is obvious to the author that this treatment of the topics of electron migration in liquids and glasses is only a snapshot of the real picture taken from his narrow little angle.

II. Experimental methods

II.A Mobility measurements

II.A.1 Introduction

A basic experiment in the study of the physical properties of excess electrons in any system is the measurement of their drift velocity under the influence of an applied electric field. In the vacuum electrons are accelerated as long as they are under the influence of the electric field and the drift velocity increases until relativistic effects set in. If the electrons interact with atoms or molecules a constant drift velocity will be obtained at a particular field strength and the drift velocity v_d increases proportional to the electric field E

$$v_d = u \cdot E \tag{1}$$

The proportionality constant u is called the mobility and is measured in $(cm^2 V^{-1} s^{-1})$ when v_d is given in $(cm\ s^{-1})$ and E in $(V\ cm^{-1})$. In electrochemistry the conductivity of one equivalent of charge carriers is usually called the mobility and is measured in $(\Omega^{-1}\ cm^{-2})$. Conversion between the two units is obtained by the Faraday constant

$$(\Omega^{-1}\ cm^{-2})/96500 = (cm^2\ V^{-1}\ s^{-1}) \tag{2}$$

Measurement of the mobility gives important information on the interaction between electrons and the atoms or molecules of the material.

A number of different methods for the determination of the mobility are available. They can be divided roughly into two groups:

(a) time of flight measurements, in which the movement of a spatial discontinuity of the charge carrier density is observed and

(b) methods in which the product of the mobility and concentration of charge carriers is measured. The following methods for the injection of excess electrons into liquids or glasses have been used.

(i) Injection from a photo-cathode immersed in the liquid.

(ii) Photoionization of solutes with a lower ionization potential than the solvent.

(iii) Ionization of the liquid or glass by high energy X-rays, γ-rays or fast electrons.

(iv) Injection of electrons from semiconductor diodes or by field emission from tips or sharp edges.

(v) Injection of electrons from the vacuum or the gas phase into the liquid.

(vi) Dissolution of an alkali metal in the liquid to yield

$$M \rightarrow M^+ + e^-_{solv}$$

The detection usually consists of the measurement of the electric current produced by the motion of those charges in an electric field. In a few cases liquid motion associated with moving charges has been applied for mobility measurements.

II.A.2 Non-polar liquids

Non-polar liquids are insulators and the observed intrinsic conductivity is mainly due to impurities. Purification of the liquid lowers the level of self-conductivity. The generation of excess charge carriers leads to an increase in conductivity. Most measurements have been carried out either with the time-of-flight method or by observing the increase in conductance produced by a known concentration of charge carriers. In the time-of-flight method one determines the time it takes a group of charge carriers to traverse a certain distance d in an electric field E. From eqn. (1) it follows for the drift time t_d in a homogeneous field $E = V/d$

$$t_d = \frac{d^2}{u \cdot V} \tag{3}$$

with V the applied voltage. Drift time measurements with charge carriers in dielectric liquids have been performed for several decades and many values have been reported, however, without identification of the species. The book by Adamczewski[15] gives an extensive coverage of these efforts.

Measurements on excess electrons in dielectric liquids were first successful with liquefied rare gases and a wealth of information has been gathered. The methods used were mainly adoption of methods from electron mobility measurements in the gas phase[16–19]. Measurements in liquid hydrocarbons became possible with the availability of better purity liquids and the first observations were reported in 1968–1969 by Minday et al.[20], by Schmidt and Allen[21], and by Tewari and Freeman[22].

Minday et al.[20,23,24] injected electrons into *n*-hexane and other liquids by illuminating a barium coated electrode of a parallel plate cell with visible light. Turning on the light led to a rise in the current until the electrons had reached the opposite electrode and a steady state was obtained. Switching off the light led to a decay of the electron current until all electrons had left the volume between the plates. Observed currents ranged from 10^{-11} to 10^{-9} A. Extreme purity of the liquids was required in order to avoid attachment of electrons to impurities and loss of efficiency of the photocathode.

Schmidt and Allen[21,25] applied a method originally developed by Hudson[26] for the measurement of electron drift velocity in argon gas to the study of excess electrons in liquid hydrocarbons. A parallel plate ionization chamber is subjected to a step function of X-rays. Ionization leads to the formation of electrons and positive ions homogeneously distributed between the plates of the measurement cell. If the mobility of the electrons is much greater than that of the positive ions then the initial rise of the ionization current is correlated to the motion of the excess electrons. Hudson derived the time-dependence of the ionization current i for the case that no loss of excess electrons due to recombination or attachment to impurities occurred. $i(t)$ is then given by

$$
i(t) = \begin{cases} 0 & \text{for } t \leq 0 \\[2mm] 2i_{max}\left(\dfrac{v_d}{d}\right)\left(t - 0.5\left(\dfrac{v_d}{d}\right)t^2\right) & \text{for } 0 \leq t \leq \dfrac{d}{v_d} \\[2mm] i_{max} & \text{for } t \geq \dfrac{d}{v_d} \end{cases} \tag{4}
$$

where v_d is the drift velocity of excess electrons, i_{max} the steady state current and d the plate separation. The electron current flows through the input resistor R of a fast amplifier and the amplified signal voltage is displayed on an oscilloscope. A general scheme of the circuit employed and a typical oscilloscopic trace are given in Fig. 1.

A similar time-of-flight method, making use of the difference in mobility of excess electrons and positive ions, has been reported by Schmidt and Bakale[27,28]. The measurement cell is irradiated with a very short pulse of high energy X-rays. During the pulse negligible drift of the charge carriers occurs so that at the end of the pulse a constant density of excess electrons and positive charge carriers exists between the plates. Since the electron mobility is greater than that of the positive charge carriers the decay of the ionization current will consist of two distinct parts separated in time. If recombination and attachment can be neglected, the decay of the ionization current consists of two straight lines from which the drift time of each carrier can be determined (Fig. 2). Identification of the excess electron is usually made by adding an electron scavenger. This leads to the transformation of a fast mobile electron into a slower ion. The fast component of the current decay then vanishes.

Another group of experimentalists determined electron mobilities from the measurement of the radiation-induced conductance during a pulse of X-rays. Conrad

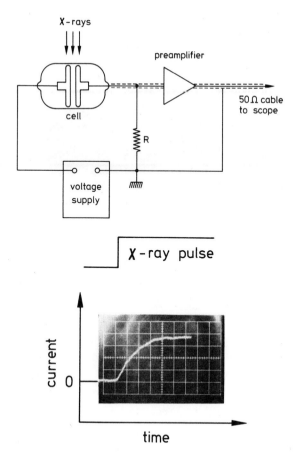

Fig. 1. a, Circuit for drift velocity measurements with the Hudson method; b, Typical rise of the electron current in tetramethylsilane under irradiation with pulsed 1.6 MeV X-rays; horizontal 0.5μ/div $E = 5$kV cm^{-1} (ref. 25).

and Silverman[29] studied the time-dependence of the ionization current after a 25 ns pulse of 2 MeV-bremsstrahlung. Since they did not use especially purified liquids the decay of the ionization current was exponential due to electron scavenging by impurities. Extrapolation back to the time of the pulse gave the radiation-induced conductance and with dose measurements and ion yield data the initial concentration n_{el} of excess electrons and positive charge carriers was obtained, $n_{el} = n_p$. Since the electron mobility u_{el} is greater than the mobility of ionic species u_p, the peak ionization current is mainly due to the electrons, i.e.

$$i_{peak} \, \alpha \, n_{el} \, u_{el}$$

It is required, naturally, that the decay of the current is long compared to the duration of the pulse. The same principle was applied by Freeman and co-workers[30-32] and by Gregg and Bakale[33].

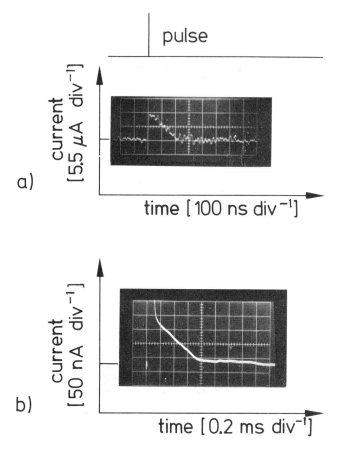

Fig. 2. a, Decay of electron current in liquid methane after irradiation with a 5 ns pulse of 15 MeV X-rays; b, decay of the ion current in liquid methane.

A microwave absorption method for the detection of excess electrons in irradiated hydrocarbons has been developed by Warman et al.[34]. For microwaves travelling through a uniform, weakly conducting medium absorption of microwave power occurs and the intensity of the microwave power after traversing a distance d is given by

$$I = I_0 \, e^{-\alpha d} \qquad (5)$$

where I_0 is the incident intensity. For small absorptions the fractional decrease in power is then

$$\frac{\Delta I}{I_0} = \alpha d \qquad (6)$$

For microwaves confined in a rectangular waveguide of broad dimension a the absorption coefficient α is given by

$$\alpha = \frac{\omega \mu_0 \, \sigma}{\left[\left(\frac{\omega \ae}{c_0}\right)^2 - \left(\frac{\pi}{a}\right)^2\right]^{1/2}} \tag{7}$$

In eqn. (7), σ is the d.c. conductivity, μ_0 is the permeability of free space, $\ae$ is the refractive index of the medium and c_0 is the velocity of light in vacuo. Substitution for α in eqn. (6) then gives

$$\frac{\Delta I}{I_0} = \frac{\omega \mu_0 \, \sigma d}{\left[\left(\frac{\omega \ae}{c_0}\right)^2 - \left(\frac{\pi}{a}\right)^2\right]^{1/2}} \tag{8}$$

Under the actual experimental conditions the medium within the waveguide is, however, not uniform but contains dielectric discontinuities at which reflections may occur. Because of this and the coherent nature of the propagating waves, interference effects occur which result in eqn. (8) being inapplicable. These effects have been theoretically and experimentally investigated by the above mentioned authors and P. P. Infelta in as yet unpublished work and they have shown that a linear relationship between $\Delta I/I_0$ and σ still holds for small absorptions i.e.

$$\frac{\Delta I}{I_0} = A\sigma \tag{9}$$

In eqn. (9) A is a function of the cell and irradiation geometry as well as the parameters given in eqn. (8).

Since the mobility of excess electrons in hydrocarbons generally exceeds the mobilities of positive and negative molecular ions the major part of the current is carried by electrons and the conductivity can be approximated by $\sigma = n_{el}u_{el}e$ where e is the electronic charge. Thus the fractional absorption of microwave power is directly proportional to the product of the electron concentration and mobility.

In Fig. 3 is shown a schematic diagram of the original experimental set-up. The

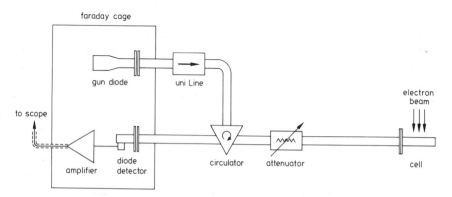

Fig. 3. Experimental set-up for the observation of microwave absorption by electrons in dielectric liquids (ref. 34).

cell consisted of a length of waveguide closed at one end with a metal plate to reflect the wave and at the other end with a vacuum tight mica "window". Electrons were produced by irradiation with pulses of 3 MeV electrons from a van de Graaff accelerator. The time response was found to be approximately 3 ns. Changes in the power level of approximately 0.1 %, corresponding to a value of $n_{el}u_{el} \simeq 10^{-8} \text{mol} 1^{-1}$ cm^2 V^{-1} s^{-1}, could be detected. The sensitivity of the method is therefore much smaller than that of d.c. methods.

Another way of producing a uniform concentration of charge carriers is by photo-ionization of solutes with a lower ionization potential than the solvent. As light source a high intensity pulse laser is employed with which ionization is achieved by two-photon absorption processes[35-37]. Beck and Thomas[37] used anthracene and pyrene as solutes, Devins and Wei[36] tetramethyl phenylene diamine (TMPD). Since anthracene and pyrene are also good electron scavengers, the photoionization current decayed in time due to recapture of the excess electrons by neutral solute molecules. As the solute is usually in excess pseudo-first-order kinetics are observed and by varying the solute concentration bimolecular rate constants for the attachment of electrons can be measured. All other mobility methods are also suitable for the measurement of electron reaction rate constants. It is only necessary that the electron lifetime in the "pure" liquid is much longer than the duration of the radiation pulse. Added scavengers shorten the lifetime.

A comparison of the various mobility methods is given in Table 1. For conductivity measurements lower purity liquids can usually be used since the sensitivities achieved so far require large doses of radiation per pulse. Time-of-flight measurements require the highest purity liquids available since only low concentrations of charge carriers can be tolerated in order to avoid space charge problems.

II.A.3 Polar liquids

Introduction of excess electrons into polar solvents can be achieved either by dissolution of alkali metals, by irradiation with high energy radiation, or by photo-ionization of suitable solutes. Since polar liquids usually have a higher self conductivity than dielectric liquids, a.c.- or pulsed d.c.-conductivity measurement techniques are applied in order to avoid electrode polarization. In liquids where e_{solv}^- is a stable entity dissolution of alkali metals of different concentrations and measurement of the conductivity provides the equivalent conductance at infinite dilution. Extensive data exist on the properties of solvated electrons in liquid ammonia, amines, and ethers and several summarizing reports have appeared on this subject[5,6,12-14]. In liquids where e_{solv}^- is not stable irradiation with a pulse of ionizing radiation produces an instantaneous concentration of solvated electrons and other charged species, resulting in an increase of the electric conductivity of the solution. Charge recombination or conversion of mobile species into less mobile species leads to a decay of the conductivity in time. If the pulse duration is short compared to the characteristic decay time of the conductivity then the maximum current during the pulse is determined by the sum of the products of yield and mobility of all charged species formed by the radiation.

TABLE 1

Comparison of time resolution and sensitivity of various mobility methods for non-polar liquids

Method of electron production	Sensitivity $(mol\,l^{-1}\,cm^2V^{-1}s^{-1})$	Detection		Time resolution (ns)	Pulse		Ref.[`]
		A	B		Length	Dose	
Photoionization	$\sim 3 \times 10^{-10}$	nu	d.c.	6.5	<20ns	—	37
High energy electrons	$\sim 10^{-8}$	nu	microwave	3	10ns	—	34
High energy electrons	$\sim 2 \times 10^{-10}$	nu	d.c.	<10	10ns	$10-10^2$ rad	85
X-ray bremsstrahlung	$\sim 10^{-15}$	t_d	d.c.	<15	10ns	mrad	28

A, value measured; nu product of mobility and concentration; t_d drift time.
B, conductivity.

The conductivity for single charged carriers is given by

$$\sigma = e\sum_i n_i \cdot u_i = \frac{i}{V_b} \cdot \frac{1}{q} \qquad (10)$$

with e the electron charge, n_i the concentration of ith type of charge carrier (cm^{-3}), u_i the mobility (cm^2 V^{-1} s^{-1}), i(A) the current flowing through the cell when a voltage V_b(V) is applied, and $1/q$ the cell constant (cm^{-1}). The current i is usually converted into a signal voltage V_s by a resistor R_s in series with the cell. The magnitude of R_s should be much smaller than V_b/i so that V_b lies across the conductance cell all the time.

The first determination of the mobility of the hydrated electron was due to Schmidt and Buck[38], who observed the transient conductivity of an alkaline aqueous solution under the influence of a 4 μs pulse of 15 MeV electrons. A differential circuit was applied to discriminate against the negative signal generated by the electrons stopped in the cell from the accelerator. The transient conductivity signal was analyzed by taking into account the concentrations and reactions of all chemical species involved in the change of the conductivity. A similar experimental set-up was described by Barker et al.[39]. Later Beck[40] combined conductivity measurement and detection of optical absorption and also determined the mobility of the hydrated electron. In all these experiments the polarity of the battery voltage was inverted with a frequency of 1–10 s^{-1} in order to avoid polarization effects on the electrodes. Vannikov et al.[41–46] extended conductivity measurements to the investigation of ketones, amines and alcohols. For liquids with higher self-conductance (10^{-5}–10^{-8} Ω^{-1} cm^{-1}) a differential method was applied where polarization effects were minimized by the use of pulsed voltages. The effects of the application of these pulses to the electrodes were compensated by a bridge method where one arm was formed by the cell and the other by a parallel circuit of C and R to simulate the capacitance and conductance of the measurement cell (Fig. 4). The high energy electron pulse was triggered while the voltage pulses were applied to the circuit. The measurement time constant of the total circuit did not exceed 0.5 μs. In all cases an increase of the conductivity was observed. The yield of the electrons and the concentration were determined by scavenger studies with anthracene. The anthracene anion exhibits a characteristic absorption from which the concentration could be determined.

II.A.4 Glasses

While in liquids radiation-produced electrons are relatively mobile and recombine with positive charge carriers, in glasses a certain fraction of the excess electrons are trapped and separated charges can be kept for long times. Increasing the temperature or illumination leads to detrapping and the electrons migrate through the material either being retrapped or recombining with a positive charge. Experimental methods for the investigation of electron migration in glasses are somewhat different from the methods applied to liquids. The three main glasses studied so far are 3-methylpentane, methyltetrahydrofuran and alkaline ice. Time-of-flight measurements and Hall effect measurements have been performed.

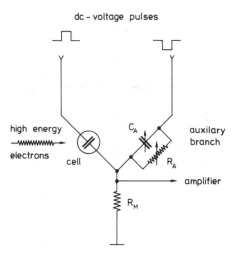

Fig. 4. Schematic circuit for measurement of radiation-induced conductivity in polar liquids (ref. 45).

Maruyama and Funabashi[47] photoionized TMPD which was deposited on one of the electrodes of a parallel plate cell ($d = 0.25$ mm). A light pulse of 1 μs duration was generated with a xenon flash lamp and the transient response of the photocurrent was measured. If no electron trapping would occur, the current–time signal should have a flat top and rise and fall times would be determined by the pulse shape. The actual photocurrent response, however, exhibited a short flat top followed by a long tail which lasted more than 30 μs. The magnitude of the photocurrent was of the order of 10^{-10} A with approximately 10^4 electrons moving across the specimen at each pulse. The quantum efficiency was estimated to 10^{-9} indicating the influence of recombination in the early stage of electron generation.

For polar glasses Kevan and co-workers developed drift mobility and Hall mobility measuring techniques[48–51]. Drift mobility was measured with the time-of-flight method. The sample had been irradiated with ^{60}Co-γ-rays to 0.2 Mrad to produce trapped electrons throughout the whole volume. The cell was illuminated with a light flash through one of the electrodes, leading to the production of some mobile electrons near this electrode. Under the influence of an external electric field these electrons moved to the opposite electrode giving rise to a transient current (Fig. 5). Only one out of 10^6 photoexcited electrons traversed the entire sample thickness. While the drift mobility is a macroscopic value, Hall effect measurements yield the true microscopic mobility which is unaffected by trapping effects. The Hall effect is due to the motion of charge carriers in crossed electric and magnetic fields. Usually the charge carrier motion occurs in a d.c. electric field, the direction of which is perpendicular to a magnetic field. However, in conductors with low density of carriers the build-up of space charge leads to a distortion of the electric field in the sample. Eisele and Kevan[50] described a double modulation method for Hall effect measurements on high impedance photoconductors which avoids these diffi-

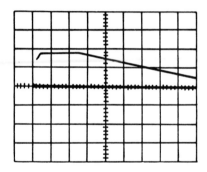

Fig. 5. Oscilloscope trace of electron current in γ-irradiated 10 M NaOH glassy ice at 77 K (ref. 48).

culties. The magnetic field is modulated by rotating the sample in the field and the electric current is modulated by intermittent illumination of the photoconductor. The Hall voltage appears as the sum and the difference of the two modulation frequencies and different error signals could be discriminated in this way. Figure 6 shows the experimental set-up. It was found that the light modulation frequency had to be 90 Hz or greater in order to avoid space charge effects. Figure 7 gives an example of Hall voltage measurements on electrons in γ-irradiated 10 M NaOH glassy ice. The rise of the Hall voltage in time seems to be due to the response time of the electronics. Prolonged illumination led to a depletion of electrons and no equilibrium Hall voltage was observed.

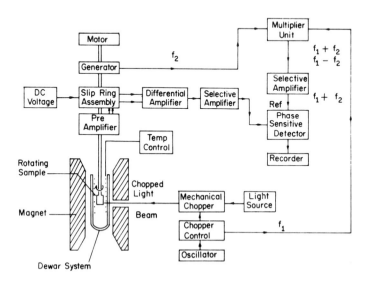

Fig. 6. Schematic for double modulation Hall mobility apparatus (ref. 50).

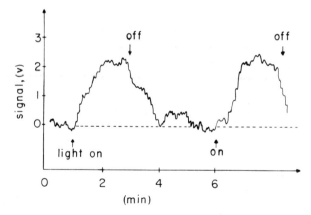

Fig. 7. Time profile of Hall voltage for electrons in γ-irradiated 10 M NaOH glassy ice (ref. 49).

The Hall mobility u_H was obtained from

$$u_H = 10^8 \cdot V_H / H\,E\,W \,(\text{cm}^2\,\text{V}^{-1}\,\text{s}^{-1}) \tag{11}$$

with H the magnetic field strength in Gauss, E the applied electric field strength, W the distance between the Hall electrodes and V_H the Hall voltage. The Hall voltage should be proportional to the magnetic field strength and the electric field strength. Figure 8 shows that this dependence is indeed observed. The accuracy of alignment of the Hall electrodes limited the application of field strength to greater than $3.5\,\text{V cm}^{-1}$.

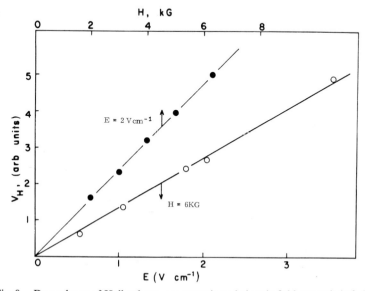

Fig. 8. Dependence of Hall voltage on magnetic and electric field strength (ref. 49).

II.B Conduction energy levels

II.B.1 Photoelectric effect

The energy which is necessary to bring an electron adiabatically from the vacuum into a liquid can be measured by means of the photoelectric effect. If a metal plate is immersed in a particular dielectric liquid, a shift of the work function occurs which is related to the energy of the electron in the conducting state V_0 by

$$V_0 = \Phi_{\text{liq}} - \Phi_{\text{vac}} \tag{12}$$

with Φ the work function of the metal by emission into the liquid or the vacuum. Investigations on hydrocarbons were carried out by Holroyd et al.[52-54] and Schiller et al.[55]. The photocurrent density j for emission into vacuum is given by Fowler's equation

$$j = \alpha A T^2 \, \text{F}(x) \tag{13}$$

where A is a universal constant, α the quantum yield, T the absolute temperature, and $\text{F}(x)$ the Fowler function with $x = (h\nu - \Phi)/k_{\text{B}} T$ and

$$\text{F}(x) = \text{e}^x - \frac{\text{e}^{2x}}{2^2} + \frac{\text{e}^{3x}}{3^2} - \frac{\text{e}^{4x}}{4^2} + \frac{\text{e}^{5x}}{5^2} - \cdots \tag{14}$$

It was found that eqn. (13) also described the current density for the emission in liquids. The experiment consists in measuring $j(\nu)/I(\nu)$ over a range of frequencies ($I(\nu)$ is a relative measure for the light intensity of the lamp in a particular frequency interval $\nu \cdots \nu + d\nu$ and it is determined with a photo multiplier). The data $j(\nu)/I(\nu)$ are plotted versus $h\nu/k_{\text{B}} T$ and the work function Φ is obtained by shifting the experimental curve in the x-direction by Δx and in the y-direction by Δy until it coincides with the general function $\text{F}(x)$. The work function is obtained from $\Delta x = \Phi/k_{\text{B}} T$. Shifts Δy represent the efficiency of electron escape from the image potential barrier which exists near the cathode[53]. Figure 9 shows some typical data obtained when the photoeffect was measured on a Zn electrode in vacuum and in neopentane.

II.B.2 Photoionization

A relative measure of V_0 in various liquids can be obtained from the variation of the ionization potential of TMPD dissolved in the particular liquid[56]. The ionization energy of TMPD in the gas phase is $I_{\text{g}} = 6.35$ eV[57]. Absorption of a photon of suitable energy leads to the formation of TMPD$^+$ and an excess electron. In the liquid the ionization energy of TMPD is modified by the energies which are necessary to transfer the positive ion and the electron into the liquid, i.e.

$$I_{\text{liq}} = I_{\text{g}} + E_{\text{p}}(\text{TMPD}^+) + V_0 - \Delta H_{\text{sol}}(\text{TMPD}) \tag{15}$$

E_{p} is the polarization energy of the positive ion, ΔH_{sol} is the heat of dissolution of TMPD. E_{p} is estimated to 1.3 eV for the hydrocarbons and since the dielectric constant varies little ($\epsilon \approx 2$ for most hydrocarbons), E_{p} is assumed to change little from one liquid to another. Shifts in the photoionization threshold, therefore, can

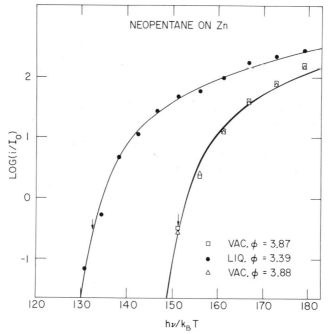

Fig. 9. Photocurrent from a Zn cathode in vacuum and neopentane vs. light energy (ref. 52).

be attributed to variations of V_0. Absolute values of V_0 require the comparison for one liquid with photoelectric values. Recently, measurements with perylene as a solute have been reported which give relative V_0 values comparable to those obtained with TMPD[58].

III. Electrons in non-polar liquids

III.A Mobilities and energy levels

The physical properties of excess electrons in liquefied rare gases have been the object of experimental and theoretical investigations for over 15 years. A great quantity of information has been accumulated about the properties of excess electrons in liquid helium and studies on other rare gases have led to a relatively detailed picture of the energetics and dynamics of excess electrons in monoatomic liquids. Several reviews and summarizing articles have appeared[6,8,9–11] so that this area can be dealt with here in a rather general way. Measurements of excess electron mobility in liquid nitrogen and hydrogen were also reported[59–61], but measurements on liquid hydrocarbons yielded contradictory results for a long time[15]. Since 1969, however, measurements of electron mobility in highly purified hydrocarbons became possible and the investigation of the physical and chemical properties of excess electrons in these simple polyatomic liquids has become a rapidly expanding field.

III.A.1 Liquefied rare gases

Measurements of the electron mobility in liquefied rare gases yielded values which fall into two groups: (a) low mobilities $u_{el} \ll 1$ and (b) high mobilities $u_{el} \gg 1$. The values for the energy of the conducting state V_0 show a parallel trend. Positive energies correspond to low mobilities while in high mobility liquids negative V_0 values were obtained. In Table 2 some typical data are compiled. In liquid helium and neon which exhibit a low electron mobility all data obtained so far are consistent with a localized electron state. (See Chap. 1, Sect.II.B). The electron is assumed to reside in a cavity the radius of which was estimated to be 14 Å for liquid He at 4.2 K. In an electric field this whole structure moves like an ionic species. A similar picture applies to liquid neon. In liquid argon, krypton and xenon the high electron mobilities

TABLE 2

Electron mobilities in the liquefied rare gases at their triple point

Liquid	Triple point temperature (K)	u_{el} $(cm^2V^{-1}s^{-1})$	Ref.	V_0 (eV)	Ref.
Xe	161	1900	62	−0.63	d,f
Kr	115	1800	62	−0.78	d,f
Ar	85	475	62	−0.33	c,f
Ne	25	1.6×10^{-3}	a	+0.5	d
^{4}He	4.2	2×10^{-2}	b	+1.05	e
^{3}He	3.0	3.7×10^{-2}	b	+0.9	d

[a] R. J. Loveland, P. G. LeComber and W. E. Spear, Phys. Lett. A, **39** (1972) 225.
[b] H. T. Davis, S. A. Rice and L. Meyer, Phys. Rev. Lett., **9** (1962) 81.
[c] J. Lekner, B. Halpern, S. A. Rice and R. Gomer, Phys. Rev., **156** (1967) 351.
[d] B. Raz and J. Jortner, in J. Jortner and N. R. Kestner (Eds.), Electrons in Fluids, Springer-Verlag, Berlin, Heidelberg, New York, 1973, p. 421, Table IV. The V_0 values were derived from spectroscopic data on Wannier impurity states originating from positive ionic cores in liquid rare gases.
[e] W. T. Sommer, Phys. Rev. Lett., **12** (1964) 271.
[f] Recent measurements of V_0 in liquid Xe, Kr and Ar by W. Tauchert and W. F. Schmidt, Z. Naturforsch. A, **30** (1975) 1085, gave somewhat different values.

indicate that the interaction between the electron and the atoms of the liquid is weak. A quasifree particle model can be applied where the magnitude of the mobility is determined by scattering processes. Further experimental evidence for the applicability of such a picture comes from the field-dependence of the drift velocity at higher field strength. It was found that in these liquids above a certain critical field the drift velocity increased proportional to $\sqrt{E}$ and became constant at even higher field strengths[62-65]. A similar behavior had been found for electrons in germanium crystals by Ryder and Shockley[66] and it was explained by Shockley on the basis of electron–phonon interactions[67].

III.A.2 Hydrocarbons and related compounds

Drift mobilities have been measured for a variety of liquids and the dependence on temperature and electric field strength provides us with important information about the transport mechanism. In addition, for many liquids V_0, the energy necessary to inject adiabatically an electron from the vacuum into the liquid, has been determined. Drift mobilities were obtained either from time-of-flight measurements, where the electron drift velocity in a constant homogeneous electric field is determined[20,23–25,28] or from the measurement of the radiation-induced conductance by taking into account the yield of charge carriers[29–32]. Time-of-flight methods are especially suited for the study of field effects. Figure 10 gives some examples of the variation of the drift velocity with the electric field strength. At low field strength the drift velocity increase is proportional to the field strength and the drift mobility is, by definition, given by

$$u_{el}(0) \equiv \frac{v_D}{E} \tag{16}$$

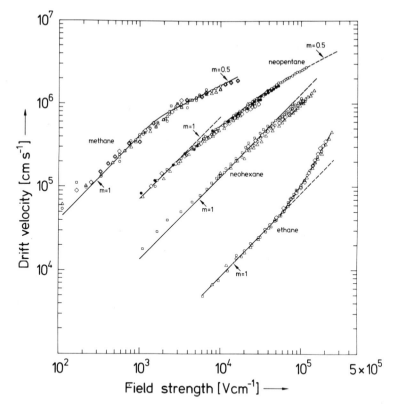

Fig. 10.   Variation of electron drift velocity with electric field strength; methane at 111 K, neopentane and neohexane at 295 K and ethane at 200 K (refs. 28 and 68).

and independent of the electric field strength. At higher field strengths deviations from this proportionality between drift velocity and field strength occur. Two types of behavior have been found. In liquids which exhibit a relatively large low field mobility $u_{el}(0)$ a decrease in the mobility is observed (cf. methane and neopentane). In liquids with a relatively small low field mobility an increase in mobility is observed above a certain field strength (e.g. ethane and propane[28,68]). In most of the liquids, however, measurements have not been carried out up to these high field strengths and only low field mobilities are reported[23-25,30-32]. Temperature studies gave positive temperature coefficients for the mobility; only in liquid methane the mobility decreased with increasing temperature[28,32] while for tetramethylsilane the mobility remained almost constant from 223 K to 293 K[25,69]. Figure 11 shows Arrhenius plots of the drift mobility in several liquids as compiled from various sources. Many more measurements have been published and for several liquids no straight lines have been found. Extrapolation of the Arrhenius plots to infinite temperature, yields for some cases, values between 10^2 and 10^3 cm^2 V^{-1} s^{-1} and on the basis of a simple trapping model it was assumed that this value would correspond to a band type motion between traps[23,24,71,72]. However, since this mobility should also depend on the temperature no straight line in the Arrhenius plot is expected and the extrapolation gives ambiguous results. Allen and Holroyd[69,70] tried to evaluate this mobility in the mobile state by calculating the probability P_m of finding the electron in the mobile state assuming that the localized states lie in a tail at the bottom of the excess electron band. The mobilities estimated in this manner, however, did not give a consistent picture.

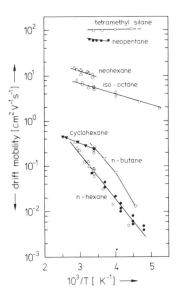

Fig. 11. Arrhenius plot of electron drift mobility in several hydrocarbons. Refs; tetramethylsilane [69]; neopentane [24]; isooctane [69]; neohexane [25]; cyclohexane [31]; n-butane [31]; *n*-hexane [24,25,69].

A more detailed investigation of excess electron transport has recently been carried out on liquid ethane[73]. Temperature and field strength dependences of the electron drift mobility were studied over a wide interval. Figure 12 shows the results obtained for the electron drift velocity as a function of the applied field. At low field strengths the drift velocity increases proportionally to the field strength. Above a certain critical field strength the drift velocity grows faster and since $u_{el} = (v_D / V_\mu)/E$ the mobility also increases. The field strength above which this deviation occurs is shifted to higher values the lower the temperature. The temperature-dependence of the low field mobility is shown in Fig. 13 as an Arrhenius plot. The relatively low electron mobility and the positive temperature coefficient seem to indicate that the electron spends most of its time in a localized state which it leaves by thermal activation. Motion occurs by jumps from one trap to another. A simple phenomenological model which yields the observed temperature and field strength dependence has been derived by Bagley[74]. Transport is assumed to occur over multiple barriers with an average height E_a separated by a distance Λ, v is the attempt frequency. The dependence of u_{el} on T and E is then given by

$$u_{el}(E, T) = \frac{\Lambda}{E} 2v \exp\left(-\frac{E_a}{k_B T}\right) \sinh\left(\frac{\Lambda e E}{2 k_B T}\right)$$ (17)

which reduces for small fields to

$$u_{el}(0, T) = \frac{\Lambda^2 e}{k_B T} v \exp\left(-\frac{E_a}{k_B T}\right)$$ (18)

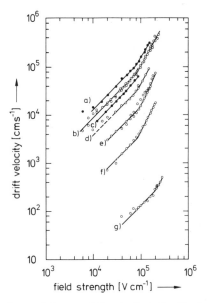

Fig. 12. Field strength-dependence of electron drift velocity in liquid ethane at various temperatures. a, 216 K; b, 197 K; c, 180 K; d. 170 K; e, 155 K; f, 142 K; g, 111 K (ref. 73).

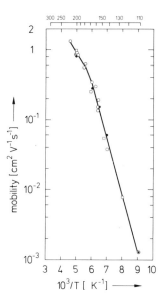

Fig. 13.   Temperature-dependence of low field electron mobility in liquid ethane (ref. 73).

A suitable equation for the field-dependence of u_{el} is obtained by combining eqns. (17) and (18) to give

$$\frac{u_{el}(E, T)}{u_{el}(0, T)} = \frac{2k_B T}{\Lambda eE} \sinh\left(\frac{\Lambda eE}{2k_B T}\right) \qquad (19)$$

This equation can be used to obtain the jumping distance as a function of temperature. Λ was found to increase with temperature from 8 Å at 111 K to 45 Å at 216 K. With these values an attempt frequency of $v = 9 \times 10^{12} \, s^{-1}$ was obtained, which has the magnitude of characteristic phonon frequencies.

In order to obtain more information on the mechanism of electron transport in liquid hydrocarbons electron drift mobilities were measured in mixtures of neopentane and n-hexane at various temperatures by Minday et al.[24,75] They found that the mobility in the mixtures as in the case of the pure components, could be expressed by the Arrhenius equation

$$u(T) = u_0 \exp\left(-E_a/k_B T\right) \qquad (20)$$

The activation energy E_a was a linear function of the composition and the mobility in the mixtures varied exponentially with the mole fraction (x_H) of n-hexane. The electron mobility u_m in the mixture was found to follow rather closely

$$u_m = u_{NP} \exp\left(\frac{E_0}{k_B T}\right) \qquad (21)$$

where the mobility subscript refers to the pure component. E_0 is the activation energy in a mixture and it is given by

$$E_0 = X_H E_H + X_{NP} E_{NP} \tag{22}$$

with E_H and E_{NP} the activation energies in hexane and neopentane. This equation was thought to describe the experimental result on mixtures of liquid methane and also liquid ethane[76], however, temperature studies and investigation of the field-dependence showed that the physical state of the excess electrons is different at high methane concentrations from that at high ethane concentrations[77]. The low field mobility did not follow eqn. (21), but at low ethane concentrations $(u_m)^{-1}$ varied linearly with the ethane concentration (cf. Fig. 14). The field-dependence of the electron drift velocity for this concentration range is shown in Fig. 15. The general behavior is maintained but the critical field E_c is shifted to higher fields as the ethane mol % increases. Since a decrease in mobility at higher field strengths is attributed in solid state physics to a warming-up of the electrons simple arguments lead to the conclusion that $u_{el} \cdot E_c = \text{const}$[67]. Figure 16 shows that this relation is fulfilled for the data of Fig. 15. More information is expected from experiments with mixtures since a continuous variation of various parameters affecting electron mobility is possible.

In the studies on the physical state of excess electrons in liquefied rare gases it was found that there exists a definite correlation between the drift mobility and the electron binding energy V_0 of the liquid. Low mobility values in helium and neon were accompanied by positive values of V_0, that means that injection from vacuum

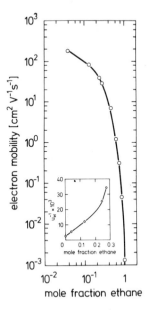

Fig. 14. Dependence of the electron mobility on the composition of the methane–ethane mixtures (ref. 77).

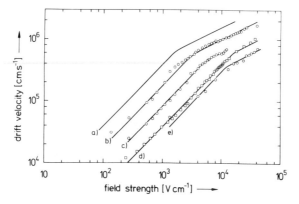

Fig. 15. Field strength dependence of electron drift velocity in methane–ethane mixtures at low ethane concentrations at $T = 111\,K$. a, pure methane; b, 4 mol % ethane; c, 13 mole % ethane; d, 24 mole % ethane, e, 27.5 mole % ethane (ref. 77).

in the liquid is an endothermic process. For liquids exhibiting high drift mobilities negative V_0 values were obtained, indicating an exothermic reaction (see Table 2). In liquid hydrocarbons the mobility values do not fall into these two groups but rather change gradually from one liquid to another. It was expected that the V_0 values would follow this behavior. Most of the V_0 data are due to Holroyd et al.[52,54,56,58] but some values have also been reported by Schiller et al.[55] In Tables 3a and 3b V_0 values are presented together with the electron drift mobility at the same temperature. As can be seen a correlation exists between V_0 and u_{el}. For liquids with a low mobility V_0 is zero or positive. Large negative values of V_0 go with high mobilities. This correlation, however, does not exist for liquid methane. Noda and Kevan[78] measured $V_0 \approx 0.0\,eV$ at $T = 109\,K$. This result was confirmed by Tauchert and

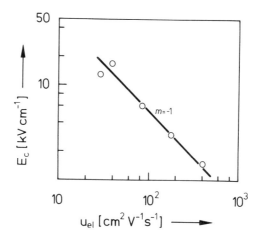

Fig. 16. Critical field E_c as a function of electron mobility u_{el} in methane–ethane mixtures for the data of Fig. 15 (ref. 77).

TABLE 3a

Ground state energies and mobilities of excess electrons in several hydrocarbons and tetramethylsilane at room temperature

Liquid	V_0 (eV)	Method	Ref.	Mobility[a] $(cm^2V^{-1}s^{-1})$
Tetramethylsilane	−0.62	PhE	52	90
	−0.59	TMPD	58	
Neopentane	−0.43	PhE	52	55, 70
	−0.38	TMPD	58	50
2,2,4 Trimethylpentane	−0.18	PhE	52	7
	−0.33	TMPD	58	
2,2-Dimethylbutane	−0.15 ±0.03	PhE	55	10
	−0.26	TMPD	58	
n-Pentane	−0.01	PhE	52	0.14
n-Hexane	+0.04	PhE	52	0.09
	+0.10	TMPD	58	
	+0.16 ±0.05	PhE	55	
n-Tetradecane	+0.21	TMPD	58	—
Cyclopentane	−0.28	PhE	52	1.1
	−0.21	TMPD	58	
Cyclohexane	+0.01	TMPD	58	0.35
				0.45
Benzene	−0.14 ±0.05	PhE	55	0.6, 0.1
Toluene	−0.22 ±0.03	PhE	55	0.56

[a]Mobility data from refs. 24, 25, 28, 31, 32 and 68.
PhE: photoelectric effect; TMPD: photoionization.

Schmidt[79]. The correlation between u_{el} and V_0 in liquefied rare gas follows from the theoretical models by Cohen and Lekner[80,81] and Springett et al.[82] since both quantities depend on the scattering length. Application of these models to liquid hydrocarbons have been attempted[72,83,84].

TABLE 3b

Ground state energies and mobilities for electrons in methane and ethane

Liquid	T(K)	V (eV)	Ref.	Mobility[a] $(cm^2V^{-1}s^{-1})$
Methane	109	0.0	78	400 ±50
	95	0.0;−0.12	79; 78(a)	
Ethane	182	+0.02	78	0.6, 0.55

[a]Mobility data from refs. 28, 31 and 68.

III.A.3 Other liquefied molecular gases
Very few data have been published on the electron mobility in liquid nitrogen, hydrogen and other liquefied molecular gases. Only low mobility values were measured and no unambiguous information on the kind of species which was observed could be obtained. Although in liquid hydrogen and deuterium the

TABLE 4

Mobilities of negative charge carriers in liquefied molecular gases

Liquid	$T(K)$	Mobility $(cm^2V^{-1}s^{-1})$		Ref.
Hydrogen	21	8.6	$\times 10^{-3}$	59
		3	$\times 10^{-2}$	60
Deuterium	21	9	$\times 10^{-3}$	60
Oxygen	77	8	$\times 10^{-3}$	60
Nitrogen	77	8	$\times 10^{-3}$	60

presence of electron scavenging impurities can probably be neglected, measurements on liquid nitrogen suffer most likely from traces of oxygen. Table 4 summarizes same data obtained so far for these systems.

III.B Reaction rates

Drift velocity measurements of excess electrons in non-polar liquids are usually obscured by the rapid reactions of electrons with impurities present in "pure" liquids. Extensive purification procedures are necessary to lower the impurity level to values which allow the measurement of drift times. If an electron scavenger is present with a concentration $[S]$ for measurements of the drift time t_d the inequality

$$t_d < (k_s[S])^{-1} \tag{23}$$

must be fulfilled. For measurements of the reaction rate constant k_s the reaction time should be shorter than the drift time

$$(k_s[S])^{-1} < t_d \tag{24}$$

In the case where a homogeneous distribution of excess electrons is generated by a short pulse of radiation the electron concentration decays exponentially in time and a pseudo-first-order rate constant k_{ap} is obtained. Variation of the scavenger concentration $[S]$ yields different values of k_{ap} and the bimolecular rate constant k_s for the reaction

$$e^- + S \xrightarrow{k_s} S^-$$

is obtained this way. Rate constant measurements were carried out either by observing the change in conductivity[28,37,76,85–88] or the change of optical absorption due to solvated electrons[89,90]. Figure 17 shows the dependence of k_{ap} on the oxygen concentration in liquid methane at $T = 109$ K. The intercept at $[O_2] = 0$ is due to a residual impurity present in the liquid. Such a linear dependence has been observed in almost all cases[91]. The bimolecular rate constant was obtained from the slope of the straight line. Allen et al.[69,70] determined rate constants by observing the decrease

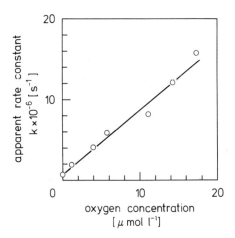

Fig. 17. Apparent rate constant $k_{ap} = k_s[S]$ for electron disappearance as a function of oxygen concentration in liquid methane at $T = 109\,\mathrm{K}$ (ref. 28).

of the steady-state electron current in the Hudson method[25] due to the addition of a scavenger.

Rate constants for quite a number of scavengers in various hydrocarbons and in tetramethylsilane have been reported. Good electron scavengers in the gas phase like SF_6, N_2O, CCl_4, O_2 etc., also react rapidly with electrons in non-polar liquids and the magnitude of the rate constants is indicative of a diffusion controlled reaction. The rate constant was thought to be proportional to the electron mobility. Measurements by Beck and Thomas[37] with biphenyl in several hydrocarbons at room temperature showed, however, that k_s varied roughly with $u_{el}^{0.5}$. Subsequent measurements with other scavengers confirmed this relation for some solutes and gave other dependencies on u_{el} for other solutes[70,91]. Some values are compiled in Table 5.

Electron transport in liquid hydrocarbons is generally described in terms of two states which the electron can occupy: (a) the extended state in which the electron is very mobile and (b) the trapped state in which the electron mobility is negligibly small. Reaction with scavengers can in principle occur in both states. Yakovlev et al.[87] interpret their experimental results on the scavenging of electrons by pyrene in n-hexane and cyclohexane with the assumption that reaction occurs while the electron is in the trapped state. From their model they obtained a relation between k_s and D_{el} (diffusion coefficient)

$$\frac{4\pi D_{el}}{k_s} = r_p^{-1} + C D_{el} \tag{25}$$

where r_p is the reaction radius and C a constant. The potential between the electron and the scavenger molecule was assumed to be given by the ion-induced dipole interaction

$$V = -\frac{\overline{\alpha}e}{2\epsilon^2 r^4} \tag{26}$$

TABLE 5

Rate constants for electron reaction with various solutes in hydrocarbons and tetramethylsilane

Liquid	Solute	T(K)	$k(M^{-1}s^{-1})$	Ref.
Methane	SF_6	110	4×10^{14}	76
	N_2O	110	8.5×10^{11}	76
	O_2	110	8.4×10^{11}	76
Ethane	SF_6	195	1.2×10^{13}	91
Propane	SF_6	195	2.2×10^{12}	91
n-Hexane	SF_6	295	1.9×10^{12}	69
	N_2O	295	1×10^{12}	69
	CCl_4	294	1.3×10^{12}	69
		293	1.2×10^{12}	a
	N_2O	293	1.5×10^{12}	90
	O_2	293	1.5×10^{11}	90
	O_2	296	2.5×10^{10}	37
	CO_2	293	1.8×10^{12}	90
	biphenyl	293	1.2×10^{12}	90
		296	7.7×10^{11}	37
	C_2H_5Br	293	1.5×10^{12}	70
Cyclohexane	SF_6	294	4×10^{12}	69
	N_2O	294	2.4×10^{12}	69
	CCl_4	294	2.7×10^{12}	69
		293	4.3×10^{12}	88
	O_2	293	1.7×10^{11}	88
		296	2.3×10^{10}	37
	biphenyl	296	2.6×10^{12}	37
		293	3.3×10^{12}	88
Neopentane	N_2O	294	2.3×10^{12}	69
	O_2	296	1.2×10^{12}	70
		296	5×10^{11}	91
	CCl_4	296	2.9×10^{13}	70
	SF_6	294	2×10^{14}	69
Isooctane	SF_6	293	5.8×10^{13}	69
	N_2O	292	9.6×10^{12}	69
	C_2H_5Br	293	5.1×10^{12}	69
Tetramethylsilane	SF_6	294	2.1×10^{14}	69
	N_2O	294	7.5×10^{11}	69

[a] J. H. Baxendale, C. Bell and P. Wardman, J. Chem. Soc., Faraday Trans. I., **69** (1973) 776.

where $\bar{\alpha}$ is the mean polarizability of the scavenger molecule, ϵ is the dielectric constant of the solvent, e is the electronic charge and r is the electron–scavenger molecule distance. The data on the temperature-dependence of u_{el} and k_s plotted according to eqn. (25) gave a straight line from which the radii of the localized states in n-hexane and cyclohexane were estimated as $r_{el} \approx 13$ Å and $r_{el} \approx 17$ Å, respectively.

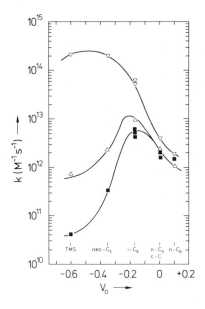

Fig. 18. Electron rate constants with three solutes at room temperature as a function of some nonpolar solvents with different V_0 values. $\triangledown\,SF_6$; $\blacksquare\,C_2H_5Br$, $\triangle\,N_2O$ (ref. 69).

Recently, Allen et al.[69] reported measurements on rate constants for electron reactions with solutes in several hydrocarbons. The values were correlated with the energy of the conducting state V_0. It was found that the rate constants exhibited maxima for particular values of V_0. Figure 18 shows three examples. The temperature-dependence of the rate constant exhibited negative or positive coefficients in different solvents dependent upon the value of V_0. The temperature-dependence was related to the temperature-dependence of V_0 which was found to increase with decreasing temperature[58]. It can then be seen from Fig. 19 that the temperature coefficient of the rate constant should be negative in solvents with V_0 values less than (more negative) the V_0 value corresponding to the maximum electron rate constant. Analogously, the temperature coefficient of the rate constant should be positive in solvents with V_0 values greater than the V_0 value corresponding to the maximum electron rate constant. This is what was observed. The reaction of electrons with N_2O has its fastest rate at $V_0(N_2O) = -0.2$ eV. In n-hexane ($V_0 = 0$) the temperature coefficient was positive, and in tetramethylsilane ($V = -0.6$ eV) it was negative. It is clear that much more exciting experimental data on electron mobility, rate constants, and energy levels can be expected in future investigations, which will lead to a more complete description of the physical state and the related chemical behaviour of excess electrons in non-polar liquids.

IV. Electron mobility in polar liquids

Investigations on the behavior of excess electrons in polar solvents ($\epsilon > 3$) have been carried out extensively for metal–ammonia solutions[5,6,12–14]. For all other

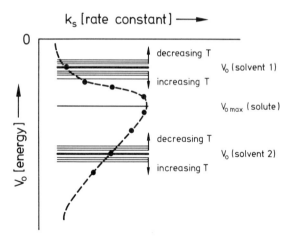

Fig. 19. The vertical bars represent energy levels for mobile electron reaction with a solute in different solvents with different V_0 values. Decreasing temperature leads to an increase in V_0 and vice versa. --- dependence of k_s on V_0.

liquids the information available is rather scarce. Mobility measurements were performed for only a few liquids and the data obtained so far are compiled in Table 6. Some relevant physical properties like the dielectric constant and the viscosity of the liquid are given but the mobility does not seem to correlate with either of these properties. In comparison with the electron mobility in hydrocarbons the values in polar liquids are in most cases much smaller and approach the mobilities for anions

TABLE 6

Electron mobility in polar liquids

Liquid	t (C)	ϵ	Viscosity $P \times 10^2$	$u_{el} \times 10^4$ $(cm^2V^{-1}s^{-1})$	Ref.
Isopropanol	20	19	2.4	5.1	44
n-Butanol	20	18	2.95	7.3/7.5	44, 45
Isobutanol	20	18	4.0	6.8	a
Isoamyl alcohol	20	15	0.40	9.7/9.5	44, 45
n-Hexanol	20	12		4.3	44
Ethylene glycol	20	39	19.9	2.8	44
Hexamethyl phosphoramide	20	30	3.6	5.5	43,45,46
Methylethyl ketone	20	21	0.43	15	45
Dimethoxyethane	−90	13	32.3	4	45
Monobutylamine	20	5	0.35	27	45
Tributylamine	20	5	1.4	20	45
Water	20	81	1	18	38
Ammonia	−60	22	0.37	100	a

[a]C. A. Kraus and E. C. Evers, Advan. Chem. Ser., **50** (1965) 1.

in these solvents (see Table 8). In liquids where the mobility of the electron and of O_2^- have been measured we notice that the electron mobility is 2–3 times greater than that of the oxygen anion. This could indicate that in these liquids the electronic motion is not purely ionic but that the electron, although strongly coupled to a cluster of solvent molecules leaves this trap to jump to another cluster of favorably oriented molecules in the neighborhood. Of special interest are the physical properties of the hydrated electron since many of its reactions have been studied[7]. Mobility values have been published by several authors[38–40] and the temperature-dependence of the mobility was obtained for the temperature range 25–50 °C (Fig. 20). At 25 °C the mobility is $\lambda_{el} = 180 \pm 5$ cm^2 Ω^{-1} and from the Nernst–Einstein relation a diffusion coefficient of $D_{el} = (4.8 \pm 0.13) \times 10^{-5}$ cm^{2}s^{-1} is calculated. The mean value of the temperature coefficient was estimated as $(2.1 \pm 0.2) \times 10^{-2}$ °C^{-1}. This temperature coefficient is comparable with that of other ions in water. Abundant information is available on the solvated electron in liquid ammonia[5,6,12–14]. Depending on the concentration of the metal roughly three regions can be distinguished: (1) dilute solutions, (2) transition range, (3) concentrated solutions.

In the dilute solutions ($C < 1$ mole % metal) which are characterized by their blue color, isolated solvated electrons and cations are formed. Studies of the conductance gave values for the equivalent conductance of these species. At -33 °C a limiting conductance of 1022 cm^2 Ω^{-1} for Na–NH$_3$ solutions was measured and the limiting conductance of NaCl–NH$_3$ solutions was found to be 309 cm^2 Ω^{-1}. Assuming the same ratio for the mobilities of Na$^+$ and Cl$^-$ in NH$_3$ as in water, the equivalent conductance of Na$^+$ in NH$_3$ was estimated as 124 cm^2 Ω^{-1} and, therefore, the mobility of the ammoniated electron was obtained[92] as 900 cm^2 Ω^{-1}. The mobility decreases with decreasing temperature and the product of viscosity and equivalent conductance stays constant; Walden's rule applies[93]. This strongly suggests that the motion of the solvated electron is of ionic nature[94]. Since the solvated electron resides in a cavity of ~ 3 Å diameter it can be thought of as an ordinary anion surrounded by a solvation shell. As the metal concentration is increased the equivalent conductance goes through a minimum and then increases at high metal

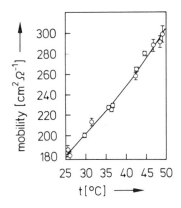

Fig. 20.   Temperature-dependence of the electron mobility in water (ref. 39).

concentration to values which are comparable to metallic conductivity. Models of the conductivity in concentrated solutions assume the existence of free electrons, the motion of which is governed by scattering by the metal ions[5,95].

Arnold and Patterson[96] proposed a model for the motion of electrons in the transition region (0.04 M–1 M) which assumes that in this concentration range, in addition to ionic transport, tunneling of electrons between metal atoms and metal ions takes place. For the calculation the system

$$Na^+ \cdots e^- \cdots Na^+$$

is considered and the wave functions of the hydrogen molecular ion are used.

The diffusion coefficient is obtained from the average jump distance $2r_s$ given by

$$\tfrac{4}{3}\pi r_s^3 = (N_0 + N_1)^{-1} \tag{27}$$

where N_0 and N_1 are concentrations of metal ions and atoms, respectively and from the jump frequency ν_T which is given by the exchange integral

$$\nu_T = \frac{2}{h} \int \psi_a \left(\frac{e^2}{\epsilon_{op} r_a} \right) \psi_b d\tau \tag{28}$$

where ϵ_{op} is the optical dielectric constant of ammonia. The diffusion coefficient is then

$$D = (2r_s)^2 \nu_T \tag{29}$$

From $u = De/k_B T$, the mobility u is calculated. It is found to increase exponentially with the third root of the concentration of metal ions and metal atoms or

$$u \propto \exp(-\text{const } r_s) \tag{30}$$

The total conductivity is the sum of ionic and tunneling contributions

$$\sigma_{tot} = \sigma_{ion} + \sigma_{tunn} \tag{31}$$

and at low metal concentrations $\sigma_{ion} \gg \sigma_{tunn}$. Above 0.1 M the tunneling contribution becomes predominant.

V. Anionic mobility

V.A Non-polar liquids

Excess electrons in non-polar liquids usually have a short lifetime due to impurities. From the rate constants for electron scavenging (Sect. III.B) one can estimate that excess electrons in n-hexane will have a lifetime of the order of microseconds if an electron scavenger with a concentration of 1 μM is present. Therefore, in many experiments designed to investigate negative charge carriers in non-polar liquids properties of anions were determined[15]. However, since in most cases the identity of the major electron scavenger was not known little can be said about the results. The uncertainty about the type and quantity of the impurity, for instance in n-hexane is reflected in Table 7 where mobility data for anions obtained by different authors

TABLE 7

Anionic mobilities in *n*-hexane*

t (°C)	$u \times 10^4$ $(cm^{-2}V^{-1}s^{-1})$	Ref.
rt	4.17	[a]
19	4.4	[b]
25	13	[c]
19	13	[d]
22	9.2	[e]
24	13	[f]
23	9.1	[g]
27	14	[h]
20	10	[i]
rt	20	[j]
20	1.5	[k]

*The identity of the anion is unknown.

[a] G. Jaffé, Ann. Phys. (Leipzig), **32** (1910) 148.

[b] I. Adamczewski, Ann. Phys., Ser. II, **8** (1937) 309.

[c] K. H. Reiss, Ann. Phys. (Leipzig), **28** (1937) 325.

[d] O. Gzowski and J. Terlecki, Acta Physica. Pol., **18** (1959) 191.

[e] O. Gzowski, Z. Phys. Chem., **221** (1962) 288.

[f] A. Hummel and A. O. Allen, J. Chem. Phys., **44** (1966) 3426.

[g] W. F. Schmidt, Z. Naturforsch. B, **23** (1968) 126.

[h] O. H. LeBlanc, J. Chem. Phys., **30** (1959) 1443.

[i] P. Chong and Y. Inuishi, Technol. Rep. Osaka Univ., **10** (1960) 545. Ref. 99.

[j] P. E. Secker and T. J. Lewis, Brit. J. Appl. Phys., **16** (1965) 1649.

[k] V. Essex and P. E. Secker, Brit. J. Appl. Phys., Ser. 2, **2** (1969) 1107.

over the decades are compiled. More data can be found in ref. 15. Investigations of the temperature dependence of anionic mobility in liquid alkanes showed positive temperature coefficients and some authors found Walden's rule $u \cdot \eta$ = const. to be fulfilled[97] while others did not[98] (η is viscosity). Measurements of the anionic drift mobility in *n*-hexane did not show any field strength-dependence up to 0.5 MV cm^{-1} [99]. Addition of a small percentage of ethyl alcohol to *n*-hexane lowered the anionic mobility from 1×10^{-3} cm^2 V^{-1} s^{-1} to 0.45×10^{-3} cm^2 V^{-1} s^{-1} [99]. Hummel[100] found that the addition of an electron scavenger (SF$_6$, N$_2$O, etc.) to cyclohexane lowered the mobility of the apparent anion by approximately a factor of two.

Very few models exist for the motion of anions or cations in dielectric liquids. In most cases the ion is considered as a hard sphere with radius r and Stokes' law is applied. Walden's rule ($u \cdot \eta =$ const.) is a result of this. The radii obtained for anions are in all cases greater than the hard core radius of the neutral molecule so that a solvation shell can be assumed. The thickness of this shell is estimated from the range of the ion-induced dipole forces[73,101–103].

V.B Polar liquids

The conductance of anions in aqueous solutions and other polar solvents has been studied for many decades and there exists an extensive literature. Short-lived anions can be observed with the pulse-radiolysis technique and in Table 8 some results are summarized. The anions were obtained by scavenging of radiation produced electrons or by injection of defined ions from specially prepared electrodes[104]. Macroscopic evidence of anion–solvent interaction is the induction of liquid motion when anionic charge carriers are injected into a liquid[105–107]. The ions move under the influence of an electric field, and through friction momentum is exchanged with the liquid. In tubes with axial ionic motion, liquid flow in the same direction is observed and a static pressure is built up which can be measured. This effect has been used to measure mobilities in liquid helium[108].

VI. Electron mobility in glasses

Experimental investigations on electronic transport in glassy matrices have been performed by a number of investigators. Measurements of transport properties like drift mobility and Hall mobility have been carried out so far on 3-methylpentane

TABLE 8

Mobility of anions obtained by electron scavenging or injection in polar solvents

Liquid	t (°C)	Anion	u_- (cm^{2}V^{-1}s^{-1})	Ref.
Isopropanol	20	O_2^-	3.3×10^{-4}	44
n-Butanol	20	O_2^-	2.6×10^{-4}	44
Isoamyl alcohol	20	O_2^-	3.9×10^{-4}	44
Hexamethyl phosphoramide	20	Anthracene$^-$	0.5×10^{-4}	43
		O_2^-	2.5×10^{-4}	
Ethanol	20	$C(NO_2)_3^-$	2.4×10^{-4}	a
		Cl^-	2.1×10^{-3}	104
Nitrobenzene	20	Cl^-	2×10^{-3}	104
Acetonitrile	20	Cl^-	2.7×10^{-3}	104
Water	22	O_2^-	6.7×10^{-4}	40

aS. A. Chaudhri and K.-D. Asmus, J. Chem. Soc. Faraday I, **68** (1972) 385.

(3MP), methyltetrahydrofuran and 10 M NaOH glassy ice. The drift mobility in 3-methylpentane was determined in the temperature range 4.2 K–77 K[47]. While the electron mobility of 0.02 cm^2 V^{-1} s^{-1} was found to be nearly independent of temperature from 4.2 K to 35 K it increased with temperature from 35 to 85 K up to a value of ≈ 0.1 cm^2 V^{-1} s^{-1}. The activation energy for this range was 0.01 eV. No dependence on the electric field strength was observed up to 40 kV cm^{-1}.

The number of electrons which took part in this drift was found to be only a fraction of the electrons originally injected into the matrix. They are probably electrons which move between shallow traps and reach the counter-electrode eventually. Another fraction will be trapped in deeper traps or become attached to impurities. Excess electrons generated by γ-irradiation are mainly localized in deep traps since their ESR signal and optical absorption can be observed for long times after their generation[109].

Although no drift experiments have been carried with 3-methylhexane (3MH) the temperature-dependence of the photocurrent in γ-irradiated samples showed a similar behavior as the drift mobility in 3MP[110]. It stayed constant between 4.2 K and 33 K and increased with temperature above this value. If the photocurrent is carried mainly by electrons this would reflect changes in the mobility.

Electron drift experiments in 2-methyltetrahydrofuran glass were reported by Huang and Kevan[111]. Excess electrons were generated by irradiation with γ-rays and photoexcited into a mobile state. As in 3MP and 3MH there was little change of the mobility between 4.2 and 36 K while above this temperature the mobility increased up to 5.5 $\times$ 10^{-2} cm^2 V^{-1} s^{-1} at 77 K. The activation energy for this range was estimated as 3.5 $\times$ 10^{-3} eV. Figure 21 shows the temperature-dependence of the drift mobility. More extensive investigations have been carried out on the transport properties of excess electrons in alkaline glassy ice by Kevan and co-workers[48-51]. They measured drift and Hall mobilities. The

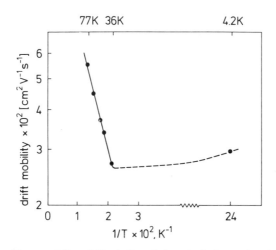

Fig. 21.   Electron drift mobility in 2-methyltetrahydrofuran glass (ref. 111).

experimental techniques have been described in Sect. II.A.4. The drift mobility data were obtained with samples which had been irradiated with ^{60}Co-γ-rays to about 0.2 Mrad. The electric field strength was varied from 300 V cm^{-1} to 20 kV cm^{-1} and measurements were carried out at 40, 77, 100, and 120 K. Figure 22 shows the experimental results. At low field strength the drift mobility was independent of the field and at higher values of the electric field strength an increase to a maximum occurred. At still higher fields the mobility decreased first with $E^{-0.62}$ and then with E^{-1} field-dependence; the higher the temperature the lower the field strength at which the mobility maximum was observed. Field-dependent mobilities are generally a sign of a perturbation of the thermal equilibrium. The electrons pick up energy from the field and a new velocity distribution results which can be correlated with a higher electron temperature. The temperature-dependence of the electron drift mobility at low field strengths showed an increase with decreasing temperature from 130–50 K. Below 50 K the drift mobility decreased with decreasing temperature whereas below 15 K a slight increase is observed again. At higher field strengths the temperature-dependence is less pronounced. The Hall mobility was measured to be (4.7 ± 1.9) cm^2 V^{-1} s^{-1} at 80 K. The low field drift mobility at the same temperature was 1.8 cm^2 V^{-1} s^{-1} which indicates that trapping effects did not influence the drift mobility to a large extent and that a microscopic mobility had been observed. The temperature-dependence of the drift mobility gave a dependence of $u \propto \exp(131/T)$ in the range from 80 to 133 K, i.e. the Hall mobility decreased with increasing temperature. The relatively large value of the electron mobility in alkaline ice and the dependence on temperature and electric field strength leads to the description of the motion by a band model (see Sect. VII).

A thermally activated mobility as has been observed in 3 MP and MTHF can be interpreted either by a trapping model, in which the electrons are thought to be localized in pre-existing physical traps and the electronic motion is treated as a thermal equilibrium between residence in traps and motion in a conduction band, or

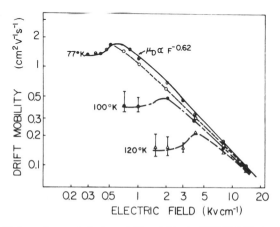

Fig. 22. Field-dependence of electron drift mobility in 10 M NaOH glassy ice (ref. 49).

by the theory of the polaron as conceived by Holstein and further developed by others[112–115], in which the electron strongly interacts with the phonons of the lattice and the motion occurs by uncorrelated phonon-assisted jumps (see Sect. VII).

VII. Models for electron transport

VII.A Non-polar liquids

An explanation of electron transport in condensed material involves three steps:

(1) the interaction of a single electron with a single atom or molecules has to be analysed;

(2) the influence of spatial order or disorder of atoms or molecules on the interaction must be taken into account and

(3) the statistics of the problem has to be solved which involves thermal equilibrium and the Boltzmann transport equation. (For more detail see Chapter 5.) Unfortunately, the complete program is difficult to work out and several simplifications or omissions are necessary in order to come to a useful interpretation of electron mobility data.

Most theoretical studies on the transport properties of excess electrons in liquids are extensions of models developed for the interpretation of electronic conduction in metals or semiconductors. However, in applying these models to non-polar liquids, certain differences in the physical properties have to be considered. First, non-polar liquids are insulators, primarily because of lack of charge carriers instead of a low mobility. Their molecules usually have ionization energies of around 10 eV. In the liquid state there is probably a reduction of this energy by 1–2 eV but excess charge carriers have to be introduced by some external agent. Second, the number of electrons which can be introduced is very small. In all experiments described electron densities seldom exceeded 10^{10} cm^{-3}. Electron–electron interactions, therefore, are small and can be neglected. Third, the effective range of the Coulomb force between an excess electron and a positive charge is determined by the Onsager length r_c of the medium defined as follows

$$r_c = \frac{e}{4\pi\epsilon\epsilon_0 k_B T} \tag{32}$$

where ϵ is the dielectric constant of the medium and $\epsilon_0 = 8.8 \times 10^{-12}$ F/m is the permitivity of free space. When r, the distance between the charges, is greater than r_c thermal motion is predominant and diffusion of positive and negative charge carriers in the bulk of the liquid occurs. Models for the physical state of excess electrons in non-polar liquids can be roughly divided into two groups:

(1) the electron is quantum-mechanically considered as an extended or delocalized state or it is treated as a free particle which interacts with the molecules of the liquid by collisions (also referred to as the quasi-free or mobile electron) and

(2) the electron is assumed to be localized either in pre-existing traps in the

liquid or by polarization interaction with the molecules of the liquid (these electrons are also referred to as solvated, trapped or localized electrons).

VII.A.1 Quasi-free electrons

The motion of quasi-free electrons in an electric field is characterized by a relatively high mobility, by a small temperature coefficient of the mobility and by a decrease of the drift mobility at higher field strength. All these properties have been observed, for instance, in liquid argon, xenon, krypton, methane, neopentane, and tetramethylsilane. A simple model which gives a classical description of the electron motion considers the electrons in thermal equilibrium with the molecules of the liquid. Due to the small electron mass m the mean velocity v_{el} is rather high. The electrons are in rapid random motion exchanging energy all the time by collisions. Application of an electric field leads to an additional velocity v_D parallel to the direction of the field. The mean free path between two subsequent collisions is Λ and the time $\tau = \Lambda / v_{el}$. The acceleration by the field is eE/m and the drift velocity is given by

$$v_D = \frac{eE}{m} \cdot \tau = \frac{eE}{m} \cdot \frac{\Lambda}{v_{el}} \tag{33}$$

At low electric field strengths

$$v_D \ll v_{el}$$

and v_{el} can be taken as the thermal velocity

$$v_{el} = v_{th} = \sqrt{2k_B T/m} \tag{34}$$

At higher field strengths the electrons can gain more energy from the field so that $v_D \ll v_{el}$ is no longer fulfilled. The energy gain between collisions is given by

$$\Delta T_f = (eE) \cdot (v_D) \cdot \tau = \frac{e^2 E^2}{m} \frac{\Lambda^2}{v_{el}^2} \tag{35}$$

The average energy loss in a collision is

$$\Delta T_c = T_{kin} \cdot f \tag{36}$$

with f the mean fractional energy loss ($f = 2m/M$ in the elastic limit, m, electron mass, M, mass of colliding particle). A stationary state is achieved when

$$\Delta T_c = \Delta T_f \tag{37}$$

or

$$\frac{e^2 E^2}{m} \cdot \frac{\Lambda^2}{v_{el}^2} = \frac{m}{2} v_{el}^2 \cdot f \tag{38}$$

that is,

$$v_{el}^4 = \frac{e^2 E^2}{m^2} \cdot \Lambda^2 \frac{2}{f} \tag{39}$$

or

$$v_{el} = \sqrt{\frac{eE}{m}} \Lambda \sqrt{2/f} \tag{40}$$

We see that in this case v_{el} increases with $\sqrt{E}$. The critical field E_c where the transition from $v_D \propto E$ to $v_D \propto \sqrt{E}$ occurs is obtained from eqns. (33), (34) and (40) as

$$E_c = \frac{v_{th}}{u} \cdot \sqrt{2/f} \quad \text{or} \quad \frac{v_D}{v_{th}} = \sqrt{f/2} \tag{41}$$

The critical field strength is inversely proportional to the mobility u and hence to Λ. The drift velocity at this field strength is $v_{th} \cdot \sqrt{2/f}$ independent of Λ.

Although such a classical model can explain qualitatively the characteristic features of electron motion in liquid methane a more detailed treatment requires the analysis of the scattering process. Schnyders et al.[116] were the first to give a theoretical description of the low field mobility in liquid argon. They considered elastic collisions only and evaluated the cross-section by the product of the energy-dependent cross-section for an isolated argon atom $Q(\epsilon)$ and the liquid structure factor $S(k)$ which is related to the radial distribution function by

$$S(k) = 1 + n \int (g(R) - 1) \, e^{ik \cdot R} \, d\mathbf{R}^3 \tag{42}$$

In the limit of thermal energies ($k = 0$) the structure factor is given by

$$S(0) = n \, k_B \, T X_{th} \tag{43}$$

where X_{th} is the isothermal compressibility of the liquid. The solution of a modified Boltzmann equation obtained in the treatment of electron motion in dilute gases was applied and yielded an equation for the drift mobility. Lekner[81] shortly afterwards treated the electron motion in liquid argon more rigorously by applying a more exact solution of the Boltzmann equation derived by Cohen and Lekner[80]. For the low field region where the drift velocity is proportional to the field strength he obtained an expression for the mobility

$$u = \frac{2}{3} \left(\frac{2}{\pi m k_B T} \right)^{1/2} \frac{e}{n \, 4\pi \, a^2 \, S(0)} \tag{44}$$

with m the electron mass, n the number density of argon atoms in the liquid and a the scattering length. Since $S(0)$ is proportional to $k_B T$ the mobility depends on the temperature as $T^{-3/2}$. The same dependence on T has been found for the mobility of electrons in non-polar crystals[117]. Davis et al. extended the Lekner theory to molecular liquids[118], where incoherent scattering has to be taken into account. Lekner's theory and the modification by Davis et al. predict that the mobility will become field-dependent for a critical field strength E_c which is given by

$$\tfrac{1}{3}(e \, E_c \, \Lambda)^2 \approx f(k_B T)^2 \, S(0) \tag{45}$$

From eqns. (43) and (44) it follows that

$$u \cdot E_c = \text{const} \tag{46}$$

which has been observed for the mobility in methane–ethane mixtures (Sect. III).

VII.A.2 Localized electrons

Low electron drift mobilities ($u < 1$ cm^2 V^{-1} s^{-1}) are usually explained by a localized electron model, where the electron is assumed to be immobilized by a trap or electron–solvent interaction for a time τ_t. Motion could occur either by tunneling to a neighboring trap or by thermal activation out of the trap into the extended state, where it spends a time τ_f before becoming localized again. The mobility of the localized electron in non-polar liquids is comparable to that of an ion since the electron moves with its "trap" in the field. This mobility is, however, still negligible compared to the mobility in the extended state. The apparent mobility u_{ap} for a model where the electron jumps from one trap to another has been derived by Frommhold[119] as

$$u_{ap} = \frac{u_f}{1 + v\tau_t} \tag{47}$$

where v is the collision frequency of trapping collisions and τ_t the mean time the electron remains in the trap. If the frequency of trapping collisions is large then

$$v\tau_t \gg 1 \quad \text{and} \quad u_{ap} \approx \frac{u_f}{v\tau_t} \tag{47a}$$

Since the release of an electron is assumed to be thermally activated the residence time τ_t depends on temperature as

$$\tau_t = \tau_0 e^{E_a/k_B T} \tag{48}$$

The value of τ_0 is approximately that of the period of vibrational oscillations about 10^{-13} s, E_a is the activation energy and it is related to the ground state energy of the trapped electron. From eqns. (48) and (47a) the apparent drift mobility is obtained

$$u_{ap} = \frac{u_f}{v\tau_0} \exp\left(-\frac{E_a}{k_B T}\right) \tag{49}$$

Minday et al.[23] used this model to explain the mobility data in liquid n-hexane and estimated a $\tau_t \approx 2 \times 10^{-10}$ s at $T = 300$ K. This model was also applied for the treatment of mobilities in mixtures of neopentane and n-hexane, where the activation energy of the mixture was assumed to be as given in eqn. (22) (p. 233). The mobility in the extended state u_f was estimated to be 150 cm^{2}V^{-1} s^{-1} for n-hexane and neopentane[71].

An experimental consequence of a trapping model is that the trap dominated drift mobility should increase with the electric field strength at higher values[74] in contrast to the quasi-free drift mobility which decreases. Without making any assumptions about the nature of the trap it can be shown that the electric field decreases τ_t by increasing the jump frequency. The mobility was found to depend on E and T as given by eqns. (17) and (18) (p. 231). This model has been applied to the mobility data of ethane and propane and Λ as a function of temperature was obtained[73,120].

A description of the electron trap in hydrocarbons can be based on the cavity model which Copeland et al. developed for polar liquids[121]. The electron is thought to be trapped in a square well potential and Davis et al.[122] estimated from the data of n-hexane that a potential well of -2.4 eV could be obtained from 6 hexane molecules. A cavity model for the localized electron in hydrocarbons was also proposed by Schiller[123]. The electron is assumed to reside in a physical cavity and the energy E_t of the system is made up from three contributions: the energy of the electron, the surface energy of the cavity and a negligible pressure-volume contribution. For the calculation of E_t a bubble of radius r was taken as a model of the cavity. The localization criterion $E_t < V_0$ can be satisfied even for values of $V_0 < 0$ for a bubble of several Ångströms radius. The probability for electron localization was calculated to be

$$P = (2\pi \, \sigma^2)^{-1/2} \int_{-\infty}^{V_0} \exp\left[-(E - E_t)^2/2\delta^2\right] dE \qquad (50)$$

where σ is an energy fluctuation parameter. Since in the trapping model the observed mobility u is related to the mobility in the extended state u_f by

$$u = u_f(1 - P) \qquad (50a)$$

the observed mobility should be proportional to $(1 - P)$. Schiller found this relation to be followed for several hydrocarbons.

VII.B Glasses

The electron drift mobility in 3-methylpentane and methyltetrahydrofuran glass show a similar temperature-dependence[111]. Above 35 K the mobility increases with temperature while it remains virtually constant between 35 K and 4.2 K. Temperature activated mobilities can be explained either by a trapping model or by a hopping model. In the trapping model the electron moves between pre-existing trapping sites (potential wells) in a conduction band. Promotion of electrons from the trapped state into the conductivity state occurs by thermal activation and the population of both levels is determined by the thermal equilibrium. The temperature-dependence of the mobility is described by an Arrhenius equation

$$u(T) = \alpha \, u_0 e^{-E_a/k_B T} \qquad (51)$$

where u_0 represents the mobility in the conducting state and $\alpha = N_c/N_t$; N_c denotes the density of states at the bottom of the conduction bond and N_t denotes the shallow trap density. No abrupt change in mobility is expected at lower temperatures.

In the hopping model[112–115] the electron is assumed to be strongly coupled with phonons of the lattice and the configuration of electron plus phonons is called a polaron. Motion can occur by uncorrelated phonon assisted lattice jumps. At lower temperatures the electron–phonon interaction becomes weaker and electronic exchange dominates. Polaron band motion occurs and Munn and Siebrand[113] have shown that there is little temperature dependence of the mobility in the transition region. Kevan[124] concludes that the data for 3MP and MTHF are best described by this hopping model.

The relatively large electron mobility in glassy ice can be rationalized by assuming a band type motion. Different scattering mechanisms can influence the motion of the electrons and thus the mobility. The main effects are

(i) scattering by the lattice or electron–phonon interaction;

(ii) scattering by defects, either neutral defects or ionized defects (Coulombic scattering) and

(iii) interaction of the electrons with each other.

This third possibility can be excluded in the present case since the electron concentrations are very small. The scattering process can be characterized by a relaxation time if the energy change of an electron in a collision is small compared to $k_B T$. In the Boltzmann transport equation approach this time is a measure of the time between two collisions and it is inversely proportional to the collision frequency. A mean free path can be defined which is related to the relaxation time by

$$\Lambda(\mathbf{k}) = \tau(\mathbf{k}) \, v\,(\mathbf{k}) \tag{52}$$

where v is the electron velocity and $\mathbf{k}$ the wave vector. Calculations of the relaxation times as a function of the wave vector are possible within the framework of the quantum theory of scattering processes. Scattering by acoustical phonons has been treated by Bardeen and Shockley[117] and the mean relaxation time was found to vary with temperature as

$$\tau_{ac} = \text{const } T^{-3/2} \tag{53}$$

Since the mobility is proportional to the relaxation time (eqn. (33)), u also decreases with increasing temperature

$$u_{el} \propto T^{-3/2}$$

For scattering by ionized centers Conwell and Weisskopf[125] obtained

$$\tau_{ion} = \text{const } T^{+3/2} \tag{54}$$

i.e. the electron mobility increases with increasing temperature. Scattering by neutral defects was calculated by Erginsoy[126] and the relaxation time was found to be independent of temperature and given by

$$\tau_{neutr} = \frac{\text{const}}{N_{neutr}} \tag{55}$$

where N_{neutr} is the concentration of scatterers.

These three scattering mechanisms are essentially elastic. Scattering on optical phonons is inelastic and the relaxation time approach is not possible. Different results for the mobility are obtained for polar and non-polar crystals. For temperatures higher than the Debye temperature ($T > \theta$) the mobility was found to vary as[127]

$$u \propto T^{-1/2}$$

while at $T < \theta$ the mobility varies with temperature as[128]

$$u \propto \exp\left(\frac{\theta}{T} - 1\right)$$

These dependencies hold for weak coupling between electrons and phonons. Stronger coupling leads to the formation of a polaron and the effective mass of the electron is increased. Low and Pines[129] derived for the mobility for $T < \theta$ the dependence

$$u \propto \exp\left(\frac{\theta}{T}\right)$$

These scattering mechanisms apply to electrons which are in thermal equilibrium with the lattice for which the drift mobility varies proportionally to the electric field strength. At higher field strength deviation from this proportionality occurs, the drift velocity increases less than linearly with field due to a deviation from the thermal equilibrium. The rate of energy gain from the field exceeds the rate of energy loss due to collisions. The effective electron temperature increases until, due to an increase of the loss rate, a stationary state is obtained again. This effect was first observed in n-Ge[66,67]. Scattering by acoustical phonons leads to a field-dependence of the mobility given by

$$u \propto E^{-1/2}$$

This mechanism is expected to be dominant at low temperatures $T < \theta$. At higher temperatures scattering by optical phonons becomes important and in non-polar materials.

$$u \propto 1/E$$

which means that the drift velocity is independent of the field strength. A detailed discussion of the effects at high electric field strength was published by Conwell[130].

The dependence of the electron drift mobility on temperature and electric field strength in alkaline glassy ice[49] could be explained with the assumption of different scattering mechanisms being active at different temperature and field strength intervals. The low field mobility is determined by scattering by acoustical and optical phonons. At higher field strength where the mobility is field-dependent, scattering by ionic species at lower temperatures and scattering by acoustical and optical phonons at higher temperatures occurs.

Most models proposed for electron transport in non-polar liquid and glasses have been developed originally for crystalline solids. Although the dependence of the drift mobility on electric field strength and temperature can be explained qualitatively, for a quantitative comparison much more information on the liquid state is necessary until a quantitative description of electron transport in liquid and glasses becomes possible.

Acknowledgements

The author wishes to thank his colleagues Dr. Bakale, U. Sowada and W. Tauchert for stimulating discussions. Support of the author's research project on electron transport in dielectric liquids by the Deutsche Forschungsgemeinschaft is gratefully acknowledged.

References

1 N. F. Mott and E. A. Davis, Electronic Processes in Non-crystalline Materials, Oxford University Press, London, 1971.

2 A. Hummel and W. F. Schmidt, Ionization of Dielectric Liquids by High Energy Radiation, Studied by Means of Electrical Conductivity Methods, HMI-Report B 117 (1971); available from Hahn-Meitner Institute, Berlin, Germany.

3 R. A. Muller, S. E. Derenzo, R. G. Smits, H. Zakland and L. W. Alvarez, UCRL Report 20 (1970) 135, available from Univ. California Radiation Lab, Berkeley, California.

4 S. E. Derenzo, R. Flagg, S. G. Louie, F. G. Mariam, T. S. Mast, A. J. Schwemin, R. G. Smits, H. Zaklad and L. W. Alvarez, CONF-72 0923–9 (1972); available from Univ. California Radiation Lab, Berkeley, California.

5 M. H. Cohen and J. C. Thompson, Advan. Phys., 17 (1968) 875.

6 R. F. Gould (Ed.), Solvated Electron, Advan. Chem. Ser. 50, Amer. Chem. Soc., Washington, D.C., 1965.

7 E. J. Hart and M. Anbar, The Hydrated Electron, Wiley-Interscience, New York, 1970.

8 J. Jortner, in G. Stein (Ed.), Radiation Chemistry of Aqueous Systems, Wiley-Interscience, New York, 1968, pp. 91–108.

9 S. A. Rice, Accounts Chem. Res., 1 (1968) 81.

10 J. Jortner, Actions Chim. Biol. Radiat. 14 (1970) 7.

11 J. Jortner, S. A. Rice and N. R. Kestner, in O. Sinanoglu (Ed.), Modern Quantum Chemistry, Part II: Interactions, Academic Press, New York, 1965, pp. 133–164.

12 G. Lepoutre and M. J. Sienko (Eds.), Metal–Ammonia Solutions, W. A. Benjamin Inc., New York, 1964.

13 J. J. Lagowski and M. J. Sienko (Eds.), Metal–Ammonia Solutions, Butterworths, London, 1969.

14 J. Jortner and N. R. Kestner (Eds.), Electrons in Fluids, Springer-Verlag, Berling, Heidelberg, New York, 1973.

15 I. Adamczewski, Ionization, Conductivity and Breakdown in Dielectric Liquids, Taylor and Francis, London, 1969.

16 L. Meyer and F. Reif, Phys. Rev., 110 (1958) 279.

17 F. Reif and L. Meyer, Phys. Rev., 119 (1960) 1164.

18 S. Cunsolo, Nuovo Cimento 21 (1961) 76.

19 L. Bruschi and M. Santini, Rev. Sci. Instrum., 41 (1970) 102.

20 R. M. Minday, L. D. Schmidt and H. T. Davis, J. Chem. Phys., 50 (1969) 1473.

21 W. F. Schmidt and A. O. Allen, J. Chem. Phys., 50 (1969) 5037.

22 P. H. Tewari and G. R. Freeman, J. Chem. Phys., 49 (1968) 4394.

23 R. M. Minday, L. D. Schmidt and H. T. Davis, J. Chem. Phys., 54 (1971) 3112.

24 R. M. Minday, L. D. Schmidt and H. T. Davis, J. Phys. Chem., 76 (1972) 442.

25 W. F. Schmidt and A. O. Allen, J. Chem. Phys., 52 (1970) 4788.

26 D. E. Hudson, USAEC Report MDDC-524 (1946); available from any U.S. AEC depository library.

27 W. F. Schmidt and G. Bakale, Chem. Phys. Lett., 17 (1972) 617.

28 G. Bakale and W. F. Schmidt, Z. Naturforsch. A, 28 (1973) 511.

29 E. E. Conrad and J. Silverman, J. Chem. Phys., 51 (1969) 450.

30 P. G. Fuochi and G. R. Freeman, J. Chem. Phys., 56 (1972) 2333.

31 J. P. Dodelet and G. R. Freeman, Can. J. Chem., **50** (1972) 2667.
32 M. Robinson and G. R. Freeman, Can. J. Chem., **52** (1974) 440.
33 E. C. Gregg and G. Bakale, Radiat. Res., **42** (1970) 13.
34 J. Warman, M. deHaas and A. Hummel, Chem. Phys. Lett., **22** (1973) 480.
35 S. S. Takada, N. E. Houser and R. C. Jarnagin, J. Chem. Phys., **54** (1971) 3195.
36 J. C. Devins and J. C. Wei, 4th International Conference on Conduction and Breakdown in Dielectric Liquids, Dublin, 1972 (unpublished).
37 G. Beck and J. K. Thomas, J. Chem. Phys., **57** (1972) 3649.
38 K. H. Schmidt and W. L. Buck, Science, **151** (1966) 70.
39 G. C. Barker, P. Fowles, D. C. Sammon and B. Stringer, Trans. Faraday Soc., **66** (1970) 1498.
40 G. Beck, Int. J. Radiat. Phys. Chem., **1** (1969) 361.
41 A. V. Vannikov, E. I. Maltsev and N. A. Bakh, Dokl. Akad. Nauk SSSR, **195** (1970) 1131.
42 A. V. Vannikov, E. I. Maltsev, Khim. Vys. Energ. **5** (1971) 174.
43 E. I. Maltsev and A. V. Vannikov, Khim. Vys. Energ. **5** (1971) 371.
44 A. V. Rudnev, A. V. Vannikov and N. A. Bakh, Khim. Vys. Energ., **6** (1972) 473.
45 A. V. Vannikov, E. I. Maltsev, V. I. Zolotarevski and A. V. Rudnev, Int. J. Radiat, Phys. Chem., **4** (1972) 135.
46 E. I. Maltsev and A. V. Vannikov, Khim. Vys. Energ. **7** (1973) 382.
47 Y. Maruyama and K. Funabashi, J. Chem. Phys., **56** (1972) 2342.
48 T. Huang, I. Eisele and L. Kevan, J. Phys. Chem., **76** (1972) 1509.
49 T. Huang, I. Eisele and L. Kevan, J. Chem. Phys., **59** (1973) 6334.
50 I. Eisele and L. Kevan, Rev. Sci. Instrum., **43** (1972) 189.
51 I. Eisele and L. Kevan, J. Chem. Phys., **55** (1971) 5407.
52 R. A. Holroyd and M. Allen, J. Chem. Phys., **54** (1971) 5014.
53 R. A. Holroyd, B. K. Dietrich and H. A. Schwarz, J. Phys. Chem., **76** (1972) 3794.
54 R. A. Holroyd and W. Tauchert, J. Chem. Phys., **60** (1974) 3715.
55 R. Schiller, Sz. Vass and J. Mandics, Int. J. Radiat. Phys. Chem., **5** (1973) 491.
56 R. A. Holroyd, J. Chem. Phys., **57** (1972) 3007.
57 G. Briegleb and J. Czekulla, Z. Elektrochem., **63** (1959) 6.
58 R. A. Holroyd and R. L. Russell, J. Phys. Chem., **78** (1974) 2128.
59 N. Zessoules, J. Brinkerhoff and A. Thomas, J. Appl. Phys., **34** (1963) 2010.
60 B. Halpern and R. Gomer, J. Chem. Phys., **51** (1969) 1031.
61 B. L. Henson, Phys. Rev. A, **135** (1964) 1002.
62 L. S. Miller, S. Howe and W. E. Spear, Phys. Rev. **166** (1968) 871.
63 D. W. Swan, Proc. Phys. Soc., London, **83** (1964) 659.
64 R. L. Williams, Can. J. Phys., **35** (1957) 134.
65 B. Halpern, J. Lekner, S. A. Rice and R. Gomer, Phys. Rev., **156** (1967) 351.
66 E. J. Ryder and W. Shockley, Phys. Rev., **81** (1951) 139.
67 W. Shockley, Bell Syst. Tech. J., **30** (1951) 990.
68 G. Bakale and W. F. Schmidt, Chem. Phys. Lett., **22** (1973) 164.
69 A. O. Allen, T. E. Gangwer and R. A. Holroyd, J. Phys. Chem., **79** (1975) 25.
70 A. O. Allen and R. A. Holroyd, J. Phys. Chem., **78** (1974) 796.
71 H. T. Davis, L. D. Schmidt and R. M. Minday, Chem. Phys. Lett., **13** (1972) 413.
72 K. Fueki, D.-F. Feng and L. Kevan, Chem. Phys. Lett. **13** (1972) 616.
73 W. F. Schmidt, G. Bakale and U. Sowada, J. Chem. Phys., **61** (1974) 5275.
74 B. G. Bagley, Solid State Commun., **8** (1970) 345.
75 R. M. Minday, L. D. Schmidt and H. T. Davis, Phys. Rev. Lett, **26** (1971) 360.
76 W. E. Schmidt, G. Bakale and W. Tauchert, Proc. 42nd Ann. Conf. on Electrical Insulation and Dielectric Phenomena, Verennes, Canada, Oct. 1973; also in 1973; Ann. Rept. Hahn-Meita u Institute, Berlin.
77 G. Bakale, W. Tauchert and W. F. Schmidt, J. Chem. Phys., (1975) in press.
78 S. Noda and L. Kevan, J. Chem. Phys., **61** (1974) 2467.
79 W. Tauchert and W. F. Schmidt, Z. Naturforsch. A., **29** (1974) 1526.
79(a) W. Tauchert and W. F. Schmidt, Proceedings of the 5th International Conference on Conduction

and Breakdown in Dielectric Liquids, Noordwijkerhout, July 28–31, 1975, p. 98, Delft University Press, 1975.

80 M. Cohen and J. Lekner, Phys. Rev., **158** (1967) 305.

81 J. Lekner, Phys. Rev., **158** (1967) 130.

82 B. E. Springett, J. Jortner and M. Cohen, J. Chem. Phys, **48** (1968) 2720.

83 R. Schiller, J. Chem. Phys., **57** (1972) 2222.

84 N. R. Kestner and J. Jortner, J. Chem, Phys., **59** (1973) 26.

85 G. Bakale,. E. C. Gregg and R. D. McCreary, J. Chem. Phys., **57** (1972) 4246.

86 G. Beck and J. K. Thomas, J. Chem. Phys., **60** (1974) 1705.

87 B. S. Yakovlev, I. A. Boriev and A. A. Balakin, Int. J. Radiat. Phys. Chem., **6** (1974) 23.

88 J. H. Baxendale, J. P. Keene and E. J. Rasburn, J. Chem. Soc. Farad. Trans. I, **70** (1974) 718.

89 J. H. Baxendale, C. Bell and P. Wardman, J. Chem. Soc., Farad. Trans. I, **69** (1973) 584.

90 J. H. Baxendale and E. J. Rasburn, J. Chem. Soc., Farad. Trans. I, **70** (1974) 705.

91 G. Bakale and W. F. Schmidt (1974), to be published.

92 D. S. Berns, in G. Lepoutre and M. J. Sienko (Eds.), Metal–Ammonia Solutions, W. A. Benjamin Inc., New York, 1964, p. 146.

93 R. R. Dewald, in J. J. Lagowski and M. J. Sienko (Eds.), Metal–Ammonia Solutions, Butterworths, London, 1969, p. 497.

94 D. S. Berns, G. Lepoutre, E. A. Bockelman and A. Patterson, Jr., J. Chem. Phys., **35** (1961) 1820.

95 J. C. Thompson, in R. F. Gould (Ed.), Solvated Electron, Advan. Chem. Ser. 50, Amer. Chem. Soc., Washington, D.C., 1965, p. 96.

96 E. Arnold and A. Patterson, in G. Lepoutre and M. J. Sienko (Eds.), Metal–Ammonia Solutions, W. A. Benjamin Inc., New York, 1964, p. 160.

97 O. Gzowski, Nature (London), **194** (1962) 173.

98 A. Hummel, A. O. Allen and F. H. Watson, J. Chem. Phys., **44** (1966) 3431.

99 P. Chong and Y. Inuishi, Technol. Rep. Osaka Univ., **10** (1960) 545.

100 A. Hummel, private communication.

101 P. E. Secker, J. Phys., D, **3** (1970) 1073.

102 M. R. Belmont and P. E. Secker, J. Phys, D, **4** (1971) 956.

103 K. de Groot, L. P. Gary and R. C. Jarnagin, J. Chem. Phys., **47** (1967) 3084.

104 P. Atten and J. P. Gosse, J. Chem. Phys., **51** (1969) 2804.

105 O. M. Stuetzer, J. Appl. Phys., **30** (1959) 984.

106 O. M. Stuetzer, J. Appl. Phys., **31** (1960) 136.

107 E. Gray and T. J. Lewis, Brit. J. Appl. Phys., Ser. 2, **2** (1969) 93.

108 B. S. Blaisse, J. M. Goldschwartz and P. C. Slagter, Cryogenics, **10** (1970) 163.

109 W. H. Hamill, in E. T. Kaiser and L. Kevan (Eds.), Radical Ions, New York, 1968, pp. 321–416.

110 T. Huang and L. Kevan, J. Amer. Chem. Soc., **95** (1973) 3122.

111 T. Huang and L. Kevan, J. Chem. Phys., **61** (1974) 4660.

112 T. Holstein, Ann. Phys. (New York) **8** (1959) 325.

113 R. W. Munn and W. Siebrand, J. Chem. Phys., **52** (1970) 6391.

114 A. L. Efros, Sov. Phys.—Solid State, **9** (1967) 901.

115 P. R. Emtage, Phys. Rev. B, **3** (1971) 2685.

116 H. Schnyders, S. A. Rice and L. Meyer, Phys. Rev., **150** (1966) 127.

117 J. Bardeen and W. Shockley, Phys. Rev., **80** (1950) 72.

118 H. T. Davis, L. D. Schmidt and R. M. Minday, Phys. Rev. A, **3** (1971) 1027.

119 L. Frommhold, Phys. Rev., **172** (1968) 118.

120 G. Bakale, U. Sowada and W. F. Schmidt, Proc. 43rd Ann. Conf. on Electrical Insulation and Dielectric Phenomena, Downingtown, Penn., USA, Oct., 1974; also in 1974 Ann. Rept. Hahn-Meitner Institute, Berlin.

121 D. A. Copeland, N. R. Kestner and J. Jortner, J. Chem. Phys., **53** (1970) 1189.

122 H. T. Davis, L. D. Schmidt and R. G. Brown, in J. Jortner and N. R. Kestner (Eds.), Electrons in Fluids, Springer-Verlag, Berlin, Heidelberg, New York, 1973, p. 393.

123 R. Schiller, J. Chem. Phys., **57** (1972) 2222.

124 L. Kevan, Int. J. Radiat. Phys. Chem., **6** (1974) 297.

125 E. Conwell and V. G. Weisskopf, Phys. Rev., **77** (1950) 388.

126 C. Erginsoy, Phys. Rev., **79** (1950) 1013.

127 F. J. Blatt, Physics of Electronic Conduction in Solids, McGraw-Hill, New York, 1968.

128 II. Fröhlich, Advan. Phys., **3** (1954) 325.

129 F. E. Low and D. Pines, Phys. Rev., **98** (1955) 414.

130 E. M. Conwell, High Field Transport in Semiconductors, Academic Press, New York, 1967.

Chapter 8

CALCULATIONS OF ANION–SOLVENT INTERACTIONS

P. SCHUSTER

I. Introduction

For a long time ion solvation has been studied extensively by both theoretical methods and experimental techniques. Nowadays, both kinds of approach have reached a high degree of accuracy and sophistication. Despite this enormous amount of information, however, no comprehensive and satisfactory theory of ion solvation can be given, due mainly to the lack of a detailed knowledge of the structure of non-simple liquids. The problems occurring in model calculations on ion solvation are two-fold. Firstly, accurate energy surfaces for ion–molecule complexes are difficult to obtain. They provide information about the most stable geometries and energies of interaction. The static energy of a complex, unfortunately, is only one side of the picture. The shape of potential surfaces for clusters consisting of an ion and many molecules, in general, indicates high flexibility, since several flat energy minima separated by low barriers are usually found. Therefore, the dynamics of the complex is the second important problem for most macroscopic properties at room temperature.

Solvation of monatomic ions has been investigated most frequently with high symmetry facilitating the construction of models. Besides monatomic ions, hydrated hydronium and hydroxide ions, $[H_3O \cdot (H_2O)_n]^+$ and $[HO \cdot (H_2O)_n]^-$, have been studied extensively, because their structures are of particular importance for any theory of autoionization or proton mobility in water. Usually, ion solvation is discussed in terms of solvation shells. We will define a solvation shell, in a very general way, as any structure surrounding the solute which differs significantly from the bulk pure solvent. In many cases different layers around the solute particle can be distinguished. In solvated ions we expect molecules in the first layer to be bound rather strongly. The symmetry of the solvated ion is thereby reflected to some extent by an ordered surrounding. The bulk of the liquid, on the other hand, has its own structure and there is no reason to believe that one ordered structure changes into the other continuously.

Frank and Wen[1] proposed a model, which distinguishes three regions around the solvated ion (Fig. 1), a more or less rigid, ordered inner solvation shell A, a region of disordered and highly mobile solvent molecules B and the bulk of the pure

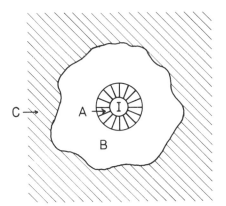

Fig. 1. Schematic model for ion solvation after Frank and Wen[1]. I represents an ion; A an ordered solvation shell; B, disordered mobile solvent, and C, bulk solvent.

solvent C. With some modifications this model has been generally accepted and seems to present an appropriate starting point for a detailed discussion of solvated monatomic ions in associated liquids[2]. Due to the lack of knowledge of molecular structures of non-simple liquids, we are not able to describe the structural details of region C, and because of the high mobility of solvent molecules in the disordered region B, any molecular description of this part is even more complicated. For the inner solvation shell A the molecules are bound much more strongly than in the bulk of the solvent, and we can assume a certain lifetime for individual equilibrium configurations. Consequently, almost all attempts to study ion solvation by molecular models are restricted to predictions of structures and properties of first solvation layers. Only a few examples discuss second or even higher layers of solvent molecules. The difficulties occurring in molecular models for regions A and C are overcome in most calculations of thermodynamic properties of ionic solutions, like solvation energies or free enthalpies, by assuming the classical laws of motion in a viscous medium and the theory of continuous dielectrics are applicable. The results of continuous models, like Born's equation for free solvation enthalpies, Einstein's derivation of partial molal viscosities, Debye's formula of rotational relaxation times, etc., are well known.

An alternative and more rigorous treatment of ion solvation is based on statistical mechanics and a priori seems to be much more appropriate. Although quite a number of papers have been published on this subject since the original derivation of the theory by McMillan and Mayer[3], progress is very slow in this field. Two more recent articles may be mentioned[4,5]. One of the most important results of many-body statistics is the pair distribution function of particles A and B, $g_{AB}(\mathbf{R})$, which represents the time averaged probability of finding a particle B at a certain distance, with a certain orientation relative to A. Furthermore, g_{AB} is an appropriate basis for comparison of theoretical and experimental results. In some cases a closely related type of correlation function, which describes the distribution of average distances between pairs of atoms in A and B, $g_{XY}(R_{XY})$, is more useful, because it can be observed in some experiments. During the last few years some progress has been made concerning the theory of many-body systems through the application of numerical techniques. These calculations begin from known or assumed two body potentials. Instead of solving the equations of statistical mechanics an ensemble of a few hundred particles is simulated in computer experiments. Two kinds of methods, molecular dynamics and Monte Carlo calculations, corresponding to constant energy or constant temperature, respectively, are generally used. Correlation functions and many other interesting properties of simple liquids, associated liquids, liquid mixtures and very recently also ionic solutions have been calculated by this approach. A brief and rather popular summary of computer experiment on liquids is recommended[6], a short general survey of solvation theories is given by Friedman[7].

During the last five years new experimental techniques have encouraged theoretical investigations. Progress in mass spectroscopy has made it possible to study ion–molecule interactions in the gas phase. Relative enthalpies and free enthalpies of complex formation for solvated ions $X(H_2O)_n$ with $n = 6$ being determined[8]. The measurement of molecular magnetic relaxation times has also made it possible to obtain some direct information on the molecular structure of ionic solutions[9]. Since most of the relevant literature on ion solvation has been treated in a recent review article[12], we shall not repeat it here.

Recent attempts to calculate structures and other properties of solvated ions may be classified into two categories: (1) calculations of energy surfaces for ion–molecule complexes and molecular associations and (2) studies of the dynamics of molecules in solvation shells. Clearly, all investigations of the latter type require the results of the first step. Therefore, we shall start our discussion on anion solvation with a brief summary of theoretical investigations of intermolecular interactions.

II. Calculation of intermolecular interactions

II.A Introduction

An extensive literature on the theory of intermolecular forces is available[12–16]. Therefore, we shall give here only a very short account of these methods as applied to ion–molecule complexes.

Even the most simple example of a molecular cluster, a complex formed from two molecules A and B leads to a very complicated energy surface of interaction

$$\Delta E_{AB} = E_{A\ldots B} - \{E_A^0 + E_B^0\} \tag{1}$$

For economic and didactic reasons, the problem of determining the energy surface and the equilibrium geometry is usually split into two sections. The energy surface between two molecules A and B at their frozen equilibrium geometries, here denoted by A_0 and B_0, is calculated. If very accurate results are desired, relaxation of molecular geometries in the complex ($A_0 \rightarrow A$ and $B_0 \rightarrow B$) is taken into account afterwards. In the first step, the resulting energy surface has six degrees of freedom (Fig. 2)

$$\Delta E_{A_0 B_0}(R) = \Delta E(R, \delta_A, \phi_A, \delta_B, \phi_B, \psi') \tag{1a}$$

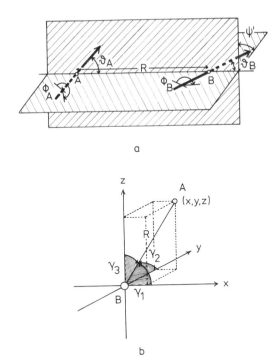

a

b

Fig. 2. Coordinate systems for the description of intermolecular interactions (a: general interaction between two molecules $R = (r, \theta_A, \phi_A, \theta_B, \phi_B, \psi')$; b: ion molecule interaction $R = (x, y, z)$ or $R = (R, \gamma_1, \gamma_2, \gamma_3)$, A represents the ion, B the molecule).

In general, one intermolecular distance and five angles are chosen as variables. The potential is simplified somewhat in the case of the interaction between two linear molecules, or between a monatomic ion and a molecule, where four or three degrees of freedom are found. In the latter case it is often an advantage to use another system of variables, which consists of the intermolecular distance R and the three Eulerian angles describing the position vector of the ion in the molecular coordinate system, $\mathbf{R} = (x, y, z)$, as shown in Fig. 2b.

$$\Delta E_{AB_0}(\mathbf{R}) = \Delta E(R, \gamma_1, \gamma_2, \gamma_3) \tag{1b}$$

$$\cos \gamma_1 = \frac{x}{R}, \cos \gamma_2 = \frac{y}{R}, \cos \gamma_3 = \frac{z}{R}$$

The three angles of course are not independent, since $\cos^2 \gamma_1 + \cos^2 \gamma_2 + \cos^2 \gamma_3 = 1$, but they are easy to obtain and more convenient for the formulae of intermolecular energies.

II.B Electrostatic models

For most purposes it is found to be appropriate to distinguish between simple and extended electrostatic models. In the simple approach molecules and ions are regarded as rigid and unpolarizable entities. The interaction between the particles is determined by their electrostatic properties. Each molecule or ion is characterized by its charge distribution, $Q(\mathbf{r})$, which in turn is determined by the charges Z_k of the N nuclei and the one particle density matrix $\rho(\mathbf{r})$ of the molecule.

$$Q(\mathbf{r}) = e_0 \left\{ \sum_{k=1}^{N} Z_k \delta(\mathbf{r} - \mathbf{R}_k) - \rho(\mathbf{r}) \right\} \tag{2}$$

$\rho(\mathbf{r})$ describes the spatial electron distribution and can be determined, in principle, by scattering techniques. In the quantum-mechanical approach $\rho(\mathbf{r})$ is obtained directly from the molecular n-electron wave function Ψ by integration

$$\rho(\mathbf{r}) = ne_0 \int \Psi^+ \Psi \, d\sigma_1 d\tau_2 d\tau_n \tag{3}$$

by σ_i we denote here the spin coordinate of electron i and by τ_i both spin and space coordinates, i.e. $d\tau_i = d\sigma_i \cdots dr_i$.

The electrostatic potential $V(\mathbf{R})$, resulting from a molecular charge distribution $Q(\mathbf{r})$ may be formulated without further difficulty, eqn. (4). In classical theories the potential is expanded as

$$V(\mathbf{R}) = \int \frac{Q(\mathbf{r})}{|\mathbf{r} - \mathbf{R}|} dr = \frac{q}{R} + \sum_k \mu_k \frac{\partial}{\partial k}\left(\frac{1}{R}\right) + \frac{1}{2!}\sum_k \sum_l Q_{kl}\frac{\partial}{\partial k}\frac{\partial}{\partial l}\left(\frac{1}{R}\right) +$$
$$+ \frac{1}{3!}\sum_k \sum_l \sum_m R_{klm}\frac{\partial}{\partial k}\frac{\partial}{\partial l}\frac{\partial}{\partial m}\left(\frac{1}{R}\right) + \cdots \tag{4}$$

$R = |\mathbf{R} - \mathbf{R}_0|$; $k = x, y$ or z, similarly for l and m (a Taylor series). $\mathbf{R}_0$ represents the origin of this expansion and q, μ, Q, R, etc. are multipole moments, the charge, the dipole moment, the second and higher moments, respectively. In most calculations the expansion is truncated after the second moment, which is usually given as the traceless quadrupole tensor (θ).

Starting from the electrostatic potential, a formally simple expression for the electrostatic energy of interaction between two particles A and B can be derived, eqn. (5). Actual calculations, nevertheless, require an appreciable effort, when A and B are

$$\Delta E_{COU}^{CL}(\mathbf{R}) = \int Q_B(\mathbf{R})V_A(\mathbf{R})d\mathbf{R} = \int Q_A(\mathbf{R})V_B(\mathbf{R})d\mathbf{R} =$$
$$= \int\int \rho_A(\mathbf{r}_1)\rho_B(\mathbf{r}_2)\frac{1}{|\mathbf{r}_1 - \mathbf{r}_2|}d\mathbf{r}_1 d\mathbf{r}_2 + \sum_{k \in A}\sum_{l \in B}\frac{Z_k Z_l}{|\mathbf{R}_k - \mathbf{R}_l|} -$$
$$- \sum_{k \in A}\int \rho_B(\mathbf{r}_1)\frac{Z_k}{|\mathbf{r}_1 - \mathbf{R}_k|}d\mathbf{r}_1 - \sum_{l \in B}\int \rho_A(\mathbf{r}_1)\frac{Z_l}{|\mathbf{r}_1 - \mathbf{R}_l|}d\mathbf{r}_1 \tag{5}$$

two polyatomic molecules. Multiple expansion for both particles A and B reduces these troubles considerably, since analytical expressions for individual terms can be given. In the case of a monatomic ion A interacting with a molecule B, the example we are mainly interested in here, one obtains

$$\Delta E_{COU}^{CL}(R) = -\frac{q_A}{R^2}\left\{\mu_x^B\cos\gamma_1 + \mu_y^B\cos\gamma_2 + \mu_z^B\cos\gamma_3\right\} -$$

$$-\frac{q_A}{3R^3}\left\{\theta_{xx}^B(3\cos^2\gamma_1 - 1) + \theta_{yy}^B(3\cos^2\gamma_2 - 1) +\right.$$

$$\left.+ \theta_{zz}^B(3\cos^2\gamma_3 - 1)\right\} + \cdots \tag{5a}$$

During the last few years the electrostatic potential has been applied successfully to give an approximate picture of associations between neutral molecules[17]. In the case of ion–molecule complexes, however, the simple electrostatic approach provides a very poor approximation. The ligands are strongly polarized in the field of the central ion (see Figs. 7 and 8). More elaborate models therefore include polarization energies (ΔE_{POL}^{CL}), which can be calculated from classical formulae. Again we obtain a fairly simple expression for the most important term in the polarization energy of interaction between a monatomic ion A and a molecule B

$$\Delta E_{POL}^{CL} = -\frac{q_A^2}{2R^4}(\alpha_{xx}^B\cos^2\gamma_1 + \alpha_{yy}^B\cos^2\gamma_2 + \alpha_{zz}^B\cos^2\gamma_3) + \cdots \tag{6}$$

In contrast to classical electrostatic and polarization energies the other contributions to the total energy of interaction can be obtained exclusively by a rigorous quantum-mechanical treatment. For practical purposes, however, empirical formulae are used in extended electrostatic theories.

The repulsive contribution to the energy of interaction between overlapping closed shell electron systems, usually called the exchange energy, is simulated by an exponential function or a high inverse power of the interatomic distance, eqns. (7a) and (7b). In the latter case

$$\Delta E_{REP} \sim A \cdot e^{-bR} \tag{7a}$$

$$\Delta F_{REP} \sim A \cdot R^{-m} \tag{7b}$$

$m = 12$ is most frequently used.

The expression for dispersion energies goes back to the original quantum-mechanical treatment by London[18] and by Slater and Kirkwood[19]. In actual calculation the approximations which use molecular polarizabilities and ionization energies as empirical parameters are taken, eqns. (8a) and (8b).

$$\Delta E_{DIS} = -\frac{3k}{2R^6} \cdot \frac{I_A \cdot I_B \cdot \alpha_A \cdot \alpha_B}{I_A + I_B} \tag{8a}$$

$$\Delta E_{DIS} = -\frac{3k'}{2R^6} \cdot \frac{\alpha_A \cdot \alpha_B}{(\alpha_A/n_A)^{1/2} + (\alpha_B/n_B)^{1/2}} \tag{8b}$$

Ionization potentials are denoted here by I_A and I_B, polarizabilities by α_A and α_B and the numbers of polarizable electrons by n_A and n_B, respectively. k and k' are conversion factors.

Three recent calculations based on extended electrostatic models may be mentioned[20-22]. The first two papers deal with ion–water complexes and will be discussed more extensively in Sect. IV. Morf and Simon[22] attempt to calculate the free enthalpies of solvation and stable coordination numbers in water, and are very successful in reproducing experimental data. The interaction of the central ion with the first layer of water molecules is calculated on a molecular basis, the interaction with the bulk liquid by Born's expression for the free enthalpies of ion solvation.

II.C Quantum-mechanical theory of intermolecular interaction

In the previous section we have seen that simple electrostatic theory is not sufficient for describing ion–molecule interactions. More elaborate extended electrostatic models, however, contain a number of empirical parameters, which have to be determined by comparison with experimental data. This deficiency degrades these models, to at best, excellent inter- or extra-polation procedures, which do not increase our knowledge of the basic principles of ion solvation. Therefore, there is substantial need for a priori investigations on ion–molecule complexes, as provided by an application of quantum mechanics to such structures.

Any quantum-mechanical treatment of intermolecular forces can be characterized as an attempt to solve the Schrödinger equation for the whole complex or supermolecule. In general, relativistic effects are neglected and the Born–Oppenheimer approximation is applied. The energy surface of the complex ABC . . . N, $E_{ABC...N'}$ represents the eigenvalue of the stationary Schrödinger equation for electron motion, which has to be calculated point by point. Two different kinds of methods are found to be appropriate, perturbation theory (PT) and molecular orbital methods (MO). Short introductions to the kind of approach and the notation used here have been presented recently[12,16]. For further details the reader is referred to some recent monographs and review articles[23-26].

II.C.1 The isolated subsystems of the complex

For all kinds of quantum-mechanical investigations on molecular associations we have to know the wave functions and energies of the isolated subsystems. At present fairly accurate ab initio calculations with electron correlation on molecules of a size up to approximately 20 electrons are within the limits of large scale computation. The agreement between calculated and experimental properties is satisfactory as far as a comparison is possible. In Table 1 calculated and experimental data for hydrogen fluoride are given as an example. Less accurate MO calculations at an ab initio level but without considering electron correlation, or using only small basis sets, show a number of systematic errors and can represent only an approximate description of molecular properties. Most importantly, for calculations on intermolecular forces, substantial errors in the calculated multipole moments and polarizabilities are introduced.

In other MO methods, CNDO, INDO, MINDO, usually called semi-empirical procedures, time-saving modifications are introduced in the calculation of the

TABLE 1

Calculated and experimental properties of the hydrogen fluoride molecule

Method[a]	Orbitals[b]	Basis set[c]	E_{tot}^{d} (au)	R_e^{e} (Å)	f^{f} mdyn Å^{-1}	μ^{g} (D)	Polarizability α_{zz}	$\alpha_{xx} = \alpha_{yy}$	$\overline{\alpha}$ (Å^3)	Ref.
CNDO/2	STO	–	–	1.00	19.1	1.86				25, 28
INDO	STO	–	–	1.01	18.6	1.98				25, 28
SCF	STO	$(2,1;1) \rightarrow [2,1;1]$	−99.479	exp.		0.88				29
SCF	STO	$(2,1;1) \rightarrow [2,1;1]$ optimized exponents, ζ	−99.536	0.922	14.4	1.44				29, 30
SCF	GTO	$(8,5,1;4,1) \rightarrow [5,4,1;3,1]$	−100.026	exp.		2.08	0.653	0.261	0.391	31
SCF	GTO	$(11,7,3;6,1) \rightarrow [7,4,3;4,1]$	−100.063	0.900	11.2	1.93	0.848	0.662	0.725	31
IEPA	GTO	$(11,7,3;6,1) \rightarrow [7,4,3;4,1]$	−100.344	0.927	9.3	1.78	0.950	0.824	0.865	31
CEPA	GTO	$(11,7,3;6,1) \rightarrow [7,4,3;4,1]$	−100.294	0.917	9.9	1.82	0.907	0.725	0.785	31
EXP			−100.527	0.917	9.7	1.82	0.960	0.720	0.800	29, 31

[a] SCF = ab initio LCAO–MO–SCF calculation, IEPA = Independent Electron Pair Approximation, CEPA = Coupled Electron Pair Approximation[24]. proximation[24,27].

[b] STO = Slater type orbitals, GTO = Gaussian type orbitals or appropriate combinations of Gaussian lobes.

[c] Throughout in this contribution the following short notation for basis sets will be used: (α, β, γ, δ/λ, μ, ν) indicates that α, s-type; β, p-type; γ, d-type; δ, f-type orbitals were applied for second row atoms and λ, s-type; μ, p-type and ν, d-type functions for H atoms. Contractions are shown in square brackets: [α', β', γ', δ'/λ', μ', ν'].

[d] Total energy, 1 au = 627.5 kcal mol^{-1} (2625 kJ mol^{-1}).

[e] Equilibrium distance, R_{HF}.

[f] Force constant.

[g] Dipole moment, $\mu = \mu_z$.

Hartree–Fock (HF) matrix elements. In order to reduce computational effort all two electron integrals besides those of the Coulomb type are omitted, eqn., (9).

$$\int \phi_a(1)\phi_b(2)\frac{1}{|\mathbf{r}_1-\mathbf{r}_2|}\phi_c(1)\phi_d(2)\,d\tau_1 d\tau_2 = (ac/bd) = \delta_{ac}\cdot\delta_{bd}\,(aa/bb) \tag{9}$$

Furthermore, overlap integrals S_{ab} are neglected, eqn. (10)

$$S_{ab} = \int \phi_a(1)\phi_b(1)\,d\tau_1 = \delta_{ab} \tag{10}$$

to be consistent with the approximation of eqn. (9). Several other approximations and empirical parameters have to be introduced in order to compensate for these simplifications. One assumption in the CNDO/2 and INDO methods will be important in the discussion. It is that the core attraction integrals are replaced by two electron Coulomb integrals[32], eqn. (11).

$$\int \phi_a(1)\phi_a(1)U_k d\tau_1 = -Z_k\int \phi_a(1)\phi_a(1)\frac{1}{|\mathbf{r}_1-\mathbf{R}_k|}d\tau_1 =$$
$$= -Z_k(aa/r_k^{-1}) \sim -Z_k(aa/bb) \tag{11}$$

b represents an orbital ϕ_b centered on atom "k"

Much larger molecules can be investigated by semi-empirical methods than by the accurate treatment but in exchange for these increased possibilities one incorporates a latent source of errors and methodical artefacts into the calculations. Further simplifications are introduced in the Extended Hückel Method (EHM)[33], where no attempt is made to calculate approximate HF matrix elements. The whole HF matrix is simulated by empirical parameters for the diagonal elements—in most cases valence state ionization potentials (VSIP) are used—and simple formulae for the off-diagonal elements (see ref. 34). EHM has been applied to a variety of problems, especially the structures and conformational equilibria of hydrocarbons[33], hydrogen bonds[35-37] and ion solvation studies[38,39]. We should remember however, that the risk of artefacts in EHM calculations is even larger than in the more common semi-empirical procedures. Some examples are mentioned here as a serious warning against an uncritical use of the method. In Extended Hückel calculations the benzene molecule is found to represent a saddle point rather than a minimum of the energy surface and to dissociate spontaneously into three acetylene molecules[40]; methyl fluoride shows no energy minimum and is unstable with respect to dissociation into CH_3^+ and F^-[41]; finally, in structures with hydrogen bonds wrong equilibrium positions for the H atoms involved in the bridges are obtained, as $H_2O-H^+ \cdots OH^-$ instead of $H_2O \cdots H-OH$[37,42].

II.C.2 Intermolecular perturbation theory

The idea of using perturbation theory for calculations on intermolecular forces is very convincing since the total energies (E_{tot}) of isolated molecules or ions are many orders of magnitude larger than the energies of interaction. The actual calculations, however, encounter so many difficulties, that only a few systems have been investigated by accurate perturbation methods. These include the systems $H_2^{+43,44}$, H_2^{43}, interactions between rare gas atoms[45], and a series of model calculations on a few electron fragments simulating intermolecular interactions in hydrogen bonded complexes[46]. No accurate studies applying perturbation methods to ion–molecule complexes have as yet been reported.

The major difficulty in intermolecular perturbation theory arises from the overlap between the wave functions of interacting particles. The stronger the systems overlap, the more terms there are which have to be included in the calculation[47] and the physical meaning of the individual terms becomes less clear. Nevertheless, perturbation theory has been very useful for two kinds of problems relating to ion–molecule interactions:

(i) the asymptotic behaviour of intermolecular energies is obtained correctly at sufficiently large distances and simple analytical formulae can be derived from quantum-mechanical perturbation theory, which become identical with the classical expressions eqns. (5), (6) and (8), in case of vanishing intermolecular overlap.

(ii) perturbation theory can be used to detect systematic errors in approximate methods like semi-empirical MO theories. We shall describe here the rather simple approach developed by Murrell et al.[48,49]. Let us regard a dimeric association

$$A + B \rightleftharpoons AB$$

The first-order contribution is usually split into two parts, which are obtained from two kinds of terms in the zeroth order wave function Ψ_{AB}^0 (12). Ψ_{AB}^0 itself is built up as a simple Hartree product of wave function of the isolated systems A and B in their

$$\Psi_{AB}^0 \sim \Psi_A^0 \cdot \Psi_B^0 - \sum_{\mu \in A} \sum_{\nu \in B} P_{\mu\nu} \{ \Psi_A^0 \cdot \Psi_B^0 \} \tag{12}$$

$P_{\mu\nu}$ exchanges the coordinates of electron μ and electron ν

ground states, Ψ_A^0 and Ψ_B^0, respectively, and a sum of terms accounting for single electron exchange between the molecules. This is certainly a serious restriction, valid only at large intermolecular distances[47]. According to their dependence on intermolecular overlap, the two first order terms can be distinguished*:

(i) ΔE_{COU}, the Coulomb energy is independent of the intermolecular overlap and converges to the classical electrostatic energy of interaction at large distances.

(ii) $\Delta E_{EX}^{(1)}$, the exchange energy represents the repulsion between two electron distributions and is roughly proportional to the square of orbital overlap, S_{ab}^2. The superscript indicates that only single electron exchange is considered.

First-order corrections to the zeroth order wave function give rise to higher order contributions to intermolecular energies. Usually, five second order terms are distinguished[48,50] three of which are especially important for the theory of intermolecular interactions:

---

*At finite intermolecular overlap this partitioning of the first order contribution is not strictly based. Other definitions for ΔE_{COU} and $\Delta E_{EX}^{(1)}$, which do not order terms according to their dependence on powers of intermolecular overlap, can be justified as well.

(i) the polarization energy, ΔE_{POL};

(ii) the dispersion energy, ΔE_{DIS}, and

(iii) the charge transfer energy, ΔE_{CHT}.

Usually the individual second-order contributions are characterized by different types of excitations into virtual orbitals of the subsystems A and B or A^+B^- and A^-B^+, respectively (see the corresponding figures in refs. 12 and 16).

ΔE_{POL} and ΔE_{DIS} do not depend explicitly on orbital overlap. At large intermolecular distances the expressions obtained from perturbation theory converge to the previously mentioned asymptotic formulae (eqns. (6) and (8a)). The polarization term, ΔE_{POL}, can be interpreted simply as the gain in energy due to rearrangement of the unperturbed electron distribution in the field of the second particle, ion or molecule. ΔE_{CHT}, is approximately proportional to the square of intermolecular orbital overlap, S_{ab}^2. It represents non-classical binding or the covalent contribution to intermolecular interaction and is caused by a net transfer of electron density from one particle to the other. Apart from energetic considerations, polarization and charge transfer can be discussed also by a careful analysis of the wave function of the complex AB. Again, partitioning of second-order energies of interaction into contributions like ΔE_{POL}, ΔE_{CHT}, etc., is not free from an arbitrariness in the case of non-vanishing intermolecular overlap.

From the previous discussion we have learned that intermolecular overlap plays a crucial role in perturbation theory. It limits the applicability of various approximations and makes a physical interpretation of individual contributions to the energy of interaction difficult or even impossible. Since overlap integrals decrease exponentially with interatomic distances a small change in R can cause a substantial change in intermolecular overlap (Fig. 3). Orbital overlap, intermolecular distances and energies of interaction of some typical complexes are compared in Table 2. We see that there is no direct correlation between these quantities. Among all ion–molecule complexes $F^- \ldots H_2O$ is exceptional with respect to orbital overlap, S_{ab}. The large numerical value of S_{ab} indicates the formation of a strong hydrogen bond, which evidently does not occur in $Cl^- \cdots H_2O$[16]. From a more empirical point of view this striking difference between the complexes $F^- \cdots H_2O$ and $Cl^- \cdots H_2O$ can be seen also in Fig. 11 where the overlap of the Van der Waals's spheres of the interacting particles at the equilibrium geometries is shown. We have to expect, therefore, that among all complexes of halide ions charge transfer effects are most important in the case of fluoride.

Partitioning of intermolecular energies by perturbation theory presents an excellent tool for testing the validity of various approximations. We shall restrict ourselves to a discussion of first-order contributions within the framework of CNDO/2 calculations. Although we shall apply perturbation theory at intermolecular distances where single electron exchange is no longer a good approximation, we shall obtain some insight into the nature of the semi-empirical approach, and the most important sources of error. In order to facilitate this approach only four electron problems are considered. Previously, we presented an analysis of the He–He and Li^+–He interaction[12,16,51]. Since we are interested exclusively in systems with non-zero electrostatic interaction, Li^+H^- will be chosen as the model system and we

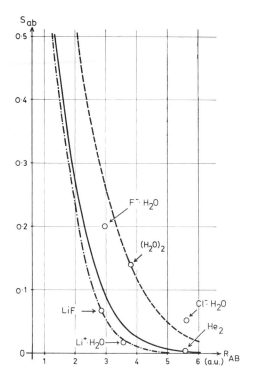

Fig. 3. Intermolecular overlap $S_{ab} = S_{ab}(R_{ab})$. Typical examples of intermolecular associations are shown as individual points ($S(1sHe, 1sHe)$: ——, $S(1sH, 2sO)$: ---- and $S(1sLi^+, 2sF^-)$: —·—·—·—·—).

TABLE 2

Overlap between valence orbitals for the two closest atoms A and B at some equilibrium geometries of molecular complexes

Complex	Atoms		R_{AB}		ΔE	S_{ab}^a
	A	B	(Å)	(au)	(kcal mol^{-1})	
He $\cdots$ He	He	He	2.96	5.59	-0.022	0.0030
He $\cdots$ HF	He	H	2.25	4.26	-0.151	0.0665
$H_2O \cdots$ HOH	O	H	2.02	3.82	-3.9	0.1097
$F^- \cdots$ HOH	F	H	1.57	2.96	-23.7	0.2045
$Cl^- \cdots$ HOH	Cl	H	2.97	5.61	-11.9	0.0554
$H_2O \cdots Li^+$	O	Li	1.89	3.57	-35.2	0.0165
$H_2O \cdots Na^+$	O	Na	2.25	4.25	-24.0	0.0061
$F^- \cdots Li^+$	F	Li	1.51	2.85	-150.0	0.0455

^{a}Overlap integrals calculated from Slater type orbitals $\phi_{ns}:\phi_{1s}$ at H, He, Li$^+$, ϕ_{2s} at O, F$^-$, Na$^+$ and ϕ_{3s} at Cl$^-$. Orbital exponents: $\zeta_H = 1.0$, $\zeta_{He} = 1.7$, $\zeta_{Li^+} = 2.7$, $\zeta_O = 2.275$, $\zeta_{F^-} = 2.425$, $\zeta_{Na^+} = 3.425$ and $\zeta_{Cl^-} = 1.916$.

shall consider its energy curve more closely here. The following expressions may be obtained for the first-order energies of interaction between two closed shell two electron systems[45,47], $|\phi_a\phi_a|$ and $|\phi_b\phi_b|$.

$$\Delta E_{COU} = 2\{2(aa/bb) - z_b(aa/r_b^{-1}) - z_a(bb/r_a^{-1})\} + \frac{z_a z_b}{|\mathbf{R}_a - \mathbf{R}_b|} \tag{13}$$

$$\Delta E_{EX}^{(1)} = -2\{(ab/ab) + S_{ab}[(aa/ab) + (ab/bb) - z_b(ab/r_b^{-1}) - z_a(ab/r_a^{-1})] - \\ - S_{ab}^2[3(aa/bb) - z_b(aa/r_b^{-1}) - z_a(bb/r_a^{-1})]\} \tag{14}$$

Definitions for the integrals are given in eqns. (9)–(11). Introducing the approximations of the CNDO/2 method into these expressions we obtain

$$\Delta E_{COU}^{CNDO/2} = 4 - 2(z_a + z_b)(aa/bb) + \frac{z_a \cdot z_b}{|\mathbf{R}_a - \mathbf{R}_b|} \tag{13a}$$

$$\Delta E_{EX}^{CNDO} = 0 \tag{14a}$$

A striking difference between ab initio and semi-empirical methods is that the exchange energy becomes zero owing to the "zero differential overlap" (ZDO) approximation (eqns. (9) and (10)). For a more detailed analysis with calculated numerical values for eqns. (13), (14) and (13a) at a number of interatomic distances R_{LiH} using simple $1s$ basis functions ϕ_a and ϕ_b are given (Fig. 4). A comparison of the energy curves obtained thereby, however, shows that the error introduced by the ZDO approximation is compensated to a large extent by a second assumption. Despite the pronounced differences in the curves for ΔE_{COU} and $\Delta E_{EX}^{(1)}$ the sum of both, the total first-order contribution to the energy of interaction, $\Delta E_{COU} + \Delta E_{EX}^{(1)}$ agrees fairly well with the CNDO/2 result

$$\Delta E_{COU} + \Delta E_{EX}^{(1)} \sim \Delta E_{COU}^{CNDO/2}$$

In fact, it is easy to trace which assumption is responsible for the error compensation in the CNDO/2 procedure, core attraction integrals in the diagonal elements of the HF operator have been replaced by negative electron–electron repulsion integrals, see eqn. (11). This modification was introduced originally[32] in order to improve calculated bond lengths in molecules which were rather unsatisfactory in the CNDO/1 treatment[52,53]. Since nuclear attraction integrals are always larger in absolute value than the corresponding two electron Coulomb integrals, eqn. (11) leads to a systematic underestimate of an attractive contribution or to a net repulsion between the two closed shell particles thus substituting the otherwise completely neglected exchange repulsion. Despite substantial errors in both contributions the total first-order energy is reproduced reasonably well by the CNDO/2 approximation in such species as $(He)_2$[16], Li^+He[12] or LiH (Fig. 4). There is no guarantee, however, that error compensation as discussed here, works equally well in other kinds of intermolecular complexes. Additionally, the introduction of empirical parameters for off-diagonal elements in the core-Hamiltonian, β_{AB}^0, provides another source of arbitrariness in semi-empirical calculations.

Figure 4 is also useful for another kind of comparison. The classical electrostatic

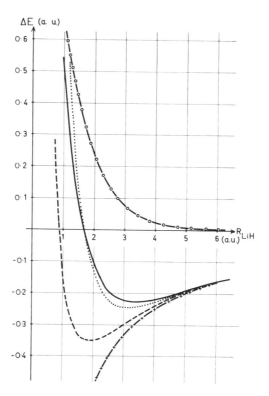

Fig. 4. Energy partitioning in LiH obtained by first-order perturbation theory (basis set: $1s(H^-)$, $\zeta = 0.7$ and $1s(Li^+)$, $\zeta = 2.7$; exact calculations: ΔE; ——, ΔE_{COU}: ------, $\Delta E_{EX}^{(1)}$: $-O-O-O$; $1/R$: $-\times-\times-\times-\times-$; CNDO/2 approximation: $\Delta E^{CNDO/2} = \Delta E_{COU}^{CNDO/2}$: · · · · · · · ·).

energy, ΔE_{COU}^{CL}, of LiH is described by a very simple analytic formula, $1/R_{LiH}$, and is shown together with the other curves. Comparison with ΔE_{COU} obtained by first-order perturbation theory demonstrates clearly that both curves start to deviate from each other around distances where orbital overlap and therefore exchange energy becomes non-negligible ($R_{LiH} \sim 5$ au $= 2.6$ Å).

II.C.3 Intermolecular MO theory

In principle intermolecular energies can be obtained in a straightforward way from MO calculations as differences in the total energies between the complex, $ABC \cdots N$, and the isolated subsystems

$$\Delta E(R) = E_{ABC\cdots N}(R) - \sum_{I=A,B,C\cdots}^{N} E_I \tag{15}$$

Actual calculations, however, are seriously restricted by the limits of numerical accuracy. Some years ago, MO calculations on ion–molecule complexes, accurate enough for a discussion of intermolecular forces were outside computational possibility. During the last decade more elaborate numerical methods have been developed and large, high speed computers have become available. Consequently, an impressive number of calculations on models for ion solvation have been performed. Although cations have been studied more extensively than anions, enough theoretical data is available for detailed hydration studies on F^-, Cl^- and OH^-.

Not unexpectedly, there are several sources of error in MO calculations, which we will summarize briefly. Let us begin with ab initio calculations on ion–molecule complexes. In this case we encounter two major sources of errors (1) neglect of electron correlation effects and (2) influence of incomplete basis sets. At present no extensive ab initio calculations including electron correlation are available for anion molecule complexes. Rough estimates for $F^- \cdots H_2O$ and $Cl^- \cdots H_2O$ at their equilibrium geometries[54] using Wigner's approximation[55] led to the conclusion that the total difference in correlation energies ΔE_{COR}, is expected to be very small, -1.4 and -0.7 kcal mol^{-1} respectively (see also Table 11). By analogy to other kinds of intermolecular associations as $(HF)_2$, where much more accurate data is available[31], we should guess that the small overall effect is due to two kinds of error compensation. Here we distinguish two types of correlation effects, intra- and intermolecular*. First of all the properties of the isolated subsystems, ions or molecules are changed to some extent by the inclusion of electron correlation. In molecules such as H_2O or HF the dipole moments decrease, and polarizabilities increase if correlation is taken into account (Table 1). Considering the asymptotic expressions for electrostatic and polarization energies, eqns. (5) and (6), we realize that the two most important intramolecular contributions of electron correlation have opposite signs and therefore compensate each other, at least in part.

The major intermolecular contribution of electron correlation to the energy of interaction can be characterized as dispersion energy and shows the expected asymptotic R^{-6} dependence as discussed previously, eqn. (8). Dispersion energy always increases the stability of molecular associations. In the case of ion–molecule interactions, however, ΔE_{DIS} is small in magnitude compared to the total energy of interaction and consequently can be neglected.

Basis set effects, on energies and other properties of ion–molecule complexes are much more difficult to discuss than correlation effects. Again we can distinguish intra- and intermolecular contributions to the energy of interaction. The former contribution results from errors due to an incorrect reproduction of molecular properties by incomplete basis sets. As we can see from calculated dipole moments of HF summarized in Table 1 the errors caused by limited basis sets can go in both directions. Only an analysis of the particular example under consideration can provide, therefore, a reliable estimate of the errors to be expected. A general type of intermolecular effect has been observed and analyzed in calculations on complexes between neutral molecules, using incomplete basis sets[16,56–58]. The basis

---

*A separation of correlation energies into intra- and intermolecular contributions is not always possible. Amongst others, electron pair methods like the independent electron pair approximation (IEPA)[24] or the coupled electron pair approximation[27] allow a partitioning of this kind.

orbitals of one subsystem improve the basis set of the other subsystem and vice versa resulting in an artificial stabilizing contribution to the energy of interaction. Similar mutual stabilization has to be expected also in calculations on ion–molecule complexes with incomplete basis sets, and was found, in fact, in a series of calculations on $Li^+ \cdots OH_2$[12,59]. In general, small basis sets overemphasize the stability of intermolecular associations.

A large number of semi-empirical calculations on ion–molecule complexes have been performed, mostly within the CNDO/2 formalism. On the whole, seven papers investigate model systems for anion solvation: $F^-(H_2O)_n$ and $Cl^-(H_2O)_n$[39,60–63] as well as $HO^-(H_2O)_n$[28,64]. Molecular geometries and energies of interaction obtained by CNDO/2 calculations on halide water complexes are compared with more reliable data in Table 3. Stable relative orientations (γ, see Fig. 11) are reproduced correctly in the case of $F^- \cdots H_2O$[63]. Absolute values for ΔE, however, are rather poor. A comparison of simple electrostatic, semi-empirical and ab initio results with the experimental data available is very disappointing for anybody who is enthusiastic about semi-empirical calculations, all values other than the CNDO/2 results agree fairly well (see also ref. 63). At the CNDO/2 minimum geometries, which deviate substantially from the ab initio results, with respect to bond lengths and intermolecular distances, the calculated energies of interaction are many times larger than the ab initio or experimental data. A remarkable improvement in CNDO/2 energies is achieved when H.F. equilibrium geometries are used as input data.

The main reason for the failure of the CNDO/2 method in predicting correct stabilities of anion–molecule complexes, stems from the fact that the parameters entering into semi-empirical calculations have been determined from neutral mole-

TABLE 3

Semi-empirical (CNDO/2), ab initio and experimental results on anion–water complexes (hydrogen bond geometry, $\alpha = 0°$).

Complex	Method of calculation	Geometry		ΔE^b (kcal mol^{-1})
		R_{OX} (Å)	R^a_{OH} (Å)	
$F^- \cdots H_2O$	CNDO/2[60,61]	2.22	1.12(1.03)	−79.8
	ab initio, SCF[70]	2.52	1.01(0.958)	−24.9
	ab initio, SCF[71]	2.55	1.00(0.950)	−22.6
	exp.[72]	–	–	−23.3
	CNDO/2[c]	2.52	1.01	−37.5
$Cl^- \cdots H_2O$	CNDO/2[60,61]	2.74	1.06(1.03)	−23.6
	ab initio, SCF[71]	3.31	0.964(0.950)	−11.4
	exp.[72]	–	–	−13.1

[a] Bond length of the HO bond involved in the hydrogen bridge. The corresponding bond length in the isolated water molecule is shown in parantheses.
[b] The experimental data correspond to ΔH at a standard state of $T = 298$ K and $p = 1$ atm.
[c] CNDO/2 calculation at the frozen SCF geometry.

cules. In fact, as we can see from the proton affinities of anions calculated by the CNDO procedure[28], relative stabilities of anions are systematically underestimated and all proton affinities are far too large in absolute values. Errors of about 240 kcal/mol^{-1} in CNDO/2 proton affinities of anions are not uncommon. Stabilization of anions by appropriate molecules with mobile hydrogen atoms like H_2O, HF, is half way to stabilization by protons and accordingly the energies of interaction obtained by CNDO calculations are again substantially too large in absolute values. This explanation is confirmed by the artificially large amount of electron density transferred from the anion to the ligand. In the following subsection we will analyze wave functions of the complex $F^- \cdots H_2O$ in some detail in order to improve our understanding of the difference between semi-empirical and ab initio calculations, and to obtain more insight into the nature of the anion–ligand bond.

This largely overemphasized charge transfer gives rise to another kind of artefact in semi-empirical calculations. In series with increasing numbers of ligands surrounding the central ion like in $F^-(H_2O)_n$, binding energies per ligand molecule decrease rather drastically with increasing n (Table 4). Again this CNDO result is not reproduced by accurate ab initio calculations or experimental data. It simply reflects the fact that charge transfer artificially introduced by the CNDO/2 procedure is not additive in complexes with varying numbers of ligands.

EHM calculations on $F^-(H_2O)_n$ and $Cl^-(H_2O)_n$ with coordination numbers $n = 4$, 6 and 8 have been performed[39]. The authors discuss preferred coordination numbers and orientations of H_2O molecules in the complex ions. They found local energy minima for these complexes at reasonable relative intermolecular distances.

TABLE 4

Semi-empirical, ab initio and experimental energies of complex formation for $X^-(H_2O)_n$

Complex	n	$\Delta E/n^a$ (kcal mol^{-1})			$\Delta H_{exp}^b/n^{72}$ (kcal mol^{-1})
		CNDO/2[39,60–63]	EHM[39]	ab initio[73]	
$F^-(H_2O)_n$	1	−79.8(2.22)		−22.8(2.54)	−23.3
	2	−59.2(2.28)		−21.8(2.55)	−20.0
	3			−20.4(2.58)	−17.9
	4	−41.0(2.36)	−15.2(2.34)	−18.4(2.62)	−16.8
	5			−16.8(2.67)	−16.1
	6	−31.6(2.43)	−13.8(2.38)		
	8	−24.9(2.52)	−20.9(2.58)		
$Cl^-(H_2O)_n$	1	−23.6(2.74)		−11.9(3.30)	−13.1
	2	−22.8(2.75)		−11.5(3.31)	−12.9
	3			−10.9(3.34)	−12.5
	4	−19.9(2.79)			−12.2
	6	−18.1(2.81)	−38.2(2.63)		
	8	−16.4(2.85)	−33.7(2.72)		

^{a}Intermolecular distances R_{OX} in Å are shown in parentheses.
^{b}Standard state $T = 298$ K, $p = 1$ atm.

Compared to more accurate data the absolute values for R_{XO} are substantially too short. The energies of complex formation, however, are rather unrealistic: H_2O molecules are bound more strongly to Cl^- than to F^-. Of course, it is a matter of taste which kind of failure we count more seriously, but in our opinion the Extended Hückel method is even less adequate in calculations on ion-molecule complexes than the semi-empirical procedures discussed above. Because of its empirical nature no systematic analysis of errors like those given previously for the CNDO/2 method is possible.

II.C.4 Analysis of molecular wave functions

Wave functions for one particular anion–molecule complex, $F^- \cdots H_2O$, are analyzed in detail. We have chosen this complex mainly for two reasons: (1) among all anion water complexes $F^- \cdots H_2O$ has been studied most frequently and very accurate calculations are available for this system and (2) from the overlap integrals shown in Table 2 we can expect charge transfer or covalent effects to be more important in this particular complex than in all other hydrated halide ions.

The major problem occurring in this kind of analysis is closely related to the concept of charge transfer between subsystems of a complex. Charge transfer between overlapping particles is not an observable quantity. Consequently, it depends always on the definitions applied, and the numbers obtained are not free from arbitrariness. Nevertheless, in chemistry, charge transfer has turned out to be a very fruitful heuristic principle. Therefore, we will use this concept here but define charge transfer in different ways. Comparison of the results obtained will show if a generalization is possible or otherwise.

Population analysis[65] is a very convenient and simple way to discuss changes in electron distributions and to calculate charge transfer in complexes. Starting from an atomic basis set (AO's) gross atomic charges can be calculated by Mulliken's procedure[65]. By summation over atoms in the individual subsystems and after adding nuclear charges: overall net charges for particles in the complex are obtained. Evidently, this procedure suggests a definition of charge transfer as the difference between net charges of the individual subsystems in the complex and in the isolated state. One source of arbitrariness in population analysis goes back to the definition of gross atomic charges, the overlap population is distributed in equal parts between the two atomic orbitals concerned, no matter the kind of shape of the overlap distribution.

A more sophisticated way to analyze the wave function of the complex starts from the one electron density function $\rho_{AB}(\mathbf{r})$ of the complex, which is obtained from the wave function by integration according to eqn. (3). ρ_{AB} gives an idea of the gross shape of the molecular system under consideration. One of the major problems encountered in using $\rho(\mathbf{r})$ for the purpose of demonstration or explanation concerns the general difficulty of how to plot a three-dimensional function. Most frequently only two-dimensional cuts are shown by means of contour lines. As an example such a cut in the molecular plane of $F^- \cdots H_2O$ is shown in Fig. 5. In systems, where a

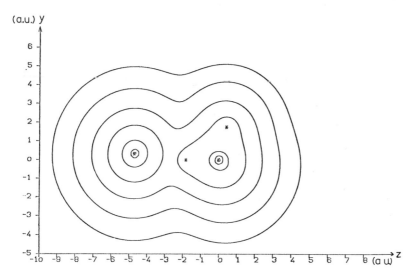

Fig. 5. Total electron density in F^-H_2O calculated in the plane of the H_2O molecule (y, z). The positions of the atoms are indicated by *. (Printed by permission of the authors G. H. F. Diercksen and W. P. Kraemer, Munich Molecular Program System, Reference Manual, Special Techn. Report, Max-Planck Institut für Physik und Astrophysik.)

principal axis can be defined*, it is appropriate to reduce the degrees of freedom by integration

$$\rho_{AB}^z(z) = \int\int_{-\infty}^{+\infty} \rho_{AB}(\mathbf{r}) \, dx dy \tag{16}$$

$\rho_{AB}^z(z)$ itself is not usually a very instructive function. The most interesting feature is provided by the sharp peaks at the nuclei of second row atoms. Examples are shown in Figs. 7 and 8.

For a more detailed investigation of the nature of the interactions between subsystems in a complex, density difference plots are much better suited than $\rho(\mathbf{r})$ itself eqn. (17)

$$\Delta\rho_{AB}(\mathbf{r}) = \rho_{AB}(\mathbf{r}) - \{\rho_A + \rho_B\} \tag{17}$$

ρ_A and ρ_B represent density functions of the isolated systems A and B, which are arranged at exactly the same geometry as in the complex. Again there is the difficulty in describing the whole three-dimensional function and therefore two-dimensional cuts are usually used for the purpose of comparison. As illustrative examples, cuts in the yz-plane $(x = 0)$ for both kinds of density functions of the complex $F^- \cdots H_2O$

---

*In linear molecules or in other structures with C_n axes $(n > 2)$ the principal axis is determined by symmetry. For other systems the axis is chosen appropriately as an axis of principal interest, e.g. along the chemical bond or along the direction of the acting intermolecular forces one wishes to study.

278

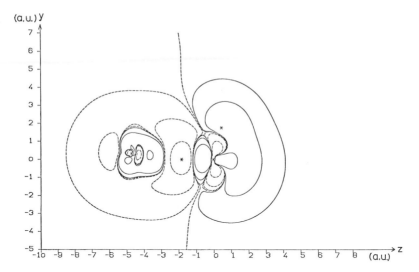

Fig. 6. Electron density difference in F^-H_2O calculated in the plane of the H_2O molecule (y, z). The positions of the atoms are indicated by * (increase in electron density: ———, decrease in electron density: – – – – –. (Printed by permission of the authors G. H. F. Diercksen and W. P. Kraemer, Munich Molecular Program System, Reference Manual, Special Techn. Report, Max-Planck Institut für Physik und Astrophysik.)

are shown in Figs. 5 and 6. The density difference plot (Fig. 6) demonstrates the decrease in electron density at the H-atom involved in the hydrogen bond $F^- \cdots HO$. At the second H-atom of the H_2O molecule, however, an opposite change in density is observed.

As described previously, a less complicated analysis of small complexes as those we are interested in here can be obtained from an integrated density difference function $\Delta\rho_{AB}^z$ shown in eqn. (18).

$$\Delta\rho_{AB}^z(z) = \int\int \Delta\rho_{AB}(\mathbf{r}) \, dx dy \qquad (18)$$

This function has previously been used to discuss polarization and charge transfer in cation molecule complexes [12,66]. Certainly, we lose information by integration over two dimensions, but the essential features of charge redistribution in the complex are reproduced correctly and can be observed easily in plots of $\Delta\rho_{AB}^z(z)$ against z.

The connecting line between F and O represents the only appropriate choice of a z-axis in $F^- \cdots H_2O$. For another kind of intermolecular association, the dimer of water, $(H_2O)_2$, is mentioned for comparison, again the direction of the hydrogen bond coinciding with the OO connection line represents an appropriate choice for the principal axis. In Figs. 7 and 8 integrated density difference plots $\Delta\rho_{AB}^z(z)$ are shown for $F^- \cdots H_2O$ and $(H_2O)_2$. Both density and density difference curves were calculated from accurate functions obtained by ab initio SCF calculations using extended GTO basis sets.

Comparing $\Delta\rho_{AB}^z(z)$ in Figs. 7 and 8 we see a number of interesting, although ex-

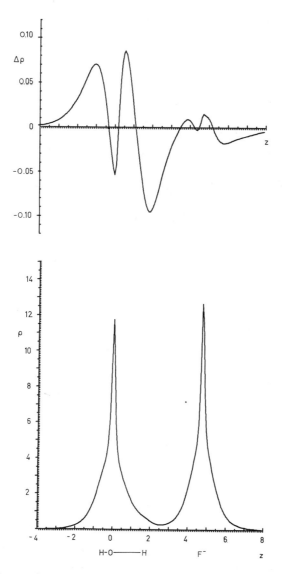

Fig. 7. Integrated electron densities (ρ) and integrated electron density differences ($\Delta\rho$) in the system F^-H_2O (ρ and $\Delta\rho$ in units of $-e_0$; z-axis in au; basis set see Table 5).

pected results. In the field of the F^- anion the H_2O molecule (B) is polarized much more strongly than in the field of a second H_2O molecule (A). Both polarization patterns are very similar in the region of the H_2O molecule (B) and differ only in amplitudes. The shape of the polarization curve itself, showing several maxima and minima, seems to be determined mainly by the internal structure of the molecule and only to a smaller extent by the polarizing external field. Both complexes show a region of reduced electron density between the interacting particles. With respect to ampli-

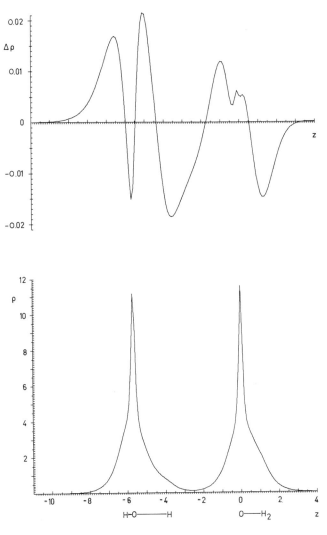

Fig. 8. Integrated electron densities (ρ) and integrated electron density differences ($\Delta\rho$) in the system $(H_2O)_2$ (ρ and $\Delta\rho$ in units of $-e_0$; z-axis in au; basis set and wave function see ref. 69).

tudes the polarization of F^- is not very different from the polarization of H_2O molecule A in $(H_2O)_2$. This result is not unexpected, if we remember that both subsystems F^- and H_2O (A) are polarized in roughly the same field resulting from an H_2O molecule (B) at exactly the same orientation.

Integral density difference curves can be used easily to calculate charge transfer (Δq) if a borderline separating the subsytems A and B has been defined eqn. (19). According to our previous integration this borderline ($z = z_0$) corresponds to a plane separating the interacting particles

$$\Delta q = \int_{z_0}^{\infty} \Delta\rho_{AB}^z(z)\,\mathrm{d}z = -\int_{-\infty}^{z_0} \Delta\rho_{AB}^z(z)\,\mathrm{d}z \tag{19}$$

Two definitions of z_0 will be applied here eqns. (20a) and (20b).

In eqn. (20a) the borderline between both subsystems A and B is drawn at that point where $\Delta\rho^z$ vanishes.

$$\Delta\rho_{AB}^z(z_0) = 0 \tag{20a}$$

and in eqn. (20b) the borderline is drawn at the point of least integral electron density between the subsystems, i.e. the point, where $\rho_{AB}^z(z)$ passes through a minimum

$$\rho_{AB}^z(z_0) = \min \;\to\; \left(\frac{\partial\rho_{AB}^z(z)}{\partial z}\right)_{z_0} = 0 \tag{20b}$$

These two definitions for z_0 lead to somewhat different numerical results concerning the amount of charge transferred (Table 5). Relative values, however, were found to be reliable in a series of different complexes[12]. The values for charge transfer calculated from a population analysis are somewhat larger, and also differ systematically if different kinds of complexes, e.g. if those with and without hydrogen bonds are compared.

We can conclude from all the kinds of analyses applied here, that in both complexes $F^- \cdots H_2O$ and $(H_2O)_2$ charge transfer is small in absolute value. From our previous considerations we expected charge transfer to be most important in $F^- \ldots H_2O$ compared to other halide water complexes yet even in this case it is smaller than $1/20$ of an electron.

TABLE 5

Charge transfer in molecular complexes obtained by different methods

Complex		Method of calculation	Geometry	Charge transfer, $\Delta q(e_0)$		
				Integration of $\Delta\rho^z$		
Electron donor	Electron acceptor			z_0: $\Delta\rho^z = 0$ eqn. 20a	z_0: $\rho^z = $ min. eqn. 20b	Population analysis[65]
H_2O	Li^+	ab initio, SCF[12,66]	SCF min.	0.0027	0.0040	0.0328
H_2O	HOH	ab initio, SCF[a]	SCF min.	0.0032	0.0079	0.011
F^-	HOH	ab initio, SCF[b]	SCF min.	0.0190	0.0378	0.0542
			CNDO min.	0.314	0.333	0.344
F^-	HOH	CNDO/2	SCF min.	0.102	0.102	0.113

[a] The wave function was taken from ref. 69.
[b] A wave function of similar quality to that used by Diercksen and Kraemer[70] was employed in this calculation. Basis set GTO: $(9,5,1/4,1) \to [6,4,1/3,1]$; $\Delta E(\text{SCF min.}) = -23.1$ kcal mol^{-1}.

282

Finally, we will use this analysis of wave functions to investigate the failure of semi-empirical methods, like CNDO/2, in the description of anion–molecule complexes. Integrated density and density difference curves calculated from CNDO/2 wave functions* of $F^- \cdots H_2O$ at the CNDO minimum and H.F. equilibrium geometries are shown in Fig. 9. Differences in $\rho^z(z)$ between CNDO and ab initio calculations mainly' result from the neglect of K-shell electrons in the semi-empirical method. At the CNDO minimum geometry, polarization of H_2O molecule as described by $\Delta\rho^z(z)$ shows some similarities with ab initio results. The amplitudes, however, are largely overemphasized, certainly as a result of the substantially too short inter-molecular distance R_{OF}. An analogous calculation starting from the CNDO wave

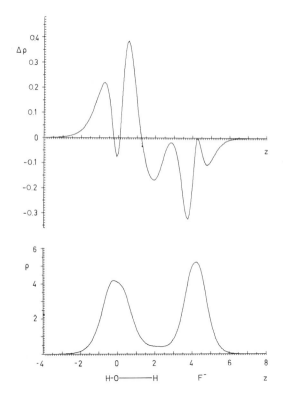

Fig. 9. Integrated CNDO/2 electron densities (ρ) and electron density differences ($\Delta\rho$) in the system F^-H_2O (ρ and $\Delta\rho$ in units of $-e_0$; z-axis in au; CNDO/2 wave function deorthogonalized by an inverse Löwdin[67,68] transformation before integration).

---

* For this purpose the CNDO basis functions, which are assumed to be orthogonal are transformed into a minimal STO basis set by an inverse Löwdin[67] orthogonalization procedure (see also ref. 66,68).

function for the H.F. equilibrium geometry leads to a similar polarization pattern as found in the ab initio calculation, with reasonable amplitudes in the outer region of the H_2O molecule. Both CNDO wave functions, however, provide a very poor description of the regions around F^- and between both subsystems. Instead of polarization of the anion a large amount of charge is transferred from F^- to H_2O and the CNDO curve, $\Delta\rho^z$, differs largely from the corresponding ab initio function. The numerical values shown in Table 5 reflect this serious error of the semi-empirical method. Again, our conclusion is not affected by the particular definition of charge transfer applied. In the case of other halide–water complexes these artefacts are less drastic. Nevertheless, unreasonably high values for charge transfer have been found in all complexes investigated to the present[60,61]. We conclude this section therefore with a serious warning against the uncritical use of semi-empirical methods, especially in fields where no careful tests of reliability have been made.

III. Partitioning of intermolecular energies

III.A Introduction

Total energies of interaction are obtained directly as differences between molecular energies within the framework of MO theory. Any partitioning of intermolecular energies, therefore, is subject to a certain set of model assumptions. Here we will discuss two cases for which different schemes of partitioning are found to be appropriate.

III.B Energy partitioning in dimeric complexes

Partitioning of MO energies in 1:1 complexes has been undertaken mainly in order to study the nature of the bond between ions and ligands, or between molecules. Furthermore, individual contributions provide a much more detailed and sensitive basis for comparison of MO results with data from other methods, like empirical electrostatic models, than total energies of interaction alone. A scheme for partitioning of intermolecular energies at the HF level, which is fairly simple and tries to come as close as possible to the perturbational approach discussed above, has been developed by several groups[76–78]. The most detailed description is that of Morokuma[78] and we will use his definitions here*.

Morokuma's analysis[78] starts from antisymmetrized wave functions of the isolated subsystems, Ψ_A^0 and Ψ_B^0, as well as from the complete Hamiltonian $\hat{H}$ of the complex AB. Four different approximate wave functions are used to describe the whole system.

---

*Another kind of energy partitioning in MO theory has been proposed by Clementi[79]. This type of partitioning called "bond energy analysis" (BEA) starts out from one center, two center and many center contributions and is found to be very useful in ion–molecule complexes[71,73,74]. Since the individual contributions of BEA are not directly related to quantities we discussed in intermolecular perturbation theory, we shall not use Clementi's technique here.

These are (1) A simple Hartree product of the two wavefunctions for the sub-systems, eqn. (21)

$$\Psi_1 = \Psi_A^0 \cdot \Psi_B^0 \tag{21}$$

(2) An antisymmetrized Hartree–Fock product of the two wave functions, eqn. (22)

$$\Psi_2 = A\{\Psi_A^0 \cdot \Psi_B^0\} \tag{22}$$

A represents an operator which takes care of antisymmetry of Ψ_2.
(3) A product of two wave functions, each one optimized in the electrostatic field of the other molecule, eqn. (23)

$$\Psi_3 = \Psi_A \cdot \Psi_B \tag{23}$$

Again Ψ_A and Ψ_B are antisymmetric and
(4) The ordinary, antisymmetric SCF wave function of the complex as obtained by the variational procedure, eqn. (24)

$$\Psi_4 = \Psi_{AB} \tag{24}$$

For all four wave functions expectation values of the energy are calculated and individual contributions to the total energy of interaction are obtained as differences

$$\Delta E_i = E_i - (E_A^0 + E_B^0); \quad E_i = \int \Psi^+ \hat{H} \Psi_i \tau; \; i = 1, 2, 3 \text{ or } 4 \tag{25}$$

Comparison with the definitions applied in intermolecular perturbation theory[48,49] shows that ΔE_1 is equivalent to the electrostatic contribution, ΔE_{COU}, defined in intermolecular perturbation theory, and becomes identical with it, when the exact solutions of the Schrödinger equations for the isolated subsystems are used.

$$\Delta E_1 = \Delta E_{COU} \tag{26}$$

ΔE_2 is the analogue to the complete first-order energy of interaction in perturbation theory including the exchange contribution. Accordingly, we define an exchange energy within the framework of MO calculations as

$$\Delta E_{EX} = \Delta E_2 - \Delta E_1 \tag{27}$$

ΔE_{EX} in eqn. (27), however, differs somewhat from the exchange contribution in Murrell's perturbational approach[48,49], since exchange is not restricted to single electrons and higher order terms in orbital overlap are included in the MO treatment.

In the trial wave function Ψ_3 the one electron MO's are optimized in an average field consisting of both the ordinary HF potential of their own molecule or ion and the electrostatic potential of the second subsystem in the complex. No exchange of electrons between the subsystems is allowed. Physically, we might interpret this contribution, ΔE_3, as a sum of electrostatic and polarization energies*

---

* Morokuma[78] originally defined ΔE_3 as a sum of Coulomb, polarization and dispersion energy. With Ψ_A and Ψ_B as single Slater determinants no correlation effects are taken into account, and hence $\Delta E_3 - \Delta E_1$ actually represents the contribution of electron polarization alone.

$$\Delta E_{POL} = \Delta E_3 - \Delta E_1 \tag{28}$$

Again, the major differences between ΔE_{POL} here and in perturbation theory consist of higher order contributions included in the MO treatment.

Finally, the remaining difference to the total SCF energy of interaction, ΔF_4, is attributed to complex stabilization as a consequence of electron delocalization or charge transfer between the subsystems.

$$\Delta E_{CHT} = \Delta E_4 - (\Delta E_{COU} + \Delta E_{EX} + \Delta E_{POL}) = \Delta E_4 + \Delta E_1 - (\Delta E_2 + \Delta E_3) \tag{29}$$

A number of higher order terms and some other second-order exchange terms[48] enter into Morokuma's charge transfer contribution. In fact, ΔE_{CHT}, defined according to eqn. (29) is a rather complex quantity. There is no doubt that it contains the stabilizing effect of charge transfer between subsystems in the complex. In actual calculations, however, pure charge transfer energy, as defined in intermolecular perturbation theory[48], is obscured somewhat by superposition with other contributions. It can happen therefore that ΔE_{CHT} becomes positive, that is repulsive, in complexes where charge transfer effects are not important (see $Li^+ \cdots OH_2$ in Table 6). Some authors[76,77] avoid these difficulties and the calculation of ΔE_3. They discuss only the sum of ΔE_{POL} and ΔE_{CHT}, which is called the delocalization energy, ΔE_{DEL}

$$\Delta E_{DEL} = \Delta E_{POL} + \Delta E_{CHT} = \Delta E_4 - \Delta E_2 \tag{30}$$

Basis effects on this kind of energy partitioning have been studied in some detail[12,59,80] and can be explained by simple semi-classical arguments[12,59], that is, as the result of basis set effects on the calculated properties of isolated subsystems. In general, extended basis sets including polarization functions have to be used if reliable values for the individual contributions are desired.

We have calculated individual contributions to the total energies of interaction according to eqns. (21)–(30)[59] for two selected ion–molecule complexes at their equilibrium geometries. We used extended basis sets similar in quality to those previously applied in the most accurate calculations available for these structures, $Li^+ \cdots OH_2$[74] and $F^- \cdots H_2O$[70,71]. The results obtained are summarized in Table 6. Additionally, energy partitioning in $(H_2O)_2$ obtained from STO–4G[75] calculations is shown[59,78]. The latter results, of course, are not as reliable as the values for $Li^+ \cdots OH_2$ and $F^- \cdots H_2O$, since they contain substantial errors due to the small basis set. Nevertheless, the differences in relative importance of individual contributions in these three complexes are large and a comparison of the numbers is possible without a serious loss in reliability. As expected, the Coulomb and polarization energies in $F^- \cdots H_2O$ and $Li^+ \cdots OH_2$ are much larger in absolute value than in $(H_2O)_2$. The attractive forces between an ion and a dipole molecule are much stronger than dipole–dipole forces and therefore lead to shorter intermolecular distances at the equilibrium geometries. Consequently, the exchange repulsion is much larger in $F^- \cdots H_2O$ or $Li^+ \cdots OH_2$ than in $(H_2O)_2$.

A more detailed comparison of the two ionic complexes shows a number of

TABLE 6

Energy partitioning in some dimeric molecular complexes obtained by ab initio LCAO–MO–SCF calculations[59]

Complex	Basis set[a] (GTO)	Geometry	Energy contributions and total energies (kcal mol^{-1})						
			ΔE_{COU}	ΔE_{EX}	ΔE_1	ΔE_{POL}	ΔE_{CHT}	ΔE_{DEL}	ΔE
$F^- \cdots H_2O$	(9,5,1;4,1) (6,4,1;3,1)	SCF-min. $\gamma = 0°$	-34.7	26.0	-8.7	-7.2	-7.2	-14.4	-23.1
$Li^+ \cdots OH_2$	(11,7,1;6,1) (4,3,1;2,1) Li(7,1) Li 3,1	SCF-min. C_{2v}-sym.	-41.7	15.3	-26.4	-12.8	5.2	-7.6	-33.9
$(H_2O)_2$	(8,4,4) (2,1;1) STO-4G[75]	SCF-min.	-7.8	9.7	1.9	-0.3	-7.9	-8.2	-6.3

[a]For $F^- \cdots H_2O$ and $Li^+ \cdots OH_2$ basis sets of almost the same quality as those used in the most accurate calculations reported in the literature[70,71,74] are employed. For $(H_2O)_2$, however, the basis set is much smaller and substantial errors in ΔE_{COU} and ΔE_{CHT} are to be expected[12,80].

interesting differences. $Li^+ \cdots OH_2$ can be regarded as a typical ion–molecule complex in the classical sense. Coulomb and polarization energies dominate and exchange repulsion provides the essential counteraction. ΔE_{CHT} is repulsive and small in comparison to the other contributions. As we have seen previously (Table 5) there is practically no charge transfer in this complex, ΔE_{CHT} represents here only the higher order "left-overs" of other contributions mentioned above. In $F^- \cdots H_2O$ we recognize features common to both ionic complexes and hydrogen bonded systems. Coulomb and polarization energies are large in absolute value, but due to the small intermolecular distance ($R_{FH} = 1.6$ Å, see Table 2) the exchange repulsion is unusually high. ΔE_{CHT} is negative and of a similar magnitude to the polarization energy indicating some covalent character in this hydrogen bond like interaction. In fact, charge transfer in $F^- \cdots H_2O$ is found to be about one order of magnitude larger than in $Li^+ \cdots OH_2$.

The most important quantity missing in Table 6 certainly is the dispersion energy. As mentioned before an estimate of ΔE_{DIS} can be obtained only from calculations which include electron correlation effects. No such calculations on anion–molecule complexes have been reported.

III.C Energy partitioning in higher clusters

The cluster expansion of many body potentials eqn. (31), can be used appropriately as a basis for energy partitioning in higher molecular aggregates, $ABC \cdots N$

$$\Delta E(R) = E_{ABC\ldots N}(R) - \sum_{I=A,B,C..}^{N} E_I =$$
$$= \sum_{I<J} V_{IJ}(R_{IJ}) + \sum_{I<J<K} V_{IJK}(R_{IJ}, R_{JK}) + \cdots \tag{31}$$

Energy surfaces based on the assumption of pairwise additivity of intermolecular forces, $\Delta E^{II}(R)$, of course, are much easier

$$\Delta E^{II}(R) = \sum_{I<J} V_{IJ}(R_{IJ}) \tag{32}$$

to obtain than the accurate energies, $\Delta E(R) \cdot V_{IJ}(R_{IJ})$ is identical with the energy surface of the dimeric complex IJ, $\Delta E_{IJ}(R)$. In statistical mechanics and in computer simulations of liquids and solutions, one of the most urgent problems concerns the relative importance of many-body forces in calculations involving higher clusters. Only in a very few cases are accurate energies available.

In the case of energy surfaces for halide–water complexes, with a first hydration layer, the importance of many body effects has been analysed by Kistenmacher et al.[73]. Their results are briefly summarized in Table 7. Many-body potentials provide a destabilizing contribution to the energies of complexes which increases substantially with increasing number of ligands in the first hydration layer. Kistenmacher et al.[73] pointed out, that many-body contributions $\Delta E^{II} - \Delta E$, in $F^-(H_2O)_n$ and $Cl^-(H_2O)_n$ are roughly proportional to the number of three-body interactions of the type $H_2O–X–H_2O$. They conclude, that this kind of repulsive interaction is respon-

TABLE 7

Deviation from pairwise additivity in near HF calculations on halide–water complexes, $X^-(H_2O)_n$

n	m^a	$F^-(H_2O)_n$			$Cl^-(H_2O)_n$		
		ΔE^{II} (kcal/mole)	ΔE (kcal/mole)	$\Delta E^{II} - \Delta E$ (kcal/mole)	ΔE^{II} (kcal/mole)	ΔE (kcal/mole)	$\Delta E^{II} - \Delta E$ (kcal/mole)
1	0	−22.8	−22.8	0	−11.9	−11.9	0
2	1	−44.4	−43.5	0.9	−23.3	−22.9	0.4
3	3	−63.8	−61.1	2.7	−34.3	−32.7	1.6
4	6	−81.0	−73.4	7.6			
5	10	−95.4	−84.1	11.3			

$^a m$ represents the number of three-body interactions of the type $H_2O–X–H_2O$ in a complex $X(H_2O)_n$, i.e. $m = n(n - 1)/2$.

sible for the major part of many-body effects. For $Li^+(OH_2)_n$ this interpretation was shown to be correct by a more detailed analysis.

Another kind of three-body interaction, encountered in structural units of the type $X^- \cdots H_2O \cdots H_2O$, has been found to determine structures and stabilities of second hydration layers. In the case of $F^-(H_2O)_2$ very accurate calculations near the HF limit are available (Table 8)[81]. Polarization and charge transfer effects lead to an increase in basicity and a decrease in acidity of the H_2O molecule bound directly to the anion. Consequently, hydrogen bonds to the lone pairs of H_2O are substantially stronger, however the hydrogen bond involving the second H-atom is much weaker than the corresponding interaction in the isolated water dimer (Table 8). In fact, a stable hydrogen bond is not formed, when a second water molecule is attached to the free H-atom in $F^- \cdots HOH$.

TABLE 8

Binding energies in near HF calculations on $F^-(H_2O)_2$[81].

Structure	Dissociation products	ΔE (kcal mol^{-1})
$HOH \cdots F^- \cdots HOH$	$F^- + 2H_2O$	−44.9
	$F^- + (H_2O)_2$	−40.1
	$[F^- \cdots HOH] + H_2O$	−20.8
$F \cdots H—O\overset{HOH}{\underset{H}{}}$	$F^- + 2H_2O$	−36.6
	$F^- + (H_2O)_2$	−31.8
	$[F^- \cdots HOH] + H_2O$	−12.5
$F^- \cdots HOH \cdots OH_2$	$F^- + 2H_2O$	−22.9
	$F^- + (H_2O)_2$	−18.2
	$[F^- \cdots HOH] + H_2O$	>0
$F^- \cdots HOH$	$F^- + H_2O$	−24.1
$(H_2O)_2$	$2H_2O$	−4.8

IV. Structures and energies of anion–molecule complexes

IV.A Introduction

From our previous discussion of various methods for calculations on anion–molecule complexes, we can expect only two approaches to produce reasonable results although they differ largely in nature and in their degree of sophistication. These are extended electrostatic models and accurate ab initio LCAO–MO calculations (see Sect. II). Almost exclusively, anion–water interactions have been studied and therefore we will summarize here the results of both kinds of calculations on $OH^-(H_2O)_n$ and halide–water complexes, $X^-(H_2O)_n$.

IV.B Electrostatic results on halide–water complexes

Before discussing in detail numerical results, we should like to give an impression of the gross shape of the energy surface for the interaction of an anion with a water molecule. We have calculated, therefore, the most important contribution to the energy of interaction, ΔE_{COU}, which also determines the angular dependence of the energy[82]. In this model calculation the anion is simulated by a negative unit charge $(-e_0)$. The electron density distribution of the H_2O molecule was calculated from an STO-3G wave function. Despite the small basis set, this wave function reproduces the dipole moment of the water molecule almost correctly, $\mu_{calc}(STO\text{-}3G) = 1.69$ D[83]*, $\mu_{exp} = 1.85$ D. We can expect therefore that the calculated energy surface, $\Delta E_{COU} = \Delta E_{COU}(R, \gamma)$ shown in Fig. 10 to come very close to the exact electrostatic potential surface, which coincides with the exact energy surface at very large intermolecular distances R. Energy curves for rotation of unit charge $(-e_0)$ around the H_2O molecule show an interesting change in shape at $R \sim 3.5$ Å. At larger distances $(R > 3.5$ Å$)$ the most stable orientation corresponds to the structure preferred by ion–dipole interaction showing C_{2v} symmetry. A decrease in R increases drastically the importance of higher terms in the multipole expansion of the ligand's potential. Finally, we end up with energy curves showing two minima at equivalent points relative to the former C_2 axis which come very close to the hydrogen bond like linear arrangement of the two atoms O and H, with the point charge $(-e_0)$ also on a straight line. An alternative explanation is that the local dipole of the OH bond determines the most stable orientation at small intermolecular distances.

Molecular geometries, energies of interaction and the individual contributions obtained by extended electrostatic calculations on halide–water complexes[20,21] are shown in Table 9, in which we see an unexpected disagreement. Equilibrium geometries are not reproduced correctly by extended electrostatic models. In all halide water complexes including $F^- \cdots H_2O$ and $Cl^- \cdots H_2O$ the planar structure with C_{2v} symmetry is predicted to be most stable, whereas more accurate calculations and experimental data, lead to a different result namely that the equilibrium structure of $F^- \cdots H_2O$ comes very close to a linear hydrogen bond like arrangement. The

---

*Johansson et al.[56] report a different value $(\mu = 1.8$ D$)$ most probably due to the molecular geometry chosen being somewhat different from the experimental one.

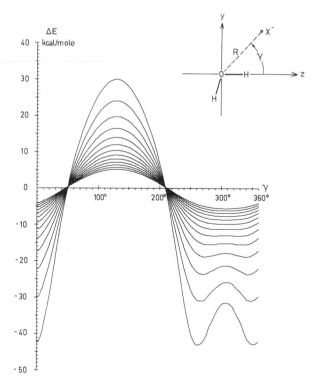

Fig. 10. Potential surface of a negative unit charge $(-e_0)$ rotating around a water molecule at different distances R. (Electron density of H_2O from an STO-3G[75] calculation, $R = 4.0$–10.0 au in steps of 0.5 au.)

failure of the extended electrostatic model calculations is rather surprising for, as we have seen previously, the exact electrostatic potential shown in Fig. 10 reproduces correctly the change in the most stable orientation of the water molecule with decreasing intermolecular distance. This fictitious discrepancy, however, reflects merely the fact that the local electrostatic potential of the OH bond is reproduced poorly by the first terms of the multipole expansion of H_2O.

Intermolecular distances $(R_{OX}-)$ obtained by the model calculations of Spears[20] are substantially shorter than those calculated by Elezier and Krindel[21] or by accurate MO methods, yet Spears' energies of interaction, agree very well with experimental data (see Table 15). This agreement, however, does not directly validate the model, because the experimental energy of formation for the $Br^- \cdots H_2O$ complex $(\Delta H_{298}(Br^- \cdots H_2O) = -12.6$ kcal mol^{-1}[72]) has been used to fit the repulsive potential.

Individual energy contributions to the total energy of interaction[20] in $F^- \cdots H_2O$ come fairly close to the values obtained previously by energy partitioning. Only the contribution of the dispersion energy seems to be unreasonably large. No direct comparison with more accurate data is possible, however, since no calculations on intermolecular correlation energies in anion–water complexes have been reported.

TABLE 9

Energies and intermolecular distances in halide–water complexes, X^-, H_2O obtained by extended electrostatic calculations [20,21]

Complex	R^a_{OX} (Å)	$\Delta E^{(1)}_{COU}$ [b]	$\Delta E^{(2)}_{COU}$	ΔE^{Cl}_{COU} [b]	ΔE_{REP}	ΔE^{Cl}_{POL}	ΔE_{DIS}	ΔE	Ref.
$F^- \cdots H_2O$	2.20	−27.9	1.4	−26.5	24.6	−11.6	−10.6	−24.0	20
	2.63 ±0.06							−17.0	21
	2.55 ±0.06							−18.5	21
$Cl^- \cdots H_2O$	2.91	−15.8	0.6	−15.2	10.2	−3.7	−5.6	−14.3	20
	3.15 ±0.08							−11.7	21
	3.10 ±0.06							−12.3	21
$Br^- \cdots H_2O$	3.11	−13.8	0.5	−13.3	8.6	−2.8	−5.0	−12.6	20
	3.42 ±0.02							− 9.8	21
	3.37 ±0.10							−10.2	21
$I^- \cdots H_2O$	3.41	−11.4	0.4	−11.1	6.5	−1.9	−4.1	−10.6	20
	3.50 ±0.03							− 9.5	21
	3.47 ±0.01							− 9.8	21

[a] In all calculations reported here a planar symmetric (C_{2v}) structure of the anion–water complex is found to be the most stable ($\gamma = 52.25°$, see Fig. 11).
[b] The electrostatic contribution to the energy of interaction is approximated here as a sum of the ion–dipole ($\Delta E^{(1)}_{COU}$). The quadrupole moment is defined relative to the center of mass (CM) of the water molecule.

As mentioned before, a rough estimate suggests there is only a small contribution from electron correlation to intermolecular energies in ion–molecule complexes[54]. As one would expect from simple physical arguments, all energy contributions decrease in absolute value in the series of halide ions ordered in increasing atomic number.

IV.C Ab initio LCAO–MO–SCF results on anion–water complexes

Owing to their somewhat different solution chemistry, results on hydrated halide and hydroxide ions will be summarized separately.

IV.C.1 Hydration of halide ions

Several ab initio SCF calculations on hydrated halide ions have been reported[54,70,71,73,81,84]. We shall restrict our discussion here to a recent series of calculations on this subject performed by Kistenmacher et al.[54,71,73].

One to one complexes have been investigated most extensively. The equilibrium structures of $F^- \cdots H_2O$ and $Cl^- \cdots H_2O$ are shown in Fig. 11. More details of the geometries are presented together with calculated energies of interaction in Table 10. In both complexes the most stable structure comes close to a linear hydrogen bond like arrangement of the three atoms O, H and X ($\gamma = 4.5°$ and $\gamma = 14.6°$ in the F^-

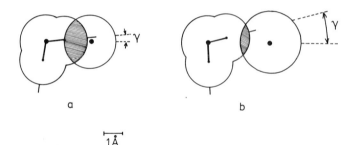

Fig. 11. Equilibrium geometries of (a) F⁻ ··· H₂O and (b) Cl⁻ ··· H₂O (ref. 71). Van der Waals'radii: $R_H = 1.2$ Å, $R_O = 1.4$ Å, $R_F = 1.4$ Å and $R_{Cl^-} = 1.9$ Å; note that γ is defined differently from Fig. 10, $\gamma \cong 360° - \gamma$.

and Cl⁻ complex, respectively). Although the Van der Waals' spheres of F and H₂O overlap strongly in F⁻ ··· H₂O, there is no problem in recognizing the water molecule. The equilibrium position of the H atom involved in the hydrogen bond is much closer to the O atom ($R_{OH} = 1.00$ Å and $R_{HF} = 1.52$ Å). A distortion of the water molecule in the field of F⁻ is detectable but by no means very strong. With respect to the HF geometry of the isolated H₂O molecule the OH bond pointing towards the F⁻ anion is elongated somewhat ($\Delta R_{OH_1} = 0.045$ Å), the second OH bond remains almost constant ($\Delta R_{OH_2} = -0.004$ Å), whereas the ∡ HOH bond angle becomes substantially smaller ($\Delta \alpha = -3.9°$) on complex formation. In

TABLE 10

Energies and equilibrium structures of halide–water complexes[71]

Complex	Geometrical constraints	$\Delta E_R{}^a$ (kcal mol⁻¹)		Intermolecular geometry		Geometry of the H₂O molecule		
			ΔE (kcal mol⁻¹)	R_{OX} (Å)	$XOH_1{}^b$ $\gamma(°)$	$R_{OH_1}{}^b$ (Å)	$R_{OH_2}{}^b$ (Å)	HOH $\alpha(°)$
F⁻···H₂O	C_{2v}–symmetry $\gamma = \alpha/2$	−16.57	−18.17	2.67	47.5	0.955	0.955	95.0
	$\gamma = 0°$	−22.60	−23.25	2.55	0.0	1.000	0.945	104.5
	planarity	−22.85	−23.53	2.51	4.5	0.995	0.946	102.7
Cl⁻···H₂O	C_{2v}–symmetry $\gamma = \alpha/2$	−10.86	−11.19	3.44	50.3	0.950	0.950	100.5
	$\gamma = 0°$	−11.36	−11.25	3.31	0.0	0.964	0.950	105.0
	planarity	−11.93	−11.92	3.31	14.6	0.961	0.946	102.4
H₂O	—	—	—	—	—	0.950	0.950	106.6

[a]Energy of interaction calculated with a rigid water molecule ($R_{OH} = 0.975$ Å, ∡ HOH = 105°) is denoted by ΔE_R, whereas ΔE represents the energy of interaction fully optimized relative to a water molecule at the HF geometry ($R_{OH} = 0.950$ Å, ∡HOH = 106.6°).
[b]H_1 denotes the hydrogen atom, which lies closer to the halogen atom X.
[c]$\gamma = 0°$ represents the hydrogen bond-like geometry with linear X, H_1 and O.

$Cl^- \cdots H_2O$ the position of the central hydrogen is even more asymmetric, $R_{OH} = 0.96$ Å and $R_{HCl} = 2.39$ Å, and the distortion of the water molecule is less pronounced than in $F^- \cdots H_2O$. Apart from the change in bond angle ($\Delta\alpha = -4.2°$) the calculated differences hardly exceed the error limits ($\Delta R_{OH_1} = 0.011$ Å, $\Delta R_{OH_2} = -0.004$ Å). Furthermore, in $F^- \cdots H_2O$ the relaxation of the H_2O molecule in the field of F^- leads to a gain in total energy of the complex ($\Delta E_R - \Delta E = 0.68$ kcal mol^{-1}), whereas in case of the heavier halide ion Cl^- the change in energy of the complex is overshadowed by the numerical errors of the individual calculations and of the interpolation procedure.

The major difference between $F^- \cdots H_2O$ and $Cl^- \cdots H_2O$ cannot be recognized from the complexes' equilibrium structures alone. In order to learn more about the energy surfaces energies have been minimized under geometrical constraints as well (Table 10). Two constraints are applied, (1) C_{2v} symmetry according to the geometry determined by ion–dipole interaction ($\gamma = \alpha/2$) and (2) linearity of the hydrogen bond X–H–O ($\gamma = 0°$). The first series of calculation were performed with a rigid water molecule. The C_{2v} complex in $F^- \cdots H_2O$ is found to be less stable than the hydrogen bond-like structure by $\Delta(\Delta E_R) = 6.3$ kcal mol^{-1}. In $Cl^- \cdots H_2O$ this difference is substantially smaller and amounts only to $\Delta(\Delta E_R) = 0.5$ kcal mol^{-1}. Relaxation of the water geometry during rotation does not change this general result, $\Delta(\Delta E) = 5.4$ kcal mol^{-1} in $F^- \cdots H_2O$ and $\Delta(\Delta E) = 0.7$ kcal mol^{-1} in $Cl^- \cdots H_2O$. Interestingly, this result is in agreement with the simple electrostatic potential surface shown in Fig. 10. The increase in intermolecular equilibrium distances leads to a change from the double minimum potential $\Delta E = \Delta E(\gamma)$ at small values of R_{XO} to a very flat energy curve with a single minimum only. Accurate HF energy surfaces[71], similarly, present an elegant illustration of this difference in binding of the water molecule to F^- and Cl^-. We can conclude, therefore, that the orientation of H_2O relative to F^- is fixed rather close to the equilibrium structure at moderate temperatures. Rotation of the water molecule from one hydrogen bond-like geometry ($\gamma = 0°$) to the other ($\gamma = \alpha$) leads to a substantial increase in energy, whereas the water molecule in the Cl^- complex is expected to be rather labile. Partial rotation from $\gamma = 0°$ to $\gamma = \alpha$, from one hydrogen bond-like arrangment $Cl-H_1-O$, to the other, $Cl-H_2-O$, does not require much more energy than thermal motion provides at room temperature ($RT \sim 0.6$ kcal mol^{-1}).

No ab initio calculations have been performed on hydrates of the heavier ions Br^- and I^-. From the shape of the electrostatic energy surface, however, we can conclude with some confidence that a similar flexibility with respect to partial rotation of the water molecule, as in case of $Cl^- \cdots H_2O$, will be found in the complexes of Br^- and I^- as well. The minimum of the energy surface in $Br^- \cdots H_2O$ and $I^- \cdots H_2O$ is expected to correspond to a structure with C_{2v} symmetry.

In order to make a comparison with experimental data Kistenmacher et al.[54] estimated the contribution of electron correlation by an approximate method due to Wigner[55]. Zero point energy corrections were calculated by conventional statistical mechanics from masses, moments of inertia and vibrational frequencies, which in turn were derived from HF equilibrium geometries and force constants. In general these corrections are rather small in comparison to the total energy of interaction

TABLE 11

Calculated and experimental ΔH_{298} values for F^-H_2O and Cl^-H_2O [54]

Complex	ΔE_{SCF} (kcal mol^{-1})	ΔE_{COR} [a] (kcal mol^{-1})	$\Delta H^0_{298} - E$ (kcal mol^{-1})	ΔH^{CALC}_{298} (kcal mol^{-1})	ΔH^{EXP}_{298} [b] (kcal mol^{-1})
$F^-\ldots H_2O$	−23.70	−1.40	2.27	−22.83	−23.3
$Cl^-\ldots H_2O$	−11.85	−0.71	1.06	−11.50	−13.1

[a] Estimated by a modified Wigner's method [55,85].
[b] Experimental enthalpies from ref. 72.

(Table 11). The agreement between calculated and experimental ΔH values is satisfactory.

Complexes of F^- and Cl^- with several water molecules have also been investigated [73,81]. As has been shown (Table 7), the assumption of pairwise additivity of intermolecular potentials is an acceptable approximation for coarse estimates on large clusters, although the errors increase substantially with increasing number of ligands. Another very important question concerns the distribution of water molecules in different hydration layers. Based on the assumption of pairwise additivity Kistenmacher et al.[73] have calculated energies for complex formation with and without the restriction that all intermolecular distances R_{OX} are equal. In the first case only one hydration layer was allowed, whereas in the latter as many layers as are advantageous to employ are formed. From Table 12 we can see that both series of energies start to deviate significantly at coordination numbers higher than 6 in the case of F^-, and 8 for Cl^-. The clusters $X(H_2O)_{10}$ show an interesting distribution of water molecules among a first and second hydration layer at the equilibrium geometry. In $F^-(H_2O)_{10}$ five water molecules are bound in the range 2.6 Å $< R_{OF} <$ 2.7 Å, and the remaining five occupy positions in a second layer, 3.8 Å $< R_{OF} <$ 4.2 Å. In the chloride complex the distribution is different, there are seven water molecules in the first layer 3.3 Å $< R_{OCl} <$ 3.6 Å and three at larger distances, 4.4 Å $< R_{OCl} <$ 4.6 Å. We shall return to the question of coordination numbers in the next section, where the dynamics of hydrated ions will be discussed. In general, the calculated energies of interaction, ΔE, agree well with the available experimental ΔH values, if both electron correlation and zero point corrections are small, as is suggested by Table 11.

IV.C.2 Hydration of the hydroxide ion

There are two reports of ab initio studies on the hydration of a hydroxide ion[86,87]. The first presents a very accurate ab initio calculation on $H_3O_2^-$, the second deals with much less extended calculations on complexes up to $H_9O_5^-$ (Table 13).

Kraemer and Diercksen[86] find the equilibrium geometry of $H_3O_2^-$ to be planar with the outer hydrogen atoms in a *trans* configuration. The O–O distance is substantially smaller than in water dimer: $R_{OO}(H_3O_2^-) = 2.51$ Å, compared to $R_{OO}(H_4O_2) = 3.00$ Å. The equilibrium position of the central hydrogen is asymmetric $R_{O_1H_1} =$

TABLE 12

Intermolecular distances and energies of formation for halide–water[73] complexes $X^-(H_2O)_n$

n	F⁻H₂Oₙ					Cl⁻(H₂O)ₙ				
	One hydration layer[a]		no restriction			One hydration layer[a]		no restriction		
	R_{OF} (Å)	ΔE^{II}[b] (kcal mol⁻¹)	ΔE^{II}[b] (kcal mol⁻¹)	ΔE (kcal mol⁻¹)	ΔH_{298}[c] (kcal mol⁻¹)	R_{OCl} (Å)	ΔE^{II}[b] (kcal mol⁻¹)	ΔE^{II}[b] (kcal mol⁻¹)	ΔE (kcal mol⁻¹)	ΔH_{298}[c] (kcal mol⁻¹)
1	2.54	-22.8	-22.8	-22.8	-23.3	3.30	-11.9	-11.9	-11.9	-13.1
2	2.55	-44.4	-44.4	-43.5	-39.9	3.31	-23.3	-23.3	-22.9	-25.8
3	2.58	-63.8	-63.8	-61.1	-53.6	3.34	-34.3	-34.3	-32.7	-37.5
4	2.62	-81.0	-81.0	-73.4	-67.1	3.37	-46.7	-46.7		-48.6
5	2.67	-95.4	-95.4	-84.1	-80.3	3.39	-55.7	-55.7		
6	2.71	-106.9	-107.2			3.42	-69.1	-69.2		
7	2.79	-115.1	-120.1			3.46	-78.6	-78.8		
8	2.89	-119.9	-131.4			3.52	-86.6	-86.7		
9	2.97	-123.8	-141.6			3.57	-91.4	-97.1		
10	3.01	-122.3	-148.5			3.59	-97.3	-104.7		

[a] All intermolecular distances are restricted to be equal or varied freely.
[b] For the definition of pairwise additivity see eqn. (32).
[c] Experimental enthalpies from ref. 72.

TABLE 13

Energies and equilibrium structure of hydrated hydroxide ions

Complex	Ref.	Basis set[a]	Symmetry	R_{OO}[b] (Å)	ΔE[c] (kcal mol^{-1})	$\Delta E_{n,n-1}$[c] (kcal mol^{-1})	$\Delta H^{\text{exp.}}_{n,n-1}$[c] (kcal mol^{-1})
$[H_3O_2]^-$	86	A	C_s	2.51	−24.3	−24.3	−22.5
	87	B	C_s	2.45	−40.7	−40.7	
$[H_5O_3]^-$	87	B	C_{2v}	$\overline{O_1O_2} = \overline{O_1O_3} = 2.53$	−70.8	−30.1	−16.4
	87	B	C_s	$\overline{O_1O_2} = \overline{O_1O_3} = 2.53$	−92.2	−21.4	
$[H_7O_4]^-$				$\overline{O_2O_4} = \overline{O_3O_4} = 2.80$			−15.1
	87	B	C_{3v}	$\overline{O_1O_2} = \overline{O_1O_3} =$ $= \overline{O_3O_4} = 2.61$	−93.9	−23.1	
	87	B	$H_7O_4^-$, C_{3v} + H_2O	$\overline{O_2O_5} = 2.70$	−112.8	−18.9	
$[H_9O_5]^-$							−14.2
	87	B	$H_7O_4^-$, C_s + H_2O	$\overline{O_1O_5} = 2.70$	−114.6	−22.2 (−20.7)	

[a]Basis set A: GTO (11,7,1/6,1) contracted to $[5,4,1/3,1]$; basis set B: STO-4/31G (8,4/4) contracted to $[3,2/2]$[89,90].
[b]The individual structures are shown in Figs. 12 and 13.
[c]ΔE represents the total energy of formation of the cluster OH$^-$(H$_2$O)$_n$ from OH$^-$ and n H$_2$O molecules. $\Delta E = E(\text{OH}^-(\text{H}_2\text{O})_n) - E(\text{OH}^-) + n\,E(\text{H}_2\text{O})$; $\Delta E_{n,n-1}$ is the energy for the addition of the "last" water molecule to the complex: $\Delta E_{n,n-1} = E(\text{OH}^-(\text{H}_2\text{O})_n) - E(\text{OH}^-\cdot(\text{H}_2\text{O})_{n-1} + E(\text{H}_2\text{O})$. Experimental reaction enthalpies are taken from ref. 88.

1.16 Å, $R_{O,H_1} = 1.36$ Å, proton motion, therefore, occurs along a double minimum potential. The calculated barrier height[86] is $\Delta E^+ = 2.7$ kcal mol^{-1}*. Compared with F$^-\cdots$H$_2$O the distortion of the water molecule bound to OH$^-$ is substantially larger ($\Delta R_{O,H_1} = 0.20$ Å) although the energy of interaction is roughly the same. Consequently, we can conclude that the symmetry of the subsystems is an important factor, causing deviation from the geometries of the isolated subunits, in the complex. As we can see from Fig. 12, overlap between OH$^-$ and H$_2$O is large in H$_3$O$_2^-$, and in this example we come close to the limits, beyond which a definition of subunits in the complex loses all meaning, e.g. in completely symmetric structures as F–H–F$^-$ it is senseless to formulate the complex as XH $\cdots$ X$^-$ or X$^-\cdots$HX. The calculated energy of interaction, $\Delta E = -24.3$ kcal mol^{-1} [86], is close to the experimentally determined enthalpy of $\Delta H_{298} = -22.5$ kcal mol^{-1} [88]. The relative importance of electron correlation and the zero point correction, however, are unknown in this case, and the agreement between the available calculated and experimental values should not be overemphasized.

The smaller basis set used by Newton and Ehrenson[87] (STO-4,31 G[89,90]) leads to a substantial error in the calculated energies of interaction (Table 13). Although all

---

*More details of the calculations on proton transfer are given in a recent review article, ref. 92.

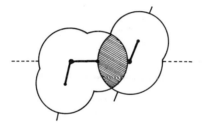

Fig. 12. Equilibrium geometry of HO$^-$H$_2$O^{86} (Van der Waals' radii: $R_H = 1.2$ Å, $R_0 = 1.4$ Å).

absolute values are too large the calculated energies reproduce the decrease in hydration enthalpies, $\Delta H_{n,n-1}$ with increasing number n, in accord with experiment. The O–O distances at the most stable geometries are somewhat smaller than in the more accurate calculations, nevertheless, the equilibrium geometry of H$_3$O$_2^-$ is nearly the same in both calculations, a planar complex with an asymmetric position of the central hydrogen is found to be the most stable. As expected, O–O distances increase in complexes with more than one water molecule bound symmetrically to a central OH$^-$. In H$_5$O$_3^-$ both H$_2$O molecules are equivalent (Fig. 13). Similarly, in H$_7$O$_4^-$ a geometry showing C_{3v} symmetry, analogous to Eigen's H$_9$O$_4^{+91}$, with three equivalent H$_2$O molecules bound to the central anion is found to be the most stable

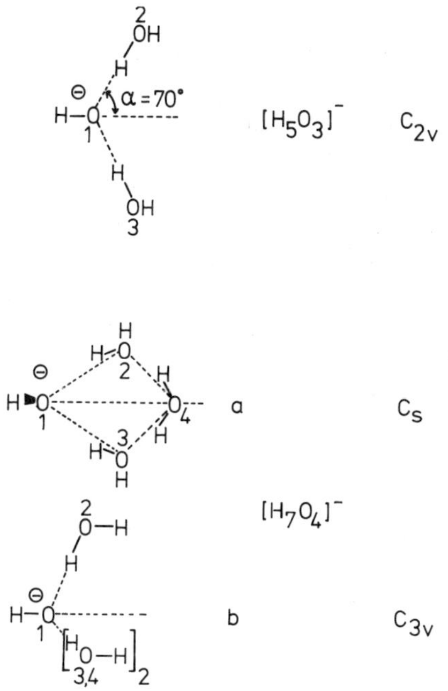

Fig. 13. Equilibrium geometries of HO$^-$(H$_2$O)$_n$^{87} with $n = 2$ and 3.

(Fig. 13b). There is, however, a second minimum in the potential surface corresponding to a cyclic structure with C_s symmetry and two types of water molecules in different layers of the hydration shell (Fig. 13a), which is only 1.7 kcal mol^{-1} higher in energy (Table 13). In $H_9O_5^-$ a fourth water molecule is bound to $H_7O_4^-$, thus starting a second hydration layer in the symmetric complex of $H_7O_4^-(C_{3v})$, since the H atom of OH$^-$ is not acidic enough to form a hydrogen bond of sufficient stability. Interestingly, the cyclic, less symmetric (C_s) central unit is found now to form the more stable complex (Table 13).

V. Dynamics of solvated anions

Three recent papers report dynamical studies on ion solvation[73,93,94], in one the method of molecular dynamics is used[93], for the others Monte Carlo techniques[73,94] are applied. As mentioned previously both methods try to simulate ensembles of molecules, or molecules and ions, under conditions which are characteristic for a liquid or an ionic solution, respectively. It is assumed that a few hundred particles are sufficient to describe a representative section of the liquid. In many calculations, periodic boundary conditions are applied, which in fact enlarge the cluster under consideration to infinity, but impose a somewhat artificial long range periodicity on the structure of the liquid.

V.A Molecular dynamical calculations

Molecular dynamics calculations can be characterized best as an attempt to integrate the classical equations of motion* for an ensemble of particles under the conditions of constant energy and constant volume. Molecular properties are introduced by the masses of the individual molecules and ions and the potential surfaces for particle interaction, which in turn represent the connecting link between energy calculations and studies of molecular dynamics. In all dynamical calculations reported until now, pairwise additivity of intermolecular potentials is assumed. Any direct introduction of three-body forces and higher interactions into the expression for the cluster potential causes a tremendous increase in numerical computation. There is one major advantage of molecular dynamics, in contrast to Monte Carlo calculations, the equations of motion are solved, and besides equilibrium properties, kinetic quantities such as diffusion coefficients and residence times, are also accessible.

The calculations of Heinzinger and Vogel[93] simulating a concentrated solution of LiCl in water (2 M) by 216 particles, 9 Li$^+$ ions, 9 Cl$^-$ ions and 198 H_2O molecules can be considered as an extension to ionic solutions of earlier molecular dynamic studies on the structure of pure water by Rahman and Stillinger[95-98]. The average

---

*In the case of rigid molecules these are coupled differential equations for translation and rotation.

temperature of the whole system was maintained at $T = 272\,K^*$. This study[93] is of a preliminary character, since the periodic box was rather small and the Coulomb part of the intermolecular potential† was cut off at a first hydration layer, i.e. between 3.5 Å and 5.5 Å depending on the nature of the interacting particles, nevertheless, a number of interesting results were obtained. All three radial pair correlation functions, obtained directly from the molecular dynamics calculation, g_{LiO}, g_{ClO} and g_{OO}, show sharp first maxima at distances $R_{\text{OLi}} = 2.1\,Å$, $R_{\text{OCl}} = 2.75\,Å$ and $R_{\text{OO}} = 2.9\,Å$. The height of the peaks decreases in this series with increasing R indicating that the strongest intermolecular force leads to the most rigid first hydration layer. A first minimum can be recognized in all three pair correlation functions. Additional maxima and minima however, are more or less overshadowed by statistical fluctuations. From radial pair distribution functions, coordination numbers in the first hydration layer can be obtained simply by integration

$$n_{\text{XO}}(R_0) = 4\pi\rho_0 \int_0^{R_0} R^2 g_{\text{XO}}(R)\,dR \qquad (33)$$

In eqn. (33), ρ_0 represents the average number density of H_2O molecules, n_{XO} is the average number of O atoms surrounding an ion X within a sphere of given radius R_0. When R_0 is taken to be equal to the distance where the first minimum of the pair correlation g_{XO} occurs, n_{XO} provides the average coordination number for the ion X. In their model calculation Heinzinger and Vogel[93] obtained the following coordination numbers, $n_{\text{Li}^+} = 5.5$ and $n_{\text{Cl}^-} = 6.0$, in rough agreement with the results from X-ray and neutron diffraction studies of LiCl solutions[11,99]. The average orientation of the water molecules in the first hydration layer corresponds to that expected from energy calculations, although substantial deviations from planarity in the $Li^+ \cdots OH_2$ subunit is found. Relative to the O atom of the H_2O molecule, Li^+ occupies one of the four corners of a tetrahedron, the remaining three corners are built up by the two H atoms of the water molecule and a hydrogen bond to a second water molecule. In the first hydration layer the three atoms Cl, H and O lie roughly on a straight line. Again, the calculated results agree well with diffraction data.

Additionally, two important dynamic quantities were calculated. The coefficients for translational and rotational diffusion of the H_2O molecule were found to be in qualitative agreement with the available experimental values (ref. 100). Considering the smallness of the periodic box and the approximations made in the pair potentials, no quantitative comparison is to be expected.

---

*Temperature equilibrium was not achieved completely. The average temperature was allowed to fluctuate about the following values: H_2O, $T_{\text{trans}} = 300\,K$ and $T_{\text{rot}} = 250\,K$; Cl^-, $T = 300\,K$ and Li^+, $T = 250\,K$.

† In this calculation a superposition of a point charge potential, $\pm e_0$ for Li^+ and Cl^-, respectively and the ST2 point charge model for the H_2O molecule[98], together with a Lennard–Jones potential was used.

V.B Monte Carlo calculations

In contrast to molecular dynamics, Monte Carlo calculations do not represent solutions of the molecular equations of motion. A set of geometrical parameters determining the configuration of the cluster is guessed by a stochastic method and the energy for this particular configuration is calculated and stored in the memory. This procedure is repeated many times such that one expects to obtain a representative distribution of cluster configurations among total energies. From these results pair correlation functions and related quantities can be obtained at given temperatures. In general, Monte Carlo calculations provide an excellent basis for a discussion of equilibrium properties of liquids. Ensembles at constant temperature can be described without further difficulty.

In both studies concerning Monte Carlo calculations on ion water complexes[73,94] a program previously applied to calculate the structure of liquid water[101,102] was used. The first calculations were restricted to single ion hydration[73] at room temperature, $F^-(H_2O)_n$ and $Cl^-(H_2O)_n$. The main idea in this investigation was to find out the influence of thermal motion on the structures and energies of hydrated ions. Central ions were placed at fixed positions, whereas positions and orientations of the water molecules were varied freely. Only single clusters were studied which means that no periodic boundary conditions were applied. The Monte Carlo calculations were based on analytical pair potentials, which were fitted to HF energy surfaces obtained in previous calculations[54,71]. Due to thermal motion less favorable configurations are occupied to some extent and consequently the average internal energy, $<E>_T$, is somewhat smaller in absolute value than at $T = 0\,K$ (Table 14), where there is no thermal energy. At room temperature the corrections due to thermal motion scarcely exceed the errors which have been introduced by the assumption of pairwise additivity of intermolecular potentials. Pair distribution functions $g_{IJ}(R_{IJ})$ and their integrals $n_{IJ}(R^0_{IJ})$ were calculated for the complexes $F^-(H_2O)_4$, $Cl^-(H_2O)_4$ and $F^-(H_2O)_{27}$. The first calculation mainly confirms that on the average the hydrogen bond like geometry with F, H and O in linear array is not disturbed significantly by

TABLE 14

Influence of thermal motion on calculated internal energies of halide–water complexes $X(H_2O)_n$[73]

n	$F^-(H_2O)_n$			$Cl^-(H_2O)_n$		
	$\Delta E_{II}{}^a$	$\langle \Delta E \rangle_{II}{}^b$ 298 K	ΔE_{SCF}	$\Delta E_{II}{}^a$	$\langle \Delta E \rangle_{II}{}^b$ 298 K	ΔE_{SCF}
	(kcal mol^{-1})	(kcal mol^{-1})	(kcal mol^{-1})	(kcal mol^{-1})	(kcal mol^{-1})	(kcal mol^{-1})
1	−22.8	−21.9	−22.8	−11.9	−10.9	−11.9
2	−44.4	−42.2	−43.5	−23.3	−21.1	−22.9
3	−63.8	−60.1	−61.1	−34.3	−30.7	−32.7
4	−81.0	−75.5	−73.4	−46.7	−39.4	

^{a} Energy calculated from two-body potentials at $T = 0$ K.
^{b} Potential energy of the cluster at $T = 298$ K.

thermal energy at $T = 298$ K, $g_{FH}(R)$ shows two well defined and sharp maxima corresponding to the four protons involved in the hydrogen bonds and the four protons further outside. As expected $g_{FO}(R)$ shows only one sharp peak. Compared to $F^-(H_2O)_4$ all maxima of pair correlation functions in $Cl^-(H_2O)_4$ are broadened and the two peaks of g_{ClH} overlap substantially, thus indicating a high flexibility with respect to partial rotation of the water molecule.

In $F^-(H_2O)_{27}$ we can expect further information, since the number of water molecules by far exceeds the coordination number in the first hydration layer. In fact, pair correlation functions and their integrals show that six water molecules surround the F^- ion in the first layer at room temperature. Again two sharp maxima of $g_{FH}(R)$ in the range of the first layer of H_2O molecules are found and we can conclude therefore that water molecules are bound tightly to F^- and rather rigidly orientated in hydrogen bond-like configurations.

In their most recent study on ion hydration, Watts et al.[94] performed a Monte Carlo calculation on a cluster containing a Li^+F^- ion pair at a fixed interionic distance and 50 water molecules. As in the calculations mentioned, which were performed on hydrated single ions[73], no periodic boundary conditions were applied. At the lower temperature investigated ($T = 298$ K), the water molecules cluster around the ion pair and do not evaporate, that is they do not fill the whole box uniformly. As expected, the tendency to evaporate is much stronger at higher temperature ($T = 500$ K), when only a few molecules remain tightly bound to the ion pair. Three interionic distances were investigated $R_{LiF} = 3.0$, 3.5 and 4.0 Å. In every case one water molecule is trapped in the region between both ions, pointing towards Li^+ with the O atom and towards F^- with one of its H atoms. With decreasing interionic distance the interstitial water molecule is pressed out remarkably from the axis connecting Li^+ and F^-. For all three interionic distances pair distribution functions g_{FH}, g_{FO}, g_{LiH} and g_{LiO} were calculated. From their integrals, coordination numbers of four for both ions in the tightly bound first layer of water molecules were derived. g_{FH} is in good agreement with a hydrogen bond-like asymmetric orientation of water molecules around F^-. At $T = 298$ K there is some evidence for the existence of a less ordered second hydration layer around the ion pair which disappears completely at higher temperatures ($T = 500$ K). The first hydration layer, however, is bound tightly at $T = 500$ K and can be easily recognized in the pair correlation diagrams.

VI. Some experimental data on anion–molecule complexes and anion solvation

So far we have discussed exclusively the result of calculations on anion–molecule complexes. In this section we will refer very briefly to some recent experimental studies which provide a basis for comparison with theory and complete our ideas on ion–molecule complexes. A somewhat more extensive summary has been given recently[12].

A few years ago the development of high pressure mass spectrometry reached a stage where it could be applied successfully to chemical problems. The technique and its applications has been summarized by Kebarle[8,103]. An impressive number of clustering equilibria of the type in eqn. (34) have been studied

$$X \cdot S_{n-1} + S \rightleftharpoons X \cdot S_n \tag{34}$$

here X represents a central ion, S a solvent molecule $X \cdot S_{n-1}$ and $X \cdot S_n$ are two clusters ions of different stoichiometry.

Based on the accessibility of suitable ion sources, single ion solvation can be studied in the high pressure mass spectrometer. Once the ion has been formed in the reaction chamber it reacts readily with solvent molecules to yield cluster ions of different size, which in turn are recorded in the mass spectrometer section of the apparatus. Provided equilibrium has been established the ratio of the intensities of cluster ions recorded is equal to their equilibrium concentrations

$$I_n/I_{n-1} = X \cdot S_n/X \cdot S_{n-1} \tag{35}$$

All important thermodynamic data for clustering equilibria can be evaluated without further difficulty, if the experiments are performed at different temperatures

$$K_{n-1,n} = \frac{[X \cdot S_n]}{[X \cdot S_{n-1}] \cdot [S]} = \frac{I_n}{I_{n-1}} \cdot \frac{1}{p_S} \tag{36}$$

p_S represents here the solvent's partial pressure in the reaction chamber. From thermodynamics

$$\Delta G^0_{n-1,n} = -RT \ln K_{n-1,n}$$

$$\Delta H^0_{n-1,n} = RT^2 \left(\frac{\partial \ln K_{n-1,n}}{\partial T} \right)_{T=T_0} \tag{38}$$

and

$$\Delta S^0_{n-1,n} = \frac{1}{T}(\Delta G^0_{n-1,n} - \Delta H^0_{n-1,n}) \tag{39}$$

Although the entropy of cluster formation provides an extremely important source of information on the detailed structure of solvation shells, ab initio theories of ion–molecule complexes are not accurate enough to give a reliable value in entropy calculations. Therefore, for comparison with calculated ΔE values ΔH^0 is most useful. Zero point corrections to the energy, $\Delta E - \Delta H^0$, are rather small and can be estimated on purely theoretical grounds with some reliability[54]. Experimental enthalpies for anion solvation in the gas phase are shown in Table 15. As has been mentioned, the agreement between gas phase enthalpies from high pressure mass spectrometry and extended ab initio LCAO–MO–SCF calculations is satisfactory. The differences in numerical values are of the same order of magnitude as the errors which have to be expected in both experimental measurements and theoretical calculations, and reach a few kcal mol^{-1} in the least favorable cases.

It is interesting to compare ΔH^0 values from gas phase clustering equilibria with single ion hydration enthalpies, ΔH^0_S (Fig. 14). In the series of hydrated halide ions the curves for $\Delta H^0_{0,1}$, $\Delta H^0_{0,2}$, $\Delta H^0_{0,4}$ and ΔH^0_S show nice similarities. In order to make a comparison we have to know in advance the differences between these quantities. Single ion solvation enthalpies describe the process shown in eqn. (40) at the limit $n \to \infty$.

$$X + S_n \rightleftharpoons X \cdot S_n \tag{40}$$

TABLE 15

Experimental solvation enthalpies $\Delta H^0_{n-1,n}$ from high pressure mass spectrometry[72,88,104-106]

Solvent molecule		$\Delta H^0_{n-1,n}{}^a$ (kcal mol^{-1}) central ion				
	n	F^-	Cl^-	Br^-	I^-	OH^-
H_2O	1	-23.3	-13.1	-12.6	-10.2	-22.5
	2	-16.6	-12.7	-12.3	-9.8	-16.4
	3	-13.7	-11.7	-11.5	-9.4	-15.1
	4	-13.5	-11.1	-10.9		-14.2
	5	-13.2				-14.1
CH_3OH	1		-14.2			
	2		-13.0			
	3		-12.3			
	4		-11.2			
	5		-10.5			
CH_3CN	1	-16.0	-13.4	-12.9	-11.9	
	2	-12.9	-12.2	-11.8	-10.5	
	3	-11.7	-10.6	-10.0	-9.3	
	4	-10.4	-6.2	-5.5		
	5	-5.3				
$(CH_3)_3COH$	1		-14.2			
C_6H_5OH	1		-19.4			
CH_3COOH	1		-21.6			
$HCOOH$	1		-37.2			

^{a}Standard state applied here: $p = 1\,atm$, $T = 298\,K$.

No single ion properties like solvation enthalpies, ΔH^0_S, or entropies, ΔS^0_S, etc. can be measured directly by thermodynamical or electrochemical methods. Only data for pairs of ions forming a salt or a neutral compound are available. Nevertheless, careful estimates based on different model assumptions have been made and led to roughly the same results[107,108]. The difference between enthalpies of solvation derived from clustering equilibria eqn. (34) by extrapolation to infinity and ΔH^0_S is represented by a sum of enthalpies for the association of solvent molecules. The enthalpies of the

$$\Delta H^0_{0,\infty} = \sum_{i=1}^{\infty} \Delta H^0_{i-1,i} = \Delta H^0_S + \sum_{i=2}^{\infty} \lambda^S_{i-1,i} \tag{41}$$

process $S_n + S \to S_{n+1}$, $\lambda^S_{n,n+1}$, converge to the heat of vaporization of the solvent, L_S, for $n \to \infty$ provided the liquid evaporates into a monomeric gas. The same standard state, of course, has to be used as a reference for all quantities in eqn. (41). Returning to our initial discussion, we have to expect both expressions for $\Delta H^0_{0,\infty}$ in eqn. (41) to grow with a constant increment for increasing $n \to \infty$. Therefore, when n is sufficiently large relative values of $\Delta H^0_{0,n}$ and ΔH^0_S should be identical in a series of central ions. Figure 14 shows that this expected parallelism in the dependence of $\Delta H^0_{0,n}$ and ΔH^0_S on the nature of the central ion is approached stepwise in the

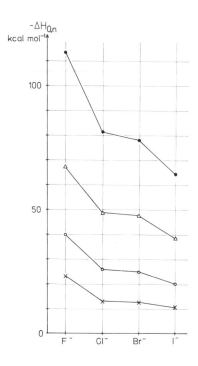

Fig. 14. Comparison of single ion solvation enthalpies $^{107}\Delta H_S^0(\cdot-\cdot-\cdot)$ and gas phase solvation enthalpies ^{72} $\Delta H_{0,n}^0$, $n = 1(X–X–X)$, $n = 2(0–0–0)$ and $n = 4(\triangle–\triangle–\triangle)$ for the series of halide ions.

series $n = 1, 2$ and 4 although it is not completely fulfilled with this small number of ligands.

From the experimental data shown in Table 15 we can obtain also some information on solvation of ions in non-aqueous solvents. Gas phase solvation enthalpies for Cl^- in CH_3OH are not very different from those in H_2O. Interestingly, the first solvation layer of Cl^- in CH_3OH is bound even more strongly than in H_2O. Not only are $\Delta H_{n-1,n}^0$ values for CH_3OH molecules larger in absolute values but also the free enthalpies, $\Delta G_{n-1,n}^0$ show the same trend. Furthermore we can see from Table 15 that the enthalpy of binding to Cl^- in one to one complexes, $\Delta H_{0,1}^0$, increases with increasing acidity of the proton in a series of organic alcohols and carboxylic acids.

Gas phase solvation in acetonitrile vapor shows a different behavior than found for water and methanol. Comparing the ΔH^0 values shown in Table 15 we recognize that the first few CH_3CN molecules are bound to the ion with comparable strengths as H_2O or CH_3OH molecules. At the values of $n = 5$ in F^- and $n = 4$ in Cl^- and Br^- a sudden drop in clustering enthalpies $\Delta H_{n-1,n}^0$, occurs which indicates that all further CH_3CN molecules are bound far less strongly than the first few inner layer molecules. This drop is very difficult to explain but one important question remains completely open, why does this drop in enthalpies occur at smaller n in the bigger anions? No theoretical investigations on solvation of anions in aprotic solvents are available but we can expect stimulating co-operation of theoreticians and experimentalists in the future.

Of all experiments which provide information about the structure of solution shells, two recent studies on ionic solutions may be mentioned as being well suited for comparison with results of dynamical calculations on ion solvation. These are the NMR relaxation studies of Hertz[9,109], and the X-ray and neutron diffraction studies, commencing from pure water[11,110], to highly concentrated solutions ($LiCl/H_2O = 3$)[11,99] which in composition approaches that of the hydrated salt $Li \cdot 2H_2O$[111].

VII. Conclusion

In this chapter we have attempted to give a brief survey of the molecular theory of anion solvation. Reliable data is not very abundant in this field. Almost exclusively it is the hydration of monatomic anions which has been studied. Two very broad fields are open for further theoretical investigations especially as experimental data is rare and sometimes difficult to obtain, these are:

(1) Solvation of polyatomic anions, a class to which the majority of the most common anions belongs. Unfortunately, they all contain a large number of electrons, which represents an almost insuperable hindrance for rigorous quantum-mechanical investigations, e.g. among the smallest polyatomic anions we find NO_2^-, CO_3^{2-} and BF_4^- with 24, 32 and 42 electrons, respectively.

(2) Solvation in non-aqueous solvents, an important field, which is just beginning to grow from infancy.

From our analysis of the calculated results on hydration of small ions three main points have been clarified and these may also be valid for other ion–molecule complexes.

(1) Accurate data can be expected exclusively from ab initio LCAO–MO–SCF calculations with extended basis sets or from very carefully conceived empirical models. Energy partitioning in MO calculations may help appreciably with the construction of such models. In ion–molecule interactions the electrostatic contribution to intermolecular energies predominates. Two other contributions are of great general importance as well, exchange repulsion reflecting more or less shape and size of the particles, and mutual polarization. Charge transfer from the anion to the molecule and covalent character of the bond between ion and ligand may well be important in some special examples, as when strong hydrogen bonds are formed, but certainly do not predominate in anion molecule complexes. The relative weight of the dispersion energy does not seem to be very large. Nevertheless, more extended investigations including electron correlation effects have to be performed before a definite conclusion can be drawn. A simplified, though realistic picture of ion–molecule interaction, would start therefore with a superposition of the three quantities mentioned above.

(2) Again with the exception of some special cases, the molecular structure of one to one ion–solvent complexes as determined by energy calculations has already provided some useful information on the actual structure predominating

in solution, since the assumption of pairwise additivity of intermolecular potentials is a reasonably good approximation to the correct energy of large clusters. Nevertheless, many body potentials represent substantial corrections and cannot be left out completely, if accurate results are desired.

(3) Calculations introducing the effects of particle motion in the complex, molecular dynamics or Monte Carlo methods, are extremely encouraging. They seem to confirm that calculations based on a few hundred particles are able to provide a reasonable description of molecular processes going on in an ionic solution or, at least, as in the case of Monte Carlo calculations, help to overcome the terrible obstacles of statistical mechanics.

Finally, we refer to the general theme of this book and shall try to give a brief comparison of common and different problems in the theory of electron and anion solvation. One of the major unsolved problems in both theories concerns our knowledge of the detailed molecular structure of polar liquids which is very poor indeed. Electrons as well as anions are embedded in a medium about which we know little of the molecular structure (for comparison see Chapter 1). In both cases we can expect strong forces ordering polar molecules in the vicinity of the dissolved particle. These ordered structures may involve one, two or even more layers of solvent molecules depending on the strength of the solute's electric field and on the changes in the liquid's structure, which have occurred in the inner layers. In principle the model applied in the theory of electron solvation (Chap. 1, Fig. 2) is very similar to that which has been applied successfully to calculate free energies of ion solvation starting from a molecular model for the first solvation layer and a continuum outside (see ref. 22). There are, however, substantial differences. In ion solvation we are dealing with an aggregation of particles of molecular size each one stable as such, and bound together by forces which are weak in comparison with ordinary chemical bonds. Therefore, the geometries of the interacting particles are changed only slightly on complex formation. The corresponding one to one complexes are stable and can be investigated in the vapor phase. Their properties tell us something about the structure of corresponding subunits in large clusters. In the case of electron solvation the one to one complexes like H_2O^- or H_3N^- to our present knowledge are not stable entities. Therefore the first stable complexes are already larger units like $(H_2O)_4^-$ or $(H_3N)_4^-$ with complicated structural details and dynamical properties, not as yet known, and hardly accessible by theory or experiment. The first attempts to solve this kind of problem have been undertaken already by energy calculations on $(H_2O)_4^-$, which have now achieved a certain degree of reliability (see Chap. 1, ref. 45). Nevertheless, we feel that even more interesting details on solvated electrons are hidden in the dynamics of these clusters, and we can expect some interesting results in the future when the methods of molecular dynamics are applied in this field.

Acknowledgements

The author expresses his gratitude to Drs. G. H. F. Diercksen and W. P. Kraemer for sending plots for F^-H_2O prior to publication and for the permission to print them in this article. We should like to thank several other groups who have sent their most recent results. Numerical calculations of our group were performed on an IBM 360/44 (Interfakultäres Rechenzentrum der Univ. Wien), an IBM 370/165 (C.I.R.C.E., Paris Sud, Orsay) and an IBM 1130 (Hochschule für Bodenkultur, Wien). The author is indebted to all of these computing centers for a generous supply of computer time. The assistance of my co-workers with whom I have engaged in many discussions on this subject, and that of Mrs. J. Dura and Mr. J. Schuster in the preparation of this article, is gratefully acknowledged.

References

1 H. S. Frank and W. Y. Wen, Discuss. Faraday Soc., **24** (1957) 133.
2 R. W. Gurney, Ionic Processes in Solution, Dover, New York, 1962; H. P. Bennetto and E. F. Caldin, J. Chem. Soc., (1971) 2198.
3 W. G. McMillan, Jr. and J. E. Mayer, J. Chem. Phys., **13** (1945) 276.
4 J. Stecki, Advan. Chem. Phys. **6** (1964) 413.
5 R. R. Dogonadze and A. A. Kornyshev, J. Chem. Soc. Faraday Trans. II., **70** (1974) 1121.
6 I. R. McDonald and K. Singer, Chem. Brit., **9** (1973) 54.
7 H. L. Friedman, Chem. Brit., **9** (1973) 300.
8 P. Kebarle, in J. F. Franklin (Ed.), Ion–Molecule Reactions, Vol. 1, Butterworths, London, 1972, pp. 315–362.
9 H. G. Hertz, in W. A. P. Luck (Ed.), Structure of Water and Aqueous Solutions, Verlag Chemie and Physik Verlag, Weinheim, 1974, pp. 439–460.
10 M. D. Zeidler, in W. A. P. Luck (Ed.), Structure of Water and Aqueous Solutions, Verlag Chemie and Physik Verlag, Weinheim, 1974, pp. 461–472.
11 A. H. Narten, in W. A. P. Luck (Ed.), Structure of Water and Aqueous Solutions, Verlag Chemie and Physik Verlag, Weinheim, 1974, pp. 345–363.
12 P. Schuster, W. Jakubetz and W. Marius, Top. Curr. Chem., **60** (1976) 1.
13 J. O. Hirschfelder, C. F. Curtiss and R. B. Bird, Molecular Theory of Gases and Liquids, Wiley, New York, 1954.
14 H. Margenau and N. R. Kestner, Theory of Intermolecular Forces, 2nd edn., Pergamon, Oxford, 1971.
15 J. O. Hirschfelder (Ed.), Intermolecular Forces, Advan. Chem. Phys., Vol. 12, Interscience, New York, 1967.
16 P. Schuster, in P. Schuster, G. Zundel and C. Sandorfy (Eds.), The Hydrogen Bond, Vol. 1, North Holland, Amsterdam, in press.
17 E. Scrocco and J. Tomasi, Top. Curr. Chem., **42** (1973) 95.
18 F. London, Z. Phys., **63** (1930) 245; Z. Phys. Chem. Abt., B, **11** (1930) 222.
19 J. C. Slater and J. G. Kirkwood, Phys. Rev., **37** (1931) 682.
20 K. G. Spears, J. Chem. Phys., **57** (1972) 1850.
21 I. Elezier and P. Krindel, J. Chem. Phys., **57** (1972) 1884.
22 W. E. Morf and W. Simon, Helv. Chim. Acta, **54** (1971) 794.
23 H. F. Schaefer III, The Electronic Structure of Atoms and Molecules, Addison-Wesley, Reading, Mass., 1972.
24 W. Kutzelnigg, Top. Curr. Chem., **41** (1973) 31.
25 J. A. Pople and D. L. Beveridge, Approximate Molecular Orbital Theory, McGraw-Hill, New York, 1970.
26 J. N. Murrell and A. J. Harget, Semi-empirical Self-Consistent Field Molecular Orbital Theory of Molecules, Wiley-Interscience, London, 1972.
27 W. Meyer, J. Chem. Phys., **58** (1973) 1017.
28 P. Schuster, Theor. Chim. Acta, **19** (1970) 212.
29 B. J. Ransil, Rev. Mod. Phys., **32** (1960) 239.
30 S. Fraga and B. J. Ransil, J. Chem. Phys., **35** (1961) 669.

308

31 H. Lischka, J. Amer. Chem. Soc., **96** (1974) 4761.

32 J. A. Pople and G. A. Segal, J. Chem. Phys., **44** (1966) 3289.

33 R. Hoffmann, J. Chem. Phys., **39** (1963) 1397.

34 M. Wolfsberg and L. Helmholtz, J. Chem. Phys., **20** (1952) 837.

35 W. Adam, A. Grimison, R. Hoffmann and C. Zuazaga de Ortiz, J. Amer. Chem. Soc., **90** (1968) 1509.

36 R. Rein, G. A. Clark and F. E. Harris, J. Mol. Struct., **2** (1968) 103.

37 A. S. N. Murthy and C. N. R. Rao, Chem. Phys. Lett., **2** (1968) 123.

38 D. A. Zhogolev, B. Kh. Bunyatyan and Yu. A. Kruglyak, Chem. Phys. Lett., **18** (1973) 135.

39 B. Kh. Bunyatyan, D. A. Zhogolev and F. Ritschl, Chem. Phys. Lett., **24** (1974) 520.

40 W. L. Bloemer and B. L. Bruner, Chem. Phys. Lett., **17** (1972) 452.

41 P. Schuster, Mh. Chemie **100** (1969) 1033.

42 P. Schuster, Int. J. Quantum Chem., **3** (1969) 851.

43 P. R. Certain and J. O. Hirschfelder, J. Chem. Phys., **52** (1970) 5992.

44 D. M. Chipman and J. O. Hirschfelder, J. Chem. Phys., **59** (1973) 2838.

45 J. N. Murrell and G. Shaw, Mol. Phys., **12** (1967) 475.

46 J. G. C. M. Van Duijneveldt-Van de Rijdt and F. B. Van Duijneveldt, J. Amer. Chem. Soc., **93** (1971) 5644.

47 J. G. C. M. Van Duijneveldt-Van de Rijdt and F. B. Van Duijneveldt, Chem. Phys. Lett., **17** (1972) 425.

48 J. N. Murrell, M. Randic and D. R. Williams, Proc. Roy. Soc. Edinburgh, Sect. A, **284** (1965) 566.

49 J. N. Murrell and G. Shaw, J. Chem. Phys., **46** (1967) 1768.

50 F. B. Van Duijneveldt and J. N. Murrell, J. Chem. Phys., **46** (1967) 1759.

51 P. Schuster, Chem. Phys. Lett. submitted for publication.

52 J. A. Pople, D. P. Santry and G. A. Segal, J. Chem. Phys., **43** (1965) S 129.

53 J. A. Pople and G. A. Segal, J. Chem. Phys., **43** (1965) S 129.

54 H. Kistenmacher, H. Popkie and E. Clementi, J. Chem. Phys., **59** (1973) 5842.

55 E. Wigner, Phys. Rev., **46** (1934) 1002.

56 A. Johansson, P. A. Kollman and S. Rothenberg, Theor. Chim. Acta, **29** (1973) 167.

57 A. Meunier, B. Levy and G. Berthier, Theor. Chim. Acta, **29** (1973) 49.

58 W. Marius, Thesis (1st degree), Univ. of Vienna, 1973.

59 A. Beyer, H. Lischka and P. Schuster, to be published.

60 H. Lischka, Th. Plesser and P. Schuster, Chem. Phys. Lett., **6** (1970) 263.

61 P. Russegger, H. Lischka and P. Schuster, Theor. Chim. Acta, **24** (1972) 191.

62 P. Cremaschi, A. Gamba and H. Simonetta, Theor. Chim. Acta, **25** (1972) 237.

63 K. Breitschwerdt and H. Kistenmacher, Chem. Phys. Lett., **14** (1972) 288.

64 M. de Paz, S. Ehrenson and L. Friedman, J. Chem. Phys., **52** (1970) 3362.

65 R. S. Mulliken, J. Chem. Phys., **23** (1955) 1833, 1841, 2338 and 2343.

66 P. Russegger and P. Schuster, Chem. Phys. Lett., **19** (1973) 254.

67 P. O. Löwdin, J. Chem. Phys., **18** (1950) 365.

68 P. Russegger, Thesis, Univ. of Vienna, 1972.

69 G. H. F. Diercksen, Theor. Chim. Acta, **21** (1971) 335.

70 G. H. F. Diercksen and W. P. Kraemer, Chem. Phys. Lett., **5** (1970) 570.

71 H. Kistenmacher, H. Popkie and E. Clementi, J. Chem. Phys., **58** (1973) 5627.

72 M. Arshadi, R. Yamdagni and P. Kebarle, J. Phys. Chem., **74** (1970) 1475.

73 H. Kistenmacher, H. Popkie and E. Clementi, J. Chem. Phys., **61** (1974) 799.

74 E. Clementi and H. Popkie, J. Chem. Phys., **57** (1972) 1077.

75 W. J. Hehre, R. F. Stewart and J. A. Pople, J. Chem. Phys., **51** (1969) 2657.

76 M. Dreyfus and A. Pullman, Theor. Chim. Acta, **19** (1970) 20.

77 P. A. Kollman and L. C. Allen, Theor. Chim. Acta, **18** (1970) 399.

78 K. Morokuma, J. Chem. Phys., **55** (1971) 1236.

79 E. Clementi, J. Chem. Phys., **46** (1967) 3842.

80 K. Morokuma, S. Iwata and W. A. Lathan, in R. Dowdel and P. Pullman (Eds.), The World of Quantum Chemistry—Proceedings of the First International Congress of Quantum Chemistry, D. Reidel, Dordrecht, Holland, 1974, pp. 277–316.

81 W. P. Kraemer and G. H. F. Diercksen, Theor. Chim. Acta, 27 (1972) 365.

82 P. Schuster and W. Marius, unpublished results.

83 M. D. Newton, W. A. Lathan, W. J. Hehre and J. A. Pople, J. Chem. Phys., 51 (1969) 3927.

84 L. Piela, Chem. Phys. Lett., 19 (1973) 134.

85 G. C. Lie and E. Clementi, J. Chem. Phys., 60 (1974) 1275.

86 W. P. Kraemer and G. H. F. Diercksen, Theor. Chim. Acta, 23 (1972) 398.

87 M. D. Newton and S. Ehrenson, J. Amer. Chem. Soc., 93 (1971) 4971.

88 M. Arshadi and P. Kebarle, J. Phys. Chem., 74 (1970) 1483.

89 R. Ditchfield, W. J. Hehre and J. A. Pople, J. Chem. Phys., 54 (1971) 724.

90 W. J. Hehre, R. Ditchfield and J. A. Pople, J. Chem. Phys., 56 (1972) 2257.

91 M. Eigen, Angew. Chemie, Int. Ed. Engl., 3 (1964) 1.

92 P. Schuster, W. Jakubetz, G. Beier, W. Meyer and B. M. Rode, in B. Pullman and E. Bergmann (Eds.), Chemical and Biochemical Reactivity, The VIth Jerusalem Symp. on Quantum Chemistry and Biochem., The Israel Academy of Sciences and Humanities, Jerusalem, 1974, pp. 257–282.

93 K. Heinzinger and P. C. Vogel, Z. Naturforsch. A, 29 (1974) 1164.

94 R. O. Watts, E. Clementi and J. Fromm, J. Chem. Phys., 61 (1974) 2550.

95 A. Rahman and F. H. Stillinger, J. Chem. Phys., 55 (1971) 3336.

96 F. H. Stillinger and A. Rahman, J. Chem. Phys., 57 (1972) 1281.

97 A. Rahman and F. H. Stillinger, J. Amer. Chem. Soc., 95 (1973) 7943.

98 F. H. Stillinger and A. Rahman, J. Chem. Phys., 60 (1974) 1545.

99 A. H. Narten, F. Vaslow and H. A. Levy, J. Chem. Phys., 58 (1973) 5017.

100 H. G. Hertz, R. Tusch and H. Versmold, Ber. Bunsenges. Phys. Chem., 76 (1971) 1177.

101 J. A. Barber and R. O. Watts, Chem. Phys. Lett., 3 (1969) 144.

102 H. Kistenmacher, H. Popkie, E. Clementi and R. O. Watts, J. Chem. Phys., 60 (1974) 4455.

103 P. Kebarle, in M. Szwarc (Ed.), Ions and Ion-Pairs in Organic Reactions, Vol. 1, Wiley-Interscience, New York, 1972, pp. 27–83.

104 R. Yamadagni and P. Kebarle, J. Amer. Chem. Soc., 93 (1971) 7139.

105 R. Yamadagni and P. Kebarle, J. Amer. Chem. Soc., 94 (1972) 2940.

106 R. Yamadagni, J. D. Payzant and P. Kebarle, Can. J. Chem., 51 (1973) 2507.

107 D. F. C. Morris, Structure and Bonding (Berlin), 4 (1968) 63.

108 D. F. C. Morris, Structure and Bonding (Berlin), 6 (1969) 157.

109 H. G. Hertz and C. Rädle, Ber. Bunsenges. Phys. Chem., 77 (1973) 21.

110 A. M. Narten, J. Chem. Phys., 56 (1972). 5681.

111 M. Falk and O. Knop, in F. Franks (Ed.), Water, Vol. 2, Plenum Press, New York, 1973, pp. 55–113.

Chapter 9

SPECTROSCOPIC STUDIES OF MONATOMIC ANION–SOLVENT INTERACTIONS

M. C. R. SYMONS

I. Ultraviolet spectroscopy

I.A Introduction

Monatomic anions have well-defined electronic spectra both in salt crystals and in fluid solutions. The onset of absorption shifts markedly to low energies on going down the Periodic Table and this effect makes iodide by far the most convenient ion to study. The lowest energy bands are almost pure Gaussian error curves and, for iodides, are not strongly dependent upon the nature of the cation. There is invariably a second intense band separated from the first by an energy approximately equal to that for the energy separation of the $2P_{3/2}$ and $2P_{1/2}$ states of the corresponding halogen atoms. These bands can loosely be described as $p \rightarrow s$ in character although the outer s-orbital is strongly modified by the surrounding medium. Additional, less intense, bands are also found in this region which can be assigned loosely to $p \rightarrow d$ transitions. The first of these bands generally lies between the two components of the $p \rightarrow s$ transition in the case of iodides (Fig. 1).

The p–s bands broaden rapidly, and their maxima shift to lower energies as the temperature is increased. When iodide ions are incorporated in other alkali halide

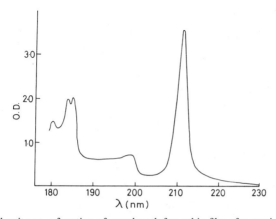

Fig. 1. Optical density as a function of wavelength for a thin film of potassium iodide at 4.2 K.

crystals, the *p–s* transitions remain well defined, but shift strongly to higher energies. For this and other reasons, these bands are best described as being localised transitions, centred around one halide ion, rather than utilising band theory. It is important that excitation within the first absorption bands does not result in photoelectric emission[1,2] or photoconductivity[3], although this does set in sharply at somewhat higher wavelengths. This corresponds to a shoulder in the UV spectra. Thus, for example, potassium iodide has its first *p–s* band at 46800 cm^{-1} and emission sets in at ca. 50000 cm^{-1}.

The most striking observation for solutions of halide ions is the marked similarity of their spectra to those of their salts (Fig. 2). In most solution studies attention has been primarily on the behaviour of iodide ions, although limited studies have shown that bromide and chloride ions behave similarly[4]. Although the absorption spectra of halide ions in solution were studied many years ago by Schiebe[5] and Lederle[6] we seem to have been the first to utilise these spectra as a method for studying ionic solvation[7–9]. These studies are rendered difficult by our lack of knowledge of the nature of the excited state[8], and this is discussed further below. Nevertheless, many significant factors regarding solvation may be extracted without invoking any detailed theory for the excited state.

Thus, for example, on adding ammonia to aqueous iodide solutions, the band maximum shifts monotonically to low energies, but the rate of shift is least in water-rich solutions and greatest in ammonia-rich solutions. This demonstrates that iodide is preferentially solvated by water molecules. It also shows that the transitions cannot involve electron transfer from iodide to one particular solvent molecule, since this would require the presence of two distinct bands for the mixed solvents. It is now common practice to describe these bands as charge-transfer to solvent (CTTS) spectra[7–9].

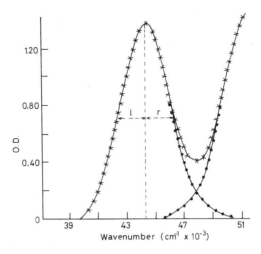

Fig. 2. The first absorption band for aqueous iodide ion (xx) together with the fit (—) of the sum of two log normal curves (●–●). Empirical width parameters are also illustrated (from ref. 14).

I.B Environmental effects on the absorption spectra for iodide ions

I.B.1 Solvents

Iodide ion spectra are in fact extremely sensitive to environmental changes. Thus, for example, on going from solvent water to hexamethylphosphoramide the first band shifts to low energies by about 6000 cm^{-1}.

In general, at a given temperature iodide ions in protic solvents exhibit high energy bands, whilst solutions in aprotic solvents exhibit low energy bands, with solutions in mixed protic–aprotic media giving intermediate bands. Recent results have been tabulated in a review by Blandamer and Fox[10], and rather than repeat this tabulation, much of the data is displayed in the form of a graph in Fig. 3.

I.B.2 Temperature

In the absence of special effects such as ion-pair formation, the shifts with temperature are remarkably linear, even over extended temperature ranges. Thus the

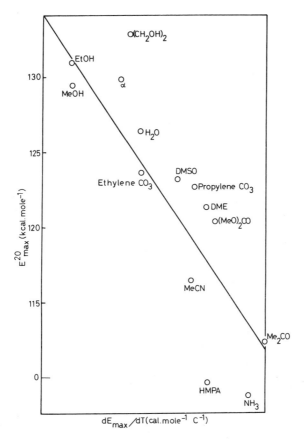

Fig. 3. Band maxima (E_{max}) for the first iodide transition at 20 °C as a function of its temperature sensitivity (dE_{max}/dT) for a range of solvents. (α is for KI crystals.)

parameter dE_{max}/dT can be taken as being as significant as E_{max} and these two are displayed together in Fig. 3. This correlation, often described as the Smith–Symons plot, shows clearly that solvents in which iodide ions display a low value of E_{max} have a large value for dE_{max}/dT, and vice versa. This means that if E_{max} values are extrapolated to near absolute zero, differences have been largely eliminated. The straight line indicated in the figure has a slope equal to the temperature selected for the data (293 K), and if all points fell on this line all solutions would have the same value for E_{max} at absolute zero.

This strongly suggests that the electron affinity of the solvent molecules is only a very minor factor in determining the energy of the transitions under consideration. (Indeed, as discussed below, when the solvent has an electron affinity that can compete favourably with the normal excited state, a new distinct band is obtained.)

We found that these results are accommodated in terms of a model in which the mean ion–solvent separation varies with temperature and strongly controls the energy of the optical transition[7-9,11]. If this is accepted, then we must conclude that good proton–donor solvents form relatively short hydrogen bonds to iodide and these only lengthen slightly on increasing the temperature. Thus fluorinated alcohols, which are the strongest acids studied, gave the highest energy transitions at room temperature, whereas ammonia, which can only form extremely weak hydrogen bonds, exhibits one of the lowest energy transitions known. Again, solvents such as methyl cyanide or ammonia, which can only form very weak hydrogen bonds, if any, move relatively far from the ion at elevated temperatures because the ion–dipole interaction is far weaker than that for hydrogen bonding.

This simple discussion ignores many factors which must contribute to the final value of E_{max} at a given temperature. These include the number of solvent molecules associated with the ion, which must to some extent control the effective solvent-shell radius in addition, of course, to changing the energy of the solvated ion in its ground state. Also, for mixed solutions of any sort, dE_{max}/dT ceases to be a significant number since preferential solvation invariably occurs, and the extent of this preference generally falls off with increasing temperature. Since protic solvents exhibit a marked preference over aprotic solvents, and since iodide has a greater value for E_{max} in the protic solvent component, this effect should add a positive increment to dE_{max}/dT for mixed protic–aprotic solvents. This is indeed observed experimentally. If a solvent has large molecules with protic and aprotic regions, one might predict that elevated temperatures would tend to favour disorientation, which would then be comparable to the effect observed for binary solvent mixtures. This may help to explain the relatively large dE_{max}/dT values exhibited by the higher alcohols such as t-butyl alcohol. 2-Cyanoethanol also fits into this category.

It is interesting that methanol or ethanol produce high energy shifts from the value of E_{max} in water, suggesting stronger interaction (H-bonding) to iodide for these solvents. This may reflect a slight increase in effective acidity for these media, or an increase in the tendency for the alcohols to donate protons to weakly solvated ions such as iodide, in preference to hydrogen bonding with other solvent molecules. (This issue is considered in depth in Sect. III.)

Equally significant is the marked curvature for the mixed water–alcohol solvent systems (Fig. 4), in a sense that suggests preferential solvation by the alcohol mole-

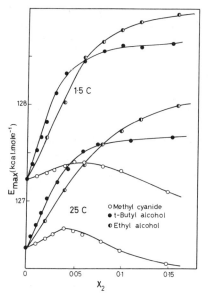

Fig. 4. Band maxima (E_{max}) for the first transition of aqueous iodide ions as a function of the mole fraction X_2 of added cosolvent at 1.5 and 25°C.

cules. For methanol this is probably a correct inference, but for *t*-butyl alcohol it is almost certainly not correct (see below) and for ethanol and other alcohols, which exhibit initial trends from the pure water value intermediate between those for methanol and *t*-butyl alcohol, both factors probably contribute.

In this connection, it is worth recalling that when methanol is added to water, there is an initial down-field shift in the O–H proton resonance value, indicating an increase in the effective hydrogen bonding. We suggest that it is this increase that is felt by the iodide, either directly, or indirectly.

The problem of comparing the effective H-bond donating power of water and an alcohol is rendered complicated by the changes in this power with the extent of H-bonding to the solvent molecule under consideration[12]. Consider the following situations for an iodide ion

316

The solvating effect in (II) will be greater than (I) because the second methanol molecule increases the H-bonding power of the first. Similarly for water, on going from (III) to (IV), enhanced interaction with I^- should result, but this can now be largely cancelled if structure (V) is formed[12].

I.B.3 Band-widths

Width changes for these Gaussian bands are another important parameter, but unfortunately these have not often been reported. The oscillator strengths for the first iodide *p–s* band are probably almost independent of environmental changes, remaining close to 0.25[11]. Thus changes in the maximum extinction coefficient (ϵ_{max}) can also be used as a measure of changes in width. The most important result is that the widths increase monotonically with increase in temperature in parallel with the shift in E_{max}. Indeed, for the temperature effect (but not the solvent effects) one can depict the shift and broadening simply in terms of an increase in the range of absorbing units present, stemming from a limiting unit having a band in the high energy region (Fig. 5). This simple model cannot be used to explain solvent effects however, since there is no correlation between band-widths and E_{max} in this sense.

Since the absorptions are governed by the Frank–Condon restriction, and since vibronic interactions in solution are too small to explain the large experimental widths (a typical half-height width being ca. 4800 cm^{-1}, compared with ca. 2000 cm^{-1} for iodide in crystals) we postulate that the width is essentially governed by the range of ion–solvent structures present at a given temperature. This range will include distributions of ion–solvent separations, relative orientations, and solvation numbers. In other words, the band-widths ($\Delta H_{1/2}$) can be taken as a measure of the precision of solvation, the narrower the band, the more precise the solvation. This accommodates the temperature effect satisfactorily.

Experimentally, widths for iodide absorption bands in different solvents do not vary extensively at a given temperature. Thus the width for the iodide band in

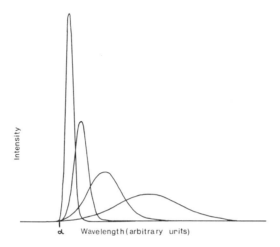

Fig. 5.   Set of curves of equal area generated for a single point (α).

$(F_3C)_2CHOH$ solution is very close to that in water. One curious and significant factor is that no marked increase in width is detected for iodide in a wide range of mixed solvents. Indeed, for certain aqueous mixtures in the water-rich region there is actually a narrowing. This result is surprising, because one would expect to find a far greater range of types of solvent shells in the mixed solvent systems, especially in the regions that display large shifts in E_{max} with changes in solvent composition. We interpret this result partially in terms of a lack of dependence of E_{max} on the electron affinity of the solvent molecules, but rather upon the mean effective radius of the solvent shell. Nevertheless, since replacement of a protic solvent molecule by an aprotic molecule leads to a relatively large shift to low energy, one would have expected a system which includes, for example, $I^-(S_1)_4(S_2)_2$, $I^-(S_1)_3(S_2)_3$, and $I^-(S_1)_2(S_2)_4$ to cover a larger range of energies than those only containing, say, $I^-(S_1)_6$ or $I^-(S_2)_6$ only. This problem is worthy of further attention.

We have previously analysed the unusually skew nature of the absorption bands for trapped and solvated electrons[13] in terms of a square-well model coupled with a harmonic distribution of radii. This highly simplified model gives a qualitatively satisfactory explanation of this asymmetry, which produces a long tail effect on the high energy side of the spectra. It is interesting that careful analysis of the absorption spectrum for I^- in water has revealed a similar but less extreme skewness (Fig. 2). (This figure, taken from ref. 14, also shows a reconstruction on the basis of a log normal fit, the skew factor ρ being 1.028.) It is perhaps worth noting that the high energy tail in Fig. 2 could be largely eliminated if the band were to be taken (arbitrarily) as pure Gaussian ($\rho = 1$) and a small extra band were assumed to be present in the 48000 cm^{-1} region. This observation means that reports of very weak bands in the region between the two major iodide bands[15] may need to be viewed with caution. This extra band (labelled C in ref. 15) has been tentatively assigned to a higher energy transition (possibly $p \rightarrow d$ in nature), and crystal data do indeed support this suggestion[16]. It is also possible that the extra feature, if real, is caused by a small concentration of ion-pairs.

I.B.4 The effect of pressure changes

Changes in pressure at constant temperature again produce linear shifts in the absorption of solvated iodide ions, with the exception of aqueous solutions[17,19]. These shifts are to high energy as the pressure increases, but there is virtually no change in the band contours under this constraint, in marked contrast with the effect of temperature changes. This rules out any description of these systems on the model implied in Fig. 5. It means that solvation is not made any more precise as pressure is increased, which firmly links the band widths to the degree of disorder of the solvation, which is controlled by temperature. We have interpreted these shifts simply in terms of a compression of the solvent cages surrounding the anions. Just the same effect is noted for I^- in, say, a KCl crystal. Very large high energy band shifts occur, relative to pure KI and this can be directly linked to the very great internal pressure experienced by the iodide occupying a chloride ion lattice site.

The most significant result is the large pressure-dependence for the typical aprotic solvent, methyl cyanide[18]. This accords with our previous concept of a large cavity

318

defined by loosely held solvent molecules, which was used to explain the low value of E_{max} and the large temperature sensitivity[7-9]. However, there is no direct relationship between E_{max} and dE_{max}/dP, since the pressure sensitivity for alcoholic solutions is considerably greater than that for aqueous solutions, except in the initial range[18,19]. Thus, for water, after an initially 'normal' shift, the pressure effect falls to a low value. It seems that the iodide ion is buffered from pressure changes in water in an unusual manner, possibly because of the three-dimensional structure exhibited by this solvent. When potassium fluoride was added this initial curvature was reduced and finally vanished, leaving the pressure-dependence equal to that for iodide in pure water in the high pressure region. In other words, the effect was just as if the KF exerted a large internal pressure at ambient external pressure. In contrast, dioxane[17] or t-butyl alcohol[18], both of which, like KF, cause E_{max} to shift to high energies, seem to extend the pressure range in which a high sensitivity is displayed! In the latter case, as discussed below, there is reason to suggest that I$^-$ ions are incorporated in the filigree of bound water molecules that encage the large co-solvent molecules. The effect of pressure may possibly be to inhibit this type of enclathration, but again, not enough is known about these systems to enable us to draw detailed conclusions from these curious trends.

I.B.5 Effect of co-solvents on aqueous iodide

Here we draw attention to the behaviour exhibited by the iodide spectrum when certain co-solvents such as dioxan, acetone or t-butyl alcohol are added to aqueous solutions (Fig. 6). This was first noted some years ago[9] and it was concluded that the sharp initial high energy shift exhibited by added dioxan in the 0.0–0.05 MF region

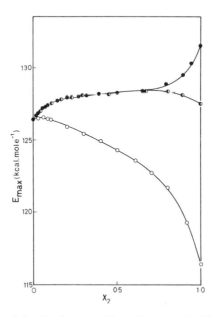

Fig. 6. Band maxima (E_{max}) for the first transition of aqueous iodide as a function of the mole fraction X_2 of added methyl cyanide (–O–) and t-butyl alcohol ($\bullet$ Na $^+$I$^-$ and $\mathbb{O}$ n-Hexyl$_4$N $^+$I$^-$).

could not signify a preferential solvation by dioxan, and hence the shift had to be understood in terms of a modification of the way in which iodide is solvated by water. The high energy shift seemed to imply an increase in the effective solvation and this was discussed in terms of dioxan molecules invading the region immediately outside the solvated ion[9]. This model has been somewhat refined to take into account the marked band-narrowing also observed during the initial shift (Fig. 6). This behaviour, suggesting an increased precision in the atmosphere around each iodide ion, was explained in terms of iodide being forced into one special type of water lattice, namely the structured water surrounding each co-solvent molecule[20]. Thus, we view normal water as offering a range of solvates to iodide ion, which are narrowed down considerably as water is induced to enclathrate large solute molecules.

I.C Models for the excited state

We have already stressed that early models involving electron transfer to one solvent molecule are unsatisfactory, provided the solvent does not have a high electron affinity. No proper quantum-mechanical treatment is yet available (to my knowledge), but two relatively crude models have been utilised as bases for discussing and qualitatively understanding the observed environmental effects. These will be described as the "delocalised"[21,22] and the "confined"[8,9] models. Since I have been the major proponent of the latter, I will outline reasons why I feel that there is a strong case for rejecting models in which the excited electron is not taken as being strongly confined.

Some arguments are based upon the marked spectral similarity between I_{solv}^- and e_{solv}^-. Thus, if it can be shown that e_{solv}^- is strongly localised within, or close to, a cavity in the solvent, then this must, surely, also be true for the excited state for I^- and related ions.

One of the most convincing arguments in favour of marked localisation comes from studies of trapped electrons in alkali hydroxide glasses[23,24]. It seems to be generally agreed that these electrons are to be found in molecular sized vacancies (possibly hydroxide ion sites[23]). The line widths of the ESR spectra for these centres was shown[23] to be markedly dependent upon coupling to protons of surrounding water molecules, but not at all upon coupling to the nuclei of nearby cations, despite the fact that, on average, a cation is most likely to be the next nearest neighbour to the trapped electron. Thus the penetration of the electron must be considerably smaller than that found for F-centres, where the electron exhibits considerable hyperfine interaction to next nearest neighbours, and even F-centres are thought to be fairly strongly confined[25-27]. A second related factor is that g-values for e_t^- remain very close to the free-spin value and are not strongly solvent dependent.

Similarly, for electrons trapped in glassy alcohols at $77 K$[28], there is a strong electron–nuclear hyperfine coupling to the OH protons which will be nearest to the electron if it is confined to a molecular-sized cavity, but the somewhat more remote C–H protons contribute very little indeed to the total linewidth. This seems to imply a relatively steep "wall" preventing major penetration of the electron away from its cavity. I find it impossible to imagine that a trapped or solvated electron could occupy a confined space in a crystal or solvent and yet that the electron in the excited state for halide ions is strongly delocalised.

It has been argued[29] that the fact that the oscillator strength for iodide in salts and in solution is close to 0.25 (for the first band) augers against the confined model. This is, surely, a false conclusion, since it was based upon a confined model which ignored the presence of the bulky iodine atom. In fact, of course, in the absence of these atoms, the oscillator strength would be high, but in its presence it falls considerably, because the transition probability depends upon the degree of overlap between the $5p$ level for iodide and the excited state orbital. In the important region of overlap this will resemble the iodine $6s$-orbital, and hence arguments based upon the use of a square-well orbital in isolation are quite invalid. In fact, it seems to me that the constancy of the oscillator strength for this transition in a range of environments suggests that this overlap must be the governing factor rather than the outermost form of the wavefunction which reflects so strongly the changes in solvation under consideration.

The reason for the confined nature of the excited state is, I believe, the Pauli exclusion principle coupled with the s-type wavefunction imposed by the central iodine atom. Thus, it is necessary to make the wavefunction of the electron orthogonal to those of the core and valence electrons of the surrounding solvent molecules. Since the lowest vacant molecular orbitals of the solvent molecules are of, relatively, very high energy in the system under consideration, this mechanism affords a high barrier to the electron and prevents a major leak of spin-density beyond the first solvent shell.

Another treatment of the iodide absorption is that of Siano and Metzler[14] who fitted their accurate band-shape data to a configurational-coordinate model, similar to those used to describe F-centres[30]. Their analysis is to some extent inconclusive, but strongly suggests, as inferred above, that coupling to normal solvent "lattice" modes is insufficient to explain the width, and that one needs to invoke the concept of a solvent cage with variable effective radius, as had been done previously.

It is interesting, in passing, to note that there are certain similarities between halide ion spectra and those of the isoelectronic rare gas atoms. Thus Brith and Ron[31] found that solid Xe exhibited a large shift to high energies for the $^3P_1 \leftarrow {}^1S_0$ and $^1P_1 \leftarrow {}^1S_0$ transitions relative to the gas phase, together with an extensive band broadening. These studies have been extended to Xe atoms in solid and liquid rare gases (Ar and Kr) by Jortner and co-workers[32,33], who confirm that these two transitions are strongly shifted to high energies, the shift being greatest in argon. These studies are of great significance in many areas, but the point I wish to make is that there is a clear qualitative analogy to results for halide ions. Since it is concluded[32,33] that the results for xenon establish strongly localised excited states dominated by an environmental "well", I feel that this once again supports the confined model for halide ions.

I.D Excited state relaxation

Relatively little is known about the way in which these excited states fall to the ground state. (Quantum yields for decomposition are very low in pure solvents.)

However, it seems that in alkali halide crystals[34], the excited state may have a long enough lifetime to permit the movement of two halide ions towards each other to give, in effect, a V_K centre (hal$_2^-$) closely associated with the excited electron. Exactly the same effect occurs for rare gas atoms in close proximity[32]. Since it has been found that the radical IOH$^-$, analogous to I$_2^-$, can be formed from aqueous iodide solutions[35,36], it seems quite probable that the excited state for iodide in water, for example, might move towards the state $[\text{I–OH}_2]^-$ prior to relaxing back to the ground state.

I.E Ion-pair formation

Ion-pairs come in a range of structures. For halide ions it is convenient to use the definitions of Griffiths and Symons[37], who considered the units (I)–(III) to be of structural significance where (I), (II) and (III) were described as contact, solvent-

$$\ominus\oplus \qquad \ominus\text{---}\,s\,\text{---}\oplus \qquad \ominus\text{---}\,ss\,\text{---}\oplus$$

$$\text{(I)} \qquad\qquad \text{(II)} \qquad\qquad \text{(III)}$$

shared and solvent-separated ion-pairs, respectively. Whilst one would expect CTTS spectra to be modified by the formation of (I) and (II), formation of (III) is unlikely to influence the absorption detectably.

The extent of spectral change is, then, very much a function of the nature of the solvent and the cation, and even formation of contact ion-pairs will not necessarily be spectroscopically very significant. One limiting case is the gas phase ion-pair, M$^+$hal$^-$, where M$^+$ is an alkali-metal cation. The iodides all exhibit the usual doublet, the first band falling in the 30800 cm^{-1} region which is almost independent of the cation. This transition is simply and satisfactorily described as a charge transfer to cation. However, the acceptor ability of the cation will be greatly reduced by co-ordinated donor solvent molecules, whilst the donor ability of the anion will be reduced by hydrogen bonding solvent molecules, and these two factors probably serve to shift any potential "charge transfer to cation" band to the high energy side of the normal CTTS absorptions. Thus it seems to be quite satisfactory to discuss iodide absorption spectra for ion-pairs in terms of modified CTTS absorptions.

Since iodide in its salts and in hydroxylic solvents has its first band in the same spectral region it is likely that contact ion-pair formation will tend to shift the absorption into this region, namely ca. 220 nm at room temperature. Similarly, contact ion-pairing with tetra alkylammonium ions should resemble aprotic solvation and give a band in the 240–250 nm region. In contrast, trialkylammonium ions, R$_3$NH$^+$, would be expected to form H-bonds to iodide and hence resemble protic solvents.

It is not so easy to generalise about solvent-shared ion-pairing. This is only expected to be spectroscopically significant for alkali-metal cations since the perturbing effects of R$_4$N$^+$ ions upon a solvent molecule linked to an anion should be too small to detect. The effect of M$^+$ on protic solvates will again be small, since one must envisage the change as a displacement, as, for example, and the indirect effect on iodide cannot be very significant. Indeed, NMR data, discussed in Sect. II, suggest that the shifts might be either to high or low energy, depending upon the nature of the cation and protic solvent.

For aprotic solvents the change is again likely to be small, provided there is no specific charge-transfer interaction with the solvent molecules. Thus consider $I^- \text{---} H_3CCN + Na^+ \rightarrow I^- \text{---} H_3CCN \text{---} Na^+$: in the absence of H-bonding to nitrogen, this should increase the (very weak) tendency of the methyl protons to H-bond to iodide, and hence cause a small high energy band-shift.

These prognostications are in fact well borne out[38,39]. Thus $R_4N^+ I^-$ ion-pairs in methylene chloride, 1,4-dioxane or tetrahydrofuran have E_{max} in the region of 41000 cm^{-1}, which is very close to the value for, say, iodide in methylcyanide (ca. 40700 cm^{-1}). However, bonding to $R_2NH_2^+$ in these solvents shifted the band markedly to high energies (ca. 45000 cm^{-1}) and bonding to sodium ion in tetrahydrofuran caused a shift of ca. 2000 cm^{-1}[38]. Both these values are well into the normal region for protic solvents, and this observation leads me towards a tentative and novel conclusion, this is that when protic solvent molecules displace aprotic molecules progressively, the major high energy shift is induced by the first molecule added, for example, for $I^-S_6 + ROH \rightarrow ROH \text{---} I^-S_5 + S$. Subsequent displacement of S by ROH is less effective because of a compensatory reduction in the H-bonding strength for a given ROH molecule as others are added. (This compensatory effect is discussed further in Sect. II.) My reason for suggesting this is the very large shift induced by one added R_3NH^+ ion. This is, of course, more acidic than normal ROH molecules, and so the shift is larger, but nevertheless, it is most significant that just one ion can have such a large effect. If this suggestion is in any sense true, then the deductions drawn in relation to the generally strongly curved plots for iodide band-shifts in mixed solvents (see Fig. 4) cannot be taken as establishing, in any quantitative sense, the occurrence of preferential solvation, since the shifts will not be linear through the series. Also, if the shift is really large for the first added protic solvent molecule, it ought to be possible to detect the growth of a distinct new band during the initial addition of the protic solvent to a solution of iodide in an aprotic solvent. We are currently looking for this effect.

Griffiths and Wijayanayake[39] have presented a systematic study of ion-pair formation which greatly extends the early work in this area[38]. Their results show that as the permittivity of alcohols falls, so ion-pairing occurs, but it is the R_4N^+ ion that causes the largest shift (to low energy) rather than alkali metal cations (to high energy). This is because the bulky alkylammonium ion displaces protic solvent molecules and replaces them, in effect, by an aprotic atmosphere. A similar situation was found in our ESR studies of ion-pairing[40]. We found that the ^{14}N hyperfine coupling for various nitrobenzene anions was modified far more by R_4N^+ ions than by alkali-metal ions in alcoholic media. The ^{14}N coupling is large in ROH solvents because of H-bonding to the nitro-group. Ions like Na$^+$ simply reinforce this effect slightly, causing small increases in $A_{iso}(^{14}N)$. However, R_4N^+ ions displace ROH molecules, thus indirectly causing a large decrease in A_{iso}.

It is interesting that the difference in the shifts caused by M$^+$ ions and R_4N^+ ions in solvents of low permittivity seems to tend towards a limiting value of ca. 1600 cm^{-1}, independent of the type of medium[39]. This may be somewhat fortuitous since, as indicated above, for some solvents it is the loss of solvation that causes the shift, whilst for others, it is the presence of the cation.

I.F Specific charge transfer complexes

Before leaving this section, reference should be made to specific interactions between halide ions and electron acceptor solvents. This phenomenon was discovered by Griffiths and Symons[37], but was not correctly diagnosed until later[41]. When tetraalkylammonium iodides are dissolved in carbon tetrachloride a very low energy band is produced (Fig. 7) in the 35000 cm^{-1} region. This is almost independent of cation and of temperature, but is rapidly lost on the addition of strongly hydrogen-bonding solvents such as methanol. The transition is, in our view, the transfer of an electron to the CCl$_4$ molecule. Chloroform and methylene chloride do not seem to form such bands, possibly because they form weak H-bonds to iodide. These studies have been greatly extended by Davis and co-workers, an example being his study of charge-transfer complex formation between halide ions and trinitrobenzene, which is, of course, a very good acceptor[42]. It is very interesting to note that the effect of the cation in these transitions is qualitatively the same as that discussed above for iodide ions in alcohols. Thus, for t-butyl alcohol the iodide–trinitrobenzene complex with Na$^+$ has a band very close to that for the solvated complex in ethanol, the sodium ion causing a very small high energy shift. In marked contrast, R$_4$N$^+$ ions cause a large low energy shift, which tends towards the value for the solvated complex in methyl cyanide[42]. I would again interpret these changes as being dominated by loss of the hydrogen-bonding solvent in the case of the R$_4$N$^+$ salts. For the sodium salt the ion pair is most likely to be type (II) rather than (I). A limiting case of charge-transfer complex formation is found with the pyridinium complexes, studied ex-

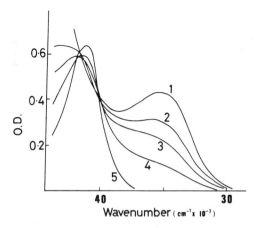

Fig. 7. Absorption spectra of (n-Hexyl)$_4$N$^+$I$^-$ in carbon tetrachloride-methyl cyanide solution, the mole fraction of methyl cyanide being (1) 0.455; (2) 0.754; (3) 0.847; (4) 0.928; (5) 1.00.

324

tensively by Kosower and co-workers[43,44], from which solvent Z-values have been derived.

I.G Polyatomic anions

Brief mention should be made of the possibility of polyatomic anions exhibiting CTTS spectra which may be useful for solvation studies. This topic has been reviewed recently by Blandamer and Fox[10]. Certainly for the protonated anions, OH^-, SH^- and NH_2^-, the absorption bands do seem to contain a strong CTTS element. However, for more complex ions such as CN^-, CNO^-, NO_2^-, NO_3^-, ClO_3^-, PO_4^{3-}, etc. the situation is far more complex because of the presence of internal transitions and the difficulty of distinguishing between them[45]. The simple theories discussed above predict a linear dependence of E_{max} for CTTS bands upon ionisation potential (IP), and this is observed in many cases (Fig. 8). However, as discussed by Fox and Hunter[46], a case can be made for a parabolic dependence, as also indicated in Fig. 8, and this effect was explained by them in terms of a theory for CTTS bands based upon Mulliken's theory for charge-transfer complexes and an acceptor orbital in the form of a shell around the ions[46]. I think that the parabolic dependence is not really established and hence that the theory is not supportable at present. It is noteworthy that a similar plot for solutions in liquid ammonia, which now includes the NH_2^-

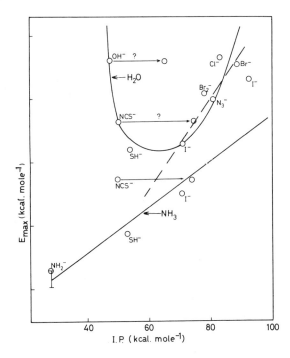

Fig. 8. E_{max} for the CTTS transitions of various anions in water and ammonia as a function of their ionisation potentials.

ion with its very low IP (ca. 28 kcal M^{-1}) is reasonably linear except for NCS^- if an IP of 49.9 kcal M^{-1} is used[46]. This value[47] is out of line with those used for $NCSe^-$ and $NCTe^-$ ions (66.0 and 59.0, respectively). If we take the linear correlation for ammonia to predict an IP of ca. 75 kcal M^{-1} for NCS^- ions this comes better into line with the series from NCO^- to $NCTe^-$ and moves the point for the water curve nicely onto the linear plot indicated in Fig. 8.

The only point now defining the parabola is that for hydroxide ions. If one uses Page's value of ca. 65 kcal M^{-1} for IP (OH^-) a point quite close to the water line is obtained. However, both OH^- and SH^- are still somewhat displaced from the main line and this may be caused by specific effects caused by the proton in these ions.

I.H Links with solvated electrons

Some years ago we drew attention to the remarkably similar way in which the spectra for iodide ions and solvated or trapped electrons varied with the nature of the medium[48]. These trends were subsequently discussed by Anbar and Hart[49], and recent extensions have provided no major exceptions to the linear correlation (Fig. 9). The slope of this line is such as to suggest that e_{solv}^- are somewhat more sensitive to environmental changes than I^-, but really they are remarkably similar, and this seems to me to be one of the strongest arguments in favour of the "solvated anion" concept of e_{solv}^-. If this species were really in some quite novel and unique form,

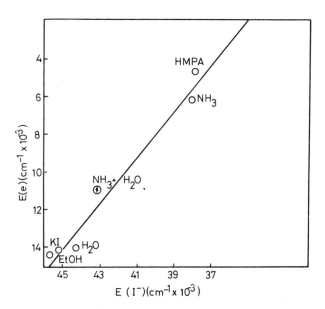

Fig. 9. Comparison of the band maxima for iodide ions with those for solvated electrons (e^-) in a range of media.

bearing no relation to the simple "electron-in-a-cavity" models, then why should its optical properties be so similar to those for I^-?

Again, the link with F-centres, of known structure, is well defined. This has extra significance, since many would argue that F-centres are quite different from e_{solv}^- or e_t^- units since the electron can be thought of as being bonded to the six cations that surround the anion vacancy. If that were true, why is there such a close link in their optical spectra?

It is worth recalling that ESR studies of metal–ammonia solutions containing added electrolytes[50] revealed g-shifts and line width increases that mainly stemmed from interaction with the anions, the shifts (to low g-values) and broadening increasing strongly on going from chloride to iodide. We attempted to explain this unexpected result in terms of a very minor contribution from species such as I^{2-}

$$e_{solv}^- + I_{solv}^- \rightleftharpoons I_{solv}^{2-}$$

This would place the unpaired electron into the first excited-state orbital discussed above. The broadening would then arise because of a rapidly modulated electron–nuclear hyperfine interaction with ^{127}I.

The g-shifts will not stem directly from this model, but provided local distortions of the solvent shell about iodide are sufficient to mix in even very small quantities of one of the $6p$-orbitals on iodide, this would be accommodated. Because of these results, we have sought to add an excess electron to halide ions in rigid glasses, but without definitive success. However, when barium sulphate doped with bromide or iodide ions was exposed to ^{60}Co γ-rays at 77 K, new species which exhibited very large hyperfine coupling to the halogen nuclei and strongly negative g-shifts were detected by ESR spectroscopy[51]. We have previously demonstrated[52] that, when anions are incorporated into ionic crystals in which the parent anion has one charge more than that of the solute, electron capture by the added anion is facile.

The reason why these anions in ammonia are more ready to interact with e_{solv}^- than the cations is, in my view, because ammonia is a good cation solvator but a very poor anion solvator. This accords with our contention that there is a large solvent cavity around halide ions in ammonia at normal temperatures. Yet another factor is that the solvent around the anions is already oriented in the required manner to stabilise e_{solv}^-, whereas the solvent coordinated to the cations will strongly oppose the close addition of e_{solv}^-. Hence when e_{solv}^- interacts with M^+ in ammonia it is more likely to do this via a type of off-centre ion-pairing.

II. Nuclear magnetic resonance spectroscopy

II.A Introduction

As with IR and Raman spectroscopy, there are two choices open to the NMR spectroscopist, in that he can study solvent nuclei or those of the ions. Here, attention is focused first on proton resonance studies of protic media, and then on studies devoted to halide ion nuclei. More comprehensive treatments have appeared relatively recently[53,54].

II.B Proton resonance for protic media

Since proton NMR absorption bands for liquids are usually extremely narrow, it might be thought that this technique would overcome all the disadvantages experienced from the excessively broad bands in the IR and Raman spectra (Sect. III). This, of course, is not the case, since these narrow bands are time averages of all the different bands contributing to the total resonance. Given that distortions, H-bond bending effects, etc. are completely averaged, one might have hoped at least to see separate features from $O-H_{free}$, $O-H_{bound}$, $O-H$---anion and HO---cation units. A hypothetical spectrum is sketched in Fig. 10. (Note, $O-H_{free}$ groups absorb at high fields: increasing the strength of H-bonding results in increasing down-field shifts. For water, by $(O-H)_{free}$ we do not imply $(H_2O)_{free}$; the water molecules involved may form up to three hydrogen bonds.)

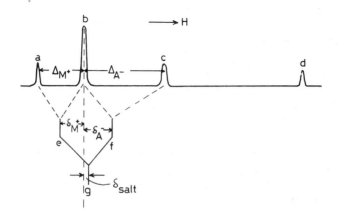

Fig. 10. Hypothetical slow exchange proton magnetic resonance spectrum for an aqueous solution of the salt M^+A^-. (a) The cation solvent shell; (b) normal water; (c) the anion solvent shell; (d) $(OH)_{free}$ protons
δ_M^+ is the shift for the average (e) of (a) and (b),
δ_A^- is the shift for the average (f) of (b) and (c), (g) is the average of (e) and (f) giving δ_{salt} the salt shift as normally measured.

II.B.1 Protic solvents in aprotic media

One method whereby this difficulty can be largely overcome is to utilise the systems considered in Sect. III.B below. As an electrolyte, $R_4N^+hal^-$, is added to a solution of, say, methanol in methylene chloride, the $O-H$ proton resonance shifts strongly down-field, generally to a limiting value that can be accurately determined and is equated to the value for the mono-solvate hal^----HOR[55,56]. Equilibrium constants derived by this method agree reasonably well with those based on the IR method, so there can be little doubt that both methods observe the same phenomenon. Furthermore, the relative shifts for Cl^-, Br^-, I^- and ClO_4^- are the same for both techniques. Unfortunately, we do not yet know the limiting shift for the fluoride ion.

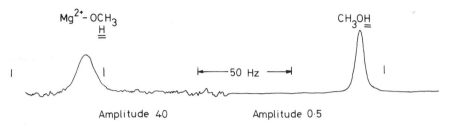

Fig. 11. Proton resonance spectrum for a methanolic solution of magnesium perchlorate showing the resolved feature for O–H protons in the solvation shell of the magnesium ions.

II.B.2 Methanol at low temperatures

Swinehart and Taube[57] discovered that by cooling methanolic solutions of magnesium salts below 0 °C a new band appeared in the PMR spectrum which narrowed on cooling further (Fig. 11). This band was shown to be caused by hydroxyl protons of methanol molecules directly bonded to the magnesium ion. This very important discovery has since been extended to a wide range of methanolic and aqueous solutions and much useful information regarding the solvation of divalent and trivalent cations has been forthcoming[58]. We have made use of this phenomenon in an alternative way[59], focusing attention upon the magnitudes of the shifts rather than on the areas of the curves, and also monitoring the shifts of the normal OH proton signals. The former shifts can be used to estimate the molal ion shifts for the cations under fast exchange conditions, and the latter can be used to obtain the molal anion shifts directly. This procedure is important since previous attempts to split "salt shifts" for the OH protons of solvents such as water or methanol into individual ion shifts were completely arbitrary, and indeed it was frequently stated that a correct separation was impossible in principle.

Of course, our method suffers several weaknesses, the major one being the assumption that the anion effects dominate the residual shift. This is not necessarily correct, since secondary cation solvation is expected to contribute to the residual shift. However, this is almost certainly a relatively small contribution, and we can at least begin to see the factors involved in ionic shifts in bulk solvents.

II.B.3 Anion solvation by water

Division into individual shifts for electrolytes in water is less readily accomplished, but an attempt has been made[60] using three unconnected methods that all give the same results within the error limits involved in obtaining the data. One method involved extrapolating from data for aluminium salts in concentrated solution for which a separate absorption was obtained for $Al(H_2O)_6^{3+}$ ions. The others were less direct, and the reader is referred to the original work for details[60].

The results for anions are displayed for water and methanol as a function of anion basicity in Figs. 12 and 13. Perhaps the most significant factor is the clear relation between the observed shift and the pK_a value. This strongly supports the picture of anion solvation by protic solvents as comprising weak hydrogen-bonding from one or more solvent molecules. Another important result is that the change in sign of the

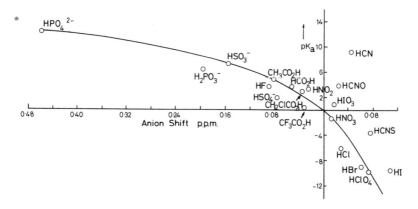

Fig. 12.　Molal shifts for aqueous anions as a function of pK_a for their conjugate acids.

shift on going from strongly basic anions (downfield shift) to non-basic anions (upfield shift) is seen to have no more significance than that of the basicity of the medium itself. In other words, if we consider the process

focusing attention on H*, then if the effective strength of the hydrogen-bond to X^- is less than that to methanol, there will be an upfield shift. Two qualifying points must be made. One is that the above equilibrium will still favour the right-hand side even if the hydrogen-bond is weaker, since the long range Coulombic force is stronger. The other factor is that the displaced methanol molecule should not be thought of as being in any sense "free"—it will simply join the remainder, the number of "free" molecules being mainly a function of the temperature.

I contend that this is a better view of anion solvation than the poorly defined concepts of "structure-making" and "structure-breaking" so often invoked to explain downfield or upfield NMR solvent shifts. In this context it is worth recalling that the iodide ion, which emerges as a strong "structure-breaker" on this classification, shows no sign of a band in the OH_{free} region in the IR or Raman spectra of its aqueous or methanolic solutions. What it does show, in accord with the NMR shifts is a band at higher frequencies than that for water polymer, which again supports the concept that the hydrogen bonds to iodide are weaker than those between solvent molecules.

Again, in accord with these ideas is the fact that the mean life time of an iodide–solvent bond is shorter than that of a solvent–solvent bond for hydroxylic solvents[61]. This reflects the fact that the bond is indeed weaker, but I can see no reason for using that fact to indicate either no solvation or indeed, negative solvation.

Before leaving the plots in Figs. 12 and 13, particular mention should be made

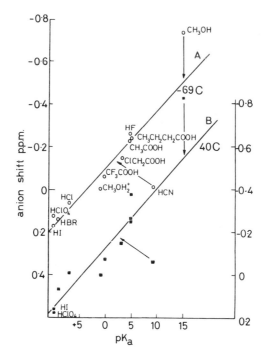

Fig. 13. Molal shifts for anions in methanolic solutions, A, at $-69\ °C$; B, at $40\ °C$, as a function of pK_a for their conjugate acids.

of the points for cyanide ions. These lie well off the correlating lines and this deviation has been satisfactorily explained in terms of strong axial solvation in which the local asymmetric field on the anion has a special effect on the proton and is comparable with the well known "ring-current" effect for aromatic compounds.

It is significant that the shift of the "bulk" solvent absorption induced by increasing the anion concentration for magnesium salts is almost exactly paralleled by a shift for the absorption caused by solvent molecules bonded to magnesium[62]. This simply means that units in which the anion is hydrogen-bonded to solvent, directly bonded to Mg^{2+}, have the same lifetime as any other anion solvate and hence these protons are similarly affected. This was true of anions such as chloride or perchlorate. However, when basic anions such as acetate were used, the magnesium solvate protons were far more strongly affected than the bulk solvent protons. Thus solvent shared ion-pairs are preferentially formed in this case, because of the increased acidity of the protons involved.

II.C ^{19}F resonance shifts for fluoride ions

Although iodide is the best anion to study by UV spectroscopy, fluoride is the best for NMR spectroscopy because of its high magnetic moment and absence of a nuclear quadrupole moment. This means that more dilute solutions can be studied than for Cl^-, Br^- or I^-, all of which have molecular quadrupole moments which can give rise to very broad lines. Many years ago we demonstrated the very marked sensitivity to environment exhibited by fluoride[63] but were unable to continue these studies because no instruments were available to us. One curious aspect of the results was the large, upfield shift from the pure water value exhibited by fluoride as methanol was added, and the surprising linearity of the shift (Fig. 14). Ethylene glycol gave a somewhat smaller shift, and ethanol still smaller. A range of solvents including isopropyl alcohol, dioxane, acetone and methylcyanide give negligibly small shifts in the mole fraction range 0–0.3, whilst formamide produced a marked shift to low-field.

In our original work we drew a comparison with our results for iodide in the CTTS region, but the gap to be bridged is, of course, very large. This is because fluoride is very strongly solvated, whilst iodide is only weakly solvated by protic solvents. Also, the iodide studies were on ca. 10^{-4} M solutions whilst the ^{19}F studies utilised 0.4 M solutions*. Nevertheless, one of the major factors thought to govern shifts for ^{19}F nuclei is the "paramagnetic" term in the shift equation. This term takes the form $\Delta/<\Delta E>$ where Δ is the sum of the squares of overlap integrals between $F^- 2p$-orbitals and orbitals on adjacent solvent molecules, and $<\Delta E>$ is the mean effec-

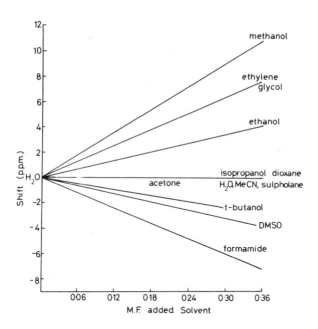

Fig. 14. ^{19}F chemical shift for aqueous fluoride ions as a function of the mole fraction of added co-solvent.

---

*The fear is sometimes expressed that the reaction

$$2F^- + H_2O \rightleftharpoons FHF^- + OH^-$$

can play an important role in studies of fluoride ion shifts. We have checked that in the work under consideration[63] this reaction made no measurable contribution to the shifts.

tive F^- excitation energy. It would seem reasonable to take ΔE for the first CTTS band as being the most significant in this expression, and this is most likely to be the most solvent sensitive term in this expression. Unfortunately, all solvents are optic ally black in the spectral region in which the CTTS bands are expected, and hence it is impossible to draw any direct correlations. However, as discussed in Sect. I, Cl^- and Br^- have optical absorption bands which have similar properties to that for I^- and hence we can guess that the F^- absorption would be similar.

Qualitatively, many trends are indeed comparable, and it may well be that changes in the ΔE^{-1} term are of major significance. Thus, relative to aqueous solutions, methanol causes a high energy shift to CTTS bands, and a large upfield shift to the ^{19}F absorption for F^-. Again, going from H_2O to D_2O causes a small high energy shift for I^- (CTTS) and a small upfield shift for F^- (NMR)[64].

We suggest that the absence of the expected downfield shift for dioxan, MeCN, etc. is only apparent, since only a small mole fraction range was covered. Strong preferential solvation by water is almost certainly the reason for this. Even iodide with these solvents shows a marked preference for water (see Sect. I) so the far more strongly bonded fluoride ion is certainly expected to exhibit a more exaggerated preference. In that case, it is most significant that the downfield shift exhibited by formamide sets in at once with no clear indication of preferential solvation by water. One factor which probably contributes to this difference is that formamide forms strong hydrogen bonds with water and hence can directly compete with F^- for available water protons. To check this concept with a solvent that also bonds strongly to water but which is unlikely to have any direct interaction with F^- we have studied

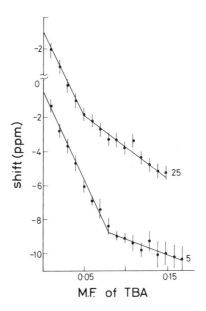

Fig. 15. Shift for $^{19}F^-$ resonance of aqueous fluoride ions as a function of the concentration of added t-butyl alcohol at 5 °C and 25 °C.

aqueous dimethyl sulphoxide solutions, and again fild a downfield shift which sets in at once (Fig. 14). This confirms that competition by "donor" solvents can be very important.

The effect of increasing the temperature of the ^{19}F resonance of aqueous potassium fluoride is then anomalous, since there is a small linear shift to high field[65] (0.034 p.p.m. deg^{-1}) rather than a shift to low field, as would be predicted by comparison with the CTTS shift for iodide (Sect. I). This is a very curious result which obviously deserves more attention.

Specific mention should perhaps be made of our studies of ^{19}F resonance from F$^-$ in aqueous t-butyl alcohol solvent systems[65]. Again, the result is curious in that there is an initial, rapid, downfield shift which levels off considerably in the 0.05–0.1 M.F. region (Fig. 15) which is opposite in sign to prediction based upon the initial high energy shift induced in the iodide absorption by t-butyl alcohol (see Sect. I). This difference is thought to be linked with the reversed temperature effect mentioned above. Thus, both the iodide CTTS shift and the ^{19}F shift induced by t-butyl alcohol can be thought of as corresponding to a cooling or an increased structuring in the immediate environment of the ion.

II.D Other halide ion resonance studies

II.D.1 Shifts

Perhaps the most significant work in this area is that of Langford and co-workers[66]. They have measured the ^{35}Cl, ^{79}Br and ^{127}I resonance in a range of pure and mixed solvents with results that key in beautifully with CTTS data for these ions. They were able to use solutions whose halide resonances were almost independent of the nature of the cations, so that ion-pair formation was not thought to be very significant. They conclude that the paramagnetic term is again the most significant variable governing these shifts and their correlation between NMR shifts

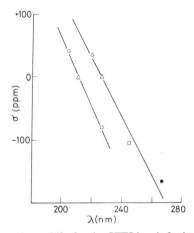

Fig. 16. A comparison of the solvent shifts for the CTTS bands for bromide and iodide ions with the NMR shifts in the same solvents (from ref. 66).

and CTTS transitions are given in Fig. 16 (see Fig. 6 in ref. 66). It would be most interesting to estimate the temperature dependence of these resonances to see if these follow the CTTS temperature shifts.

One striking result for their chloride ion studies is the very marked narrowing of the resonance in methyl cyanide compared with water. This surely indicates once more the far weaker solvent interaction in the latter solvent, such that nuclear relaxation via coupling between the quadrupolar nucleus and fluctuations in the immediate electric field gradients is greatly reduced.

Equally interesting are the mixed solvent trends: thus for water–methyl cyanide solvents, there is a very marked curvature indicating preferential solvation by water, which becomes more marked on going from I^- to Cl^- (Fig. 17). The trend for I^- accords fairly well with that for the I^- CTTS data, and the increasing selectivity fits in nicely with our earlier results for F^- discussed above. In contrast, however, dimethylsulphoxide induces an almost linear shift to lowfield as a function of mole fraction. Since these two solvents are expected to interact somewhat similarly with halide ions this is a most curious contrast. Stengle et al.[66] stress that the water–methyl cyanide interaction (HOH---NCMe) is weaker than that for dimethyl sulphoxide (HOH---OSMe), which means that the latter solvent can compete far more effectively for water molecules than can the former.

Deverell and Richards' study[67] of Cl^-, Br^- and I^- in aqueous electrolyte solutions is less readily explained in terms of CTTS excited states since the corresponding values are not all known. It is clear that anions have a major effect on the halide shifts in fairly concentrated solutions, with the heavier halide ions giving downfield shifts on the lighter and vice versa. This means anion–anion contacts are occurring and it agrees with the limited information available for the I^- CTTS band, since chloride produces a high energy shift which is greater than that for bromide, whilst, so far as can be judged, iodide itself causes a small low energy shift. This agrees well with chloride causing a larger upfield NMR shift of the ^{127}I resonance than bromide, with iodide itself causing a downfield shift.

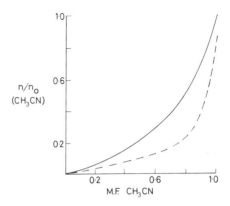

Fig. 17. Relative shifts (n/n_0) for aqueous bromide and iodide ions as a function of added methyl cyanide——I^- for NMR and CTTS shifts, – – – Br^- for NMR shifts (from ref. 66).

II.D.2 Line width studies

There have been many reports of relaxation studies for the halide ions[52], mostly for Cl^-, Br^- and I^- which relax via their quadrupole interaction. Line width studies for fluoride ion suffer from the broadening effect caused by FHF^- formation and care must be taken to eliminate this. Hertz and Rädle[68] have recently presented an interesting study of the relaxation rate $1/T_1$ for potassium fluoride (1 M solution) in D_2O. In H_2O there is an added contribution from the $^{19}F^-$–^1H interaction (the ^{19}F–^2H interaction is small and can be neglected). From these measurements they have calculated a value for the distance of closest approach for the proton. From similar studies using $D_2^{17}O$ there is a dipole–dipole interaction between ^{19}F and ^{17}O and hence a value for the distance of closest approach for these nuclei was calculated. The results are only compatible with a model in which there is one strong, linear, hydrogen bond per water molecule. This is, of course, in no sense surprising since it fits in with all current concepts and with X-ray data for salt hydrates[69].

Of the many studies of the other halide ions, attention is called to an interesting comparison between aqueous alkali metal halide solutions and aqueous halides having various ammonium, phosphonium and sulphonium cations[70]. In all cases the organic cations cause a line broadening, which increases as the number of carbon atoms increases. Curiously, when H_2O was replaced by D_2O there was a ca. 20% width increase, which led to the suggestion that the phenomenon is caused by anion–water interactions which are enhanced by the organic cations which act as structure makers. Other theories, based on contact ion-pairing or micelle formation were rejected.

III. Infrared and Raman spectroscopy

III.A Introduction

Despite the ready availability of instruments, IR and Raman spectroscopic techniques have not been extensively applied to the study of ionic solvation. Two quite different approaches can be used, one being to study the spectra of the ions themselves and the other to study the effect of electrolytes on the spectra of solvent molecules. The former method is, of course, useless for monatomic anions, but can be used most effectively to study such ions as NO_3^- or ClO_4^-[71]. The latter, however, suffers from the fact that, for hydroxylic solvents, it is usually impossible to differentiate between the separate effects of cations and anions. This should not be so, since there is no time averaging as in NMR spectroscopy (see Sect. II) but arises experimentally because shifts are small and lines are very broad relative to these shifts.

III.B Protic solvents in aprotic solutions

The systems to be considered comprise solvents like carbon tetrachloride or methylene chloride containing protic solvents in low concentration, to which R_4N^+ hal^- salts are added progressively. Initially, narrow O–H stretching fre-

336

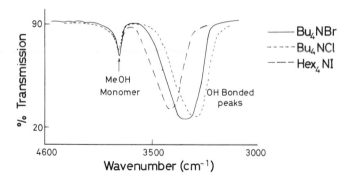

Fig. 18. Infrared spectra for solutions of tetrabutyl ammonium halides in carbon tetrachloride containing methanol in low concentration, showing the O–H stretching band for methanol monomers and MeO–H–––hal⁻ units.

quencies from monomeric ROH molecules can be detected (possibly together with broader bands from dimers or oligomers). As the salt is added, the absorbance for the monomer band falls, whilst a new, broader band, shifted to low energy by $\delta E(X^-)$ grows in (Fig. 18). This band can safely be assigned to the unit hal⁻---HOR, since its frequency is independent of R in R_4N^+. As expected, δE increases as the anion radius falls, with the apparent exception of F⁻, if we use the data of Allerhand and Schleyer[72]. However, we have repeated and extended these studies, and find that fluoride ions actually induce a greater shift than that of chloride ions (Fig. 19).

Some attempts have been made to add a second solvent molecule to the anion, but it is difficult to distinguish such effects from those of solvent–solvent interactions. However, we believe that this can be done, at least for chlorides, and the clear implication of the data is that a new band, shifted slightly back towards the unbonded OH line, is produced. In other words, there is a slight, but real, fall in the strength of the first H-bond as a second protic solvent is added to the ion. This result unfortunately makes it difficult to extrapolate from these results to an expectation for solutions in bulk protic solvents.

III.C The fundamental O–H stretch

This region is most unsatisfactory to work with because of the great breadth of the bands. As can be seen from Fig. 20, the absorption profile for fairly concentrated salt solutions in water show only minor changes, which are almost impossible to monitor in any quantitative manner. These bands can be narrowed somewhat by

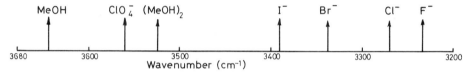

Fig. 19. Display of infrared data for dilute solutions of methanol in carbon tetrachloride containing various tetraalkyl ammonium salts.

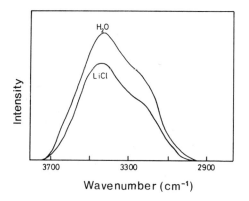

Fig. 20.   Infrared absorption spectra for water and aqueous lithium chloride.

using HOD in D_2O (or, conversely, by studying the O–D stretch for HOD in H_2O) since this replaces the two coupled frequencies for H_2O (v_1 and v_3) and also removes extensive coupling with H-bonded neighbour molecules. This is not enough to help much. However, in a few specific cases, electrolytes will generate new bands, as is the case for a range of perchlorates in water or alcohols[73–76] (Fig. 21). Since the new band, clearly associated with the presence of perchlorate ions, occurs in the region expected for non-bonded O–H groups (O–H_{free}) it has been concluded[73] that perchlorate and related ions simply act as water structure-breakers, whilst

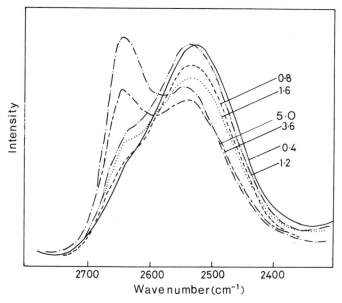

Fig. 21.   HOD in H_2O (O–D stretch) containing various concentrations of sodium perchlorate (molalities indicated).

remaining themselves unsolvated. We have strongly opposed this view, and on the basis, in particular, of NMR studies, we prefer the concept that ClO_4^- ions are very weakly H-bonded, probably symmetrically to four water (or ROH) molecules[74].

One way in which we are exploiting this result is to study the behaviour of the remaining absorption, after subtraction of that for solvent associated with perchlorate ions, as a function of the cations used. Only shifts can be seen, but at least the spectral changes can only be ascribed to the cations! This enables us to say, conclusively, that the small shifts to high energy and concomitant band enhancements observed in the Raman spectra of solutions of chlorides, bromides or iodides (increasing in that order) are largely caused by the anions. Iodide in particular causes a large spectral enhancement in the Raman spectrum (Fig. 22). This provides us with a link to the work outlined in Sect I. Thus the Raman spectrum utilises excited electronic states, and the relatively low-lying CTTS state for iodide, involving as it must the surrounding solvent molecules, serves to increase the oscillator strengths of those solvent molecules directly bonded to the anion. This effect falls off steadily for bromide and chloride, and is in fact likely to be reversed for fluoride since this ion has its CTTS absorption (if any) at higher energies than the onset of absorption by the solvent itself.

From these results we can conclude firmly that the strength of the H-bonding interaction between water (or methanol) increases in the order $ClO_4^- \sim BF_4^- < I^- < Br^- < Cl^- \leq H_2O \; (< F^-)$, which is the same order that appears from the studies in inert solvents. However, except for ClO_4^- it is not yet possible to derive quantitative data.

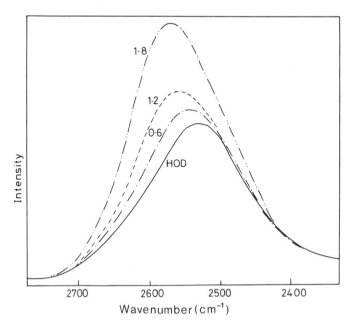

Fig. 22. Raman spectrum in the O–D stretching region for HOD in H_2O, and solutions of NaI (molalities indicated).

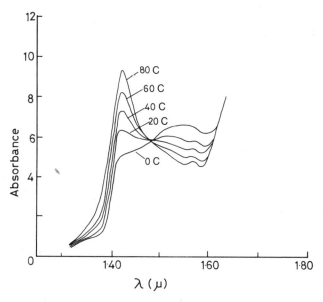

Fig. 23. First overtone (O–H stretch) absorption for HOD in D_2O as a function of temperature, showing the growth of the (O–H)$_{free}$ band in the 1.4 μ region.

III.D The O–H stretch overtones

In some respects there is a real gain in studying the first or second overtones for the O–H stretching frequency of water or alcohols in the near-IR region. This is because different selection rules now operate. All components are of course greatly reduced in intensity, but the bands for strongly H-bonded O–H are reduced more than those for O–H$_{free}$ and hence it is possible to detect O–H$_{free}$ groups in relatively low concentration (Fig. 23). This is also facilitated by the fact that the separation between O–H$_{free}$ and O–H$_{bound}$ bands is increased in the overtone region.

For this reason, most studies in this spectral region have been centred on the factors that cause water or alcohols to acquire or to lose O–H$_{free}$ groups. Some theories of water structure require the presence of free H_2O rather than just O–H$_{free}$ groups, but it has been possible to use near-IR results to establish that the bands under consideration are certainly not due to free water molecules[77]. Our own results on perchlorate solutions in this region show a difference in E_{max} for the peak assigned to O–H$_{free}$ and that assigned to water weakly H-bonded to ClO_4^- ions. Unfortunately, resolution is not sufficient to give any clear measure of the concentration of O–H$_{free}$ groups in the presence of ClO_4^- ions. Other information is, in general, similar to that obtained from the fundamental region.

IV. Conclusions

No attempt has been made to produce a full coverage of the literature in this field. Instead, pertinent results have been quoted which serve to build up a picture of the nature of anionic solvation. All these results point to a picture of anion solvation which involves, for solvents such as water or alcohols, strong H-bonding to solvent molecules, the number of solvent molecules being poorly defined, but surely in the region of four. The strength of this interaction falls on going from F^- to I^- and hence the lifetime of a given A^----S unit falls in the same sense. The strength of the H-bonding to anions is about equal to that between bonded solvent molecules for chloride, being greater for fluoride and weaker for bromide or iodide.

Aprotic solvents such as MeCN interact more weakly and approach the anions less closely, giving rise to a larger, more deformable, solvent shell. Weakly H-bonding solvents such as ammonia behave similarly.

Mixed protic–aprotic solvents generally display preferential solvation by the protic medium but this is reduced greatly if the aprotic solvent is strongly basic since it then actively competes with the anion for hydrogen-bonds.

References

1 E. A. Taft and H. R. Philipp, J. Phys. Chem. Solids, **3** (1957) 1.
2 H. Hartmann, Z. Elektrochem., **61** (1957) 908.
3 J. E. Eby, K. J. Teegarden and D. B. Dutton, Phys. Rev., **116** (1959) 1099.
4 M. J. Blandamer, T. R. Griffiths, L. Shields and M. C. R. Symons, Trans. Faraday Soc., **60** (1964) 1524.
5 G. Scheibe, Z. Electrochem., **35** (1929) 701.
6 E. Lederle, Z. Physik Chem., Abt. B, **10** (1930) 121.
7 M. Smith and M. C. R. Symons, J. Chem. Phys., **25** (1956) 1074.
8 M. Smith and M. C. R. Symons, Trans. Faraday Soc., **54** (1958) 338, 346.
9 M. Smith and M. C. R. Symons, Discuss. Faraday Soc., **24** (1957) 206.
10 M. J. Blandamer and M. F. Fox, Chem. Rev., **70** (1970) 59.
11 T. R. Griffiths and M. C. R. Symons, Trans. Faraday Soc., **56** (1960) 1125.
12 M. C. R. Symons, Nature (London), **239** (1972) 257.
13 M. J. Blandamer, L. Shields and M. C. R. Symons, J. Chem. Soc., (1965) 3759.
14 D. B. Siano and D. E. Metzler, J. Chem. Soc. Faraday II, **68** (1972) 2042.
15 M. F. Fox and E. Hayon, Chem. Phys. Lett., **14** (1972) 442.
16 K. Teegarden and G. Baldini, Phys. Rev., **155** (1967) 896.
17 M. J. Blandamer, T. E. Gough and M. C. R. Symons, Trans. Faraday Soc., **59** (1963) 1748.
18 M. J. Blandamer, T. E. Gough and M. C. R. Symons, Trans. Faraday Soc., **62** (1966) 296.
19 M. J. Blandamer and T. R. Burdett, J. Chem. Soc. Faraday II, **68** (1972) 577.
20 M. J. Blandamer, M. C. R. Symons and M. J. Wootten, Chem. Commun., (1970) 147.
21 G. Stein and A. Treinin, Trans. Faraday Soc., **55** (1959) 1036.
22 R. Platzman and J. Franck, Z. Phys., **138** (1954) 411.
23 M. J. Blandamer, L. Shields and M. C. R. Symons, J. Chem. Soc., (1964) 4352.
24 J. Jortner and B. Sharf, J. Chem. Phys., **37** (1962) 2506.
25 T. A. Claxton, D. J. Greenslade. K. D. J. Root and M. C. R. Symons, Trans. Faraday Soc., **62** (1966) 2050.
26 G. Feher, Phys. Rev., **105** (1957) 1122.
27 B. S. Gourary and F. J. Adrian, Solid State Phys., **10** (1958) 127.
28 M. J. Blandamer, L. Shields and M. C. R. Symons, J. Chem. Soc., (1965) 1127.
29 J. Jortner and A. Treinin, Trans. Faraday Soc., **58** (1962) 1503.
30 G. A. Russell and C. C. Klick, Phys. Rev., **101** (1956) 1473.
31 M. Brith and A. Ron, J. Chem. Phys., **50** (1969) 3045.
32 B. Raz and J. Jortner, Proc. Roy. Soc. Edinburgh, Sect. A, **317** (1970) 113.
33 A. Gedanken, B. Raz and J. Jortner, J. Chem. Phys., **58** (1973) 1178.

34 M. J. Marrons, F. W. Patten and M. N. Kabler, Phys. Rev. Lett., **31** (1973) 467.

35 I. Marov and M. C. R. Symons, J. Chem. Soc., A, (1971) 201.

36 I. S. Ginns and M. C. R. Symons, J. Chem. Soc. Dalton, (1972) 143.

37 T. R. Griffiths and M. C. R. Symons, Mol. Phys., **3** (1960) 174.

38 M. J. Blandamer, T. E. Gough and M. C. R. Symons, Trans. Faraday Soc., **62** (1966) 286.

39 T. R. Griffiths and R. H. Wijayanayake, Trans. Faraday Soc., **66** (1970) 1563.

40 J. Oakes, J. Slater and M. C. R. Symons, Trans. Faraday Soc., **66** (1970) 546.

41 M. J. Blandamer, T. E. Gough and M. C. R. Symons, Trans. Faraday Soc., **62** (1966) 301.

42 K. M. C. Davis, J. Chem. Soc., B, (1969) 1020.

43 E. M. Kosower, J. Amer. Chem. Soc., **80** (1958) 3253, 3261.

44 E. M. Kosower and G-S. Wu, J. Amer. Chem. Soc., **83** (1961) 3142.

45 E. Gusarsky and A. Treinin, J. Phys. Chem., **69** (1965) 2176.

46 M. F. Fox and T. F. Hunter, Nature (London), **223** (1969) 177.

47 F. M. Page, Advan. Chem., **36** (1962) 68.

48 M. J. Blandamer, R. Catterall, L. Shields and M. C. R. Symons, J. Chem. Soc., (1964) 4357.

49 M. Anbar and E. J. Hart, J. Phys. Chem., **69** (1965) 1244.

50 R. Catterall and M. C. R. Symons, J. Chem. Soc., (1964) 4342.

51 K. V. S. Rao and M. C. R. Symons, Chem. Commun., (1971) 268.

52 R. S. Eachus and M. C. R. Symons, J. Chem. Soc., (1968) 790.

53 C. Deverell, Progr. NMR Spectrosc., **4** (1969) 235.

54 J. F. Hinton and E. S. Amis, Chem. Rev., **67** (1967) 367.

55 R. D. Green and J. S. Martin, J. Amer. Chem. Soc., **90** (1968) 3659.

56 S. Ormondroyd, E. A. Philpott and M. C. R. Symons, Trans. Faraday Soc., **67** (1971) 1253.

57 J. W. Swinehart and H. Taube, J. Chem. Phys., **37** (1962) 1579.

58 See, for example, N. A. Matwiyoff and H. Taube, J. Amer. Chem. Soc., **90** (1968) 2796; A. Fratiello, in J. O. Eduards (Ed.), Inorganic Reaction Mechanisms, Wiley, New York, 1972, p. 57.

59 R. N. Butler and M. C. R. Symons, Trans. Faraday Soc., **65** (1969) 945.

60 J. Davies, S. Ormondroyd and M. C. R. Symons, Trans. Faraday Soc., **67** (1971) 3465.

61 O. Y. Samoilov, Structure of Aqueous Electrolyte Solutions and Hydration of Ions, Consultants Bureau, New York, 1965; see also H. G. Hertz, in F. Franks (Ed.), Water, A Comprehensive Treatise, Vol. 3, Plenum Press, London, 1973, p. 301.

62 R. N. Butler, J. Davies and M. C. R. Symons, Trans. Faraday Soc., **66** (1970) 2426.

63 A. Carrington, F. Dravnicks and M. C. R. Symons, Mol. Phys., **3** (1960) 174.

64 C. Deverell, K. Schaumburg and H. T. Bernstein, J. Chem. Phys., **49** (1968) 1276.

65 R. G. Anderson and M. C. R. Symons, Trans. Faraday Soc., **65** (1969) 2550.

66 T. R. Stengle, Y-C. E. Pan and C. H. Langford, J. Amer. Chem. Soc., **94** (1972) 9037.

67 C. Deverell and R. E. Richards, Mol. Phys., **16** (1969) 421.

68 H. G. Hertz and C. Rädle, Ber. Bunsenges. Phys. Chem., **77** (1973) 531.

69 R. K. McMullan and G. A. Jeffrey, J. Chem. Phys., **31** (1959) 1231; D. Feil and G. A. Jeffrey, J. Chem. Phys., **35** (1961) 1863.

70 H. Wennerstrom, B. Lindman and S. Forsen, J. Phys. Chem., **75** (1971) 2936.

71 D. E. Irish, A. R. Davis and R. A. Plane, J. Chem. Phys., **50** (1969) 2262.

72 A. Allerhand and P. Schleyer, J. Amer. Chem. Soc., **85** (1963) 1233.

73 G. E. Walrafen, J. Chem. Phys., **52** (1970) 4176.

74 D. M. Adams, M. J. Blandamer, M. C. R. Symons and D. Waddington, Trans. Faraday Soc., **67** (1971) 611.

75 L. J. Bellamy, M. J. Blandamer, M. C. R. Symons and D. Waddington, Trans. Faraday Soc., **67** (1971) 3435.

76 G. Brink and M. Falk, Can. J. Chem., **48** (1970) 2096.

77 L. M. Kleiss, H. A. Strobel and M. C. R. Symons, Spectrochim. Acta, Part A, **29** (1973) 829.

AUTHOR INDEX

Numbers in parentheses following the page number are reference numbers. The page number where the full reference is given follows the parenthesized reference number.

345

Edwards, P.P., 61 (63) 88; 62, 85 (66) 88; 62 (67) 88; 68, 70 (108) 88; 78 (175) 90
Efros, A.L., 247, 252 (114) 257
Eggarter, T.P., 8 (15) 40; 8 (17) 41
Egland, R.J., 46 (3) 86; 61, 62, 85 (65) 88; 69 (142) 89
Ehrenson, S., 274 (64) 308; 294, 296, 297 (87) 309
Eiben, K., 62 (68) 88; 68 (105) 88; 68 (122) 89
Eigen, M., 163 (56) 173
Eisele, I., 180 (48) 209; 223, 245 (48) 255; 223, 245, 254 (49) 255; 223, 245 (50) 255; 223, 245 (51) 255
Ekstrom, A., 66 (87) 88; 140, 141 (3) 171; 143 (14) 172; 145 (20) 172
Eliezer, I., 264, 289, 290, 291 (21) 307
Emtage, P.R., 247, 252 (115) 258
Erginsoy, C., 253 (126) 257
Ershov, B.G., 68 (109) 88; 68 (113) 89; 68 (117) 89; 68 (119) 89; 68 (129) 89; 68 (130) 89; 68 (131) 89; 68 (132) 89; 68 (133) 89; 68 (135) 89; 68 (136) 89; 68 (137) 89; 68 (139) 89; 68 (141) 89; 85 (189) 90; 104, 105 (41) 114; 177 (32) 209; 181, 184 (54) 209; 183 (58) 209; 184, 185 (59) 209; 185 (70) 210

Falk, M., 305 (111) 309; 337 (76) 341
Fano, U., 205 (111) 211
Farhataziz, 28, 29 (79) 42; 33 (90) 42; 143 (17) 172
Feher, G., 319 (26) 340; 48 (7) 86; 49 (19) 87
Feng, D.-F., 16, 18, 19 (47) 41; 18 (49) 41; 19 (53) 41; 23 (70) 42; 27 (74) 42; 27 (78) 42; 145 (22) 172; 230, 235 (72) 256
Finkelstein, G., 29 (85) 42; 38 (102) 43
Fischer-Hjalmars, I., 191, 193 (78) 210
Flagg, R., 213 (4) 254
Flanigan, M.C., 164 (57) 173
Flynn, G.J., 95 (20) 114; 131 (29) 138
Foner, S.N., 82 (180) 90
Forsen, S., 335 (70) 341
Fowler, W.B., 107, 109 (47) 114; 107, 108, 109 (48) 114
Fowles, P., 222, 241 (39) 255
Fox, M.F., 313, 324 (10) 340; 317 (15) 340; 324, 325 (46) 341
Fraga, S., 266 (30) 307
Franck, J., 319 (22) 340
Frank, H.S., 259 (1) 307
Franklin, J.F., 261 (8) 307
Freed, S., 84 (183) 90

Freeman, G.R., 27 (75) 42; 176 (22) 209; 215 (22) 255; 217, 229, 230 (30) 255; 217, 229, 230 (31) 255; 217, 229, 230 (32) 255
Freudenthal, A.M., 152 (27) 172
Friauf, R.J., 85 (190) 90
Friedman, H.L., 261 (7) 307; 274 (64) 308
Fröhlich, H., 153 (30) 172; 253 (128) 257
Fromm, J., 298, 300, 301 (94) 309
Frommhold, L., 162 (50) 173; 250 (119) 257
Fueki, K., 16, 18, 19 (47) 41; 18 (49) 41; 23 (70) 42; 27 (74) 42; 27 (78) 42; 38 (102) 43; 66 (91) 88; 66 (103) 88; 104, 105 (41) 114; 142, 145 (12) 172; 145 (22) 172; 230, 235 (72) 256
Fukui, K., 13 (40) 41
Funabashi, K., 23, 33 (71) 42; 33 (71) 42; 140, 146 (5) 171; 223, 245 (47) 255
Fuochi, P.G., 217, 229, 230 (30) 255

Gaathon, A., 18, 19, 21, 22, 29, 36, 37 (51) 41; 19 (54) 41; 35 (93) 42; 36 (95) 43
Galkin, A.A., 49 (22) 87
Gallivan, J.B., 145 (18) 172
Gamba, A., 274, 275 (62) 308
Gangwer, T.E., 230, 236, 239 (69) 256
Garg, S.K., 157 (36) 172
Garstens, M.A., 49 (24) 87
Gary, L.P., 244 (103) 256
Gauthier, M., 176, 191 (25) 209
Gedanken, A., 320 (33) 340
Geiringer, H., 152 (27) 172
Gel'danskii, V.I., 164, 167 (61) 173
Getoff, N., 207 (113) 211
Gilles, L., 27 (76) 42; 95, 102 (22) 114
Gillis, H.A., 94, 95 (21) 114; 95, 107, 113 (26) 114; 66 (80) 88; 142 (9) 171; 156 (35) 172
Ginns, I.S., 321 (36) 341
Gires, F., 100 (37) 114
Girina, E.L., 181, 184 (54) 209
Glarum, S.H., 134 (35) 138
Glazunov, P.Ya., 68 (117) 89; 68 (119) 89
Gold, M., 85 (155) 89
Goldansky, V.I., 181, 184, 185 (53) 209; 184 (62) 210
Golden, S., 84 (156) 89
Goldsborough, J.P., 82 (181) 90
Goldschwartz, J.M., 244 (108) 256
Gomer, R., 227 (60) 255; 228 (65) 255
Gormley, J.A., 207 (113) 211
Gosse, J.P., 244 (104) 256
Goto, H., 66 (91) 88
Gough, T.E., 317, 318 (17) 340; 317, 318 (18) 340; 322 (38) 341; 323 (41) 341

Gould, R.F., 213, 214, 220, 227, 239, 241 (6) 254
Gourary, B.S., 47 (4) 86; 319 (27) 340
Graceffa, P., 72 (153) 89
Grand, D., 93, 107 (11) 113; 176, 191 (25) 209; 180 (46) 209
Gray, E., 244 (107) 256
Gray, L.P., 13 (39) 41
Green, R.D., 327 (55) 341
Greenslade, D.J., 319 (25) 340
Gregg, E.E., 217 (33) 255; 236 (85) 256
Grie, M., 66 (94) 88
Griffiths, T.R., 312 (4) 340; 314, 315 (11) 340; 321, 323 (37) 341; 322, 323 (39) 341
Grimison, A., 267 (35) 308
Grindberg, O.Ya., 68 (129) 89
Grossweiner, L.I., 176, 193, 194 (24) 209
Gurney, R.W., 260 (2) 307
Gusarsky, E., 324 (45) 341
Gzowski, O., 243 (97) 256

Hager, S.L., 40 (107) 43; 93, 106 (14) 113
Hahne, S., 35 (92) 42
Hallada, C., 75 (161) 90
Halndamer, M.J., 66 (85) 88
Halpern, B., 227 (60) 255; 228 (65) 255
Hamill, W.H., 72 (146) 89; 245 (109) 256; 140 (2) 171; 145 (18) 172; 175, 201 (1) 208; 175, 201 (2) 208; 175, 201 (4) 208; 175, 201 (6) 208
Hamlet, P., 93 (15) 114; 180 (49) 209
Hamm, R.N., 205 (112) 211
Hansen, E.M., 28, 29 (79) 42
Harget, A.J., 265 (26) 307
Harris, F.E., 267 (36) 308
Harrison, H.R., 8 (19) 41
Harrocks, D.L., 106 (46) 114
Hart, E.J., 91 (3) 113; 94 (18) 114; 141 (7) 171; 142 (11) 172; 213, 241 (7) 254; 325 (49) 341
Hartmann, H., 312 (2) 340
Hase, H., 65, 66 (77) 88; 66 (100) 88; 95 (24) 114; 104, 105 (41) 114; 142, 143, 145 (12) 172; 143 (17) 172; 176, 179, 192, 204 (21) 209; 178 (34) 209; 179 (40) 209; 179 (41) 209; 179, 186 (42) 209; 181, 183 (55) 209; 185, 187 (66) 210; 204 (110) 211
Hasen, E.E., 77, 84 (164) 90
Hasing, J., 117 (5) 137
Hayashi, K., 66 (92) 88; 66 (94) 88
Hayon, E., 66 (81) 88; 66 (93) 88; 317 (15) 340
Hehre, W.J., 285, 290 (75) 308; 289 (83) 309; 296 (89) 309; 296 (90) 309
Heinzinger, K., 298, 299 (93) 309

Heller, J.M., Jr., 205 (112) 211
Helmholtz, L., 267 (34) 308
Henglein, A., 191 (180) 210
Henriksen, T., 68 (120) 89
Henson, B.L., 227 (61) 255
Hentz, R.R., 28, 29 (79) 42; 33 (90) 42; 141 (8) 171; 143, 158 (16) 172; 143 (17) 172; 159 (40) 172; 162 (46) 172
Hernandez, J.P., 8 (18) 41
Hertz, H.G., 261, 305 (9) 307; 299 (100) 309; 305 (109) 309; 335 (68) 341
Hertz, P., 169 (68) 173
Higashimura, T., 65, 66 (75) 88; 95 (16) 114; 95 (24) 114; 142, 145 (10) 172; 142, 143, 144, 145 (12) 172; 176, 179, 192, 204 (21) 209; 179 (39) 209; 179 (40) 209; 179, 186 (42) 209; 185, 187 (66) 210; 204 (110) 211
Hinton, J.F., 326 (54) 341
Hiroike, K., 14 (43) 41
Hirschfelder, J.O., 261 (13) 307; 261 (15) 307; 268 (43) 308; 268 (44) 308
Ho, K.K., 106 (45) 114; 179 (37) 209
Hochanadel, C., 207 (114) 211
Hoffman, R., 267 (33) 308; 267 (35) 308
Hohn, C.H., 51 (34) 87
Hoijtink, G.J., 72 (147) 89; 132 (34) 138
Holroyd, R.A., 226, 234 (52) 255; 226 (53) 255; 226, 234 (54) 255; 226, 234 (56) 255; 227, 234, 239 (58) 255; 230, 236, 239 (69) 256; 230, 236, 237 (70) 256
Holstein, T., 146 (23) 172; 247, 252 (112) 256
Holteng, J.A., 191, 195, 196 (74) 210
Holzworth, N.A.W., 9 (24) 41
Honna's, P.I., 195, 196 (86) 210
Horani, M., 63, 64 (74) 88
Horii, H., 179 (43) 209
Horsfield, A., 61 (55) 87
Hotop, H., 135 (37) 138
Houser, N.E., 220 (35) 255
Howat, G., 13, 37, 40 (42) 41; 21 (59) 42
Howe, S., 228 (62) 255
Huang, J.J., 196 (94) 210
Huang, T., 223, 245 (48) 255; 223, 244, 254 (49) 255; 245 (110) 256; 245, 251 (111) 256
Hudson, D.E., 216 (26) 255
Huff, L., 100 (38) 114
Hughes, T.R., Jr., 51, 53, 59 (37) 87
Hummel, A., 213 (2) 254; 218 (34) 255; 243 (98) 256; 243 (100) 256
Hunt, J.W., 27 (76) 42; 95, 102 (22) 114; 96 (30) 114; 156 (31) 172; 175, 200 (8) 208; 175, 200 (9) 208; 175, 177, 200 (10) 208; 175, 200, 201, 205 (11) 208; 175, 200, 202 (12) 208; 175, 200, 203 (13) 209

354

Walker, D.C., 21 (58) 41; 40 (106) 43; 66 (80)
  88; 91, 94 (2) 113; 93, 95, 98, 99, 100, 101,
  102, 103, 104, 105, 106, 107, 110, 111,
  112 (13) 113; 93, 97 (12) 113; 94 (17) 114;
  94, 95 (21) 114; 95 (20) 114; 96 (31) 114;
  97, 99, 100, 101, 102, 106, 107, 109 (32)
  114; 98, 99, 102, 107, 108, 109 (34) 114;
  105, 107, 112 (42) 114; 110, 112 (49) 114;
  112, 113 (50) 114; 112 (51) 114; 131 (29)
  138; 131 (32) 138; 180 (50) 209; 200, 201,
  203 (104) 210
Wallace, S.C., 98, 99 (35) 114; 200, 201, 203
  (104) 210
Walrafen, G.E., 337 (73) 341
Warashina, T., 176, 179, 192, 204 (21) 209;
  179 (39) 209; 179 (41) 209; 204 (110) 211;
  142, 145 (10) 172; 142, 145 (12) 172; 143
  (17) 172
Wardman, P., 27 (77) 42; 66 (78) 88; 72 (149)
  89; 95, 113 (19) 114; 156, 157 (32) 172;
  159 (40) 172; 177 (28) 209; 177 (29) 209;
  236 (89) 256
Warf, J.C., 75 (162) 90
Warman, J., 218 (34) 255
Watanabe, T., 13 (37) 41
Watson, F.H., 243 (98) 256
Watts, R.O., 298, 300, 301 (94) 309; 300 (101)
  309; 300 (102) 309
Waugh, J.S., 52 (40) 87; 77, 84 (164) 90
Webster, B.C., 13, 37, 40 (42) 41; 19 (52) 41;
  21 (59) 42
Wei, J.C., 220 (36) 255
Weiss, J.J., 68 (121) 89; 68 (125) 89
Weisskopf, V.G., 253 (125) 257
Weissman, M., 13 (41) 41; 29 (85) 42
Wen, W.Y., 259 (1) 307
Wennerstrom, H., 335 (70) 341
Weyl, W., 141 (6) 171
Wigner, E., 273, 293, 294 (55) 308
Wijayanayake, R.H., 322, 323 (39) 341
Willard, J.E., 40 (107) 43; 66 (87) 88; 85 (191)
  90; 93, 106 (14) 113; 104, 105 (41) 114;
  140 (1) 171; 145 (20) 172; 176, 184 (20) 209
Williams, D.R., 268, 284, 285 (48) 308

Williams, F.I.B., 10 (27) 41; 46 (2) 86; 66 (86) 88;
  66 (93) 88; 66 (97) 88; 66 (98) 88; 66 (99)
  88; 145 (18) 127
Williams, R.L., 228 (64) 255
Wilson, E.G., 12, 20, 22 (34) 41
Wolff, R.K., 96 (30) 114; 156 (31) 172; 175,
  200 (9) 208; 175, 179, 200 (10) 208; 175,
  200, 201, 205 (11) 208; 175, 200, 202 (12)
  208; 175, 200, 203 (13) 209
Wolfsberg, M., 267 (34) 308
Wooten, M.J., 319 (20) 340
Wu, G.-S., 324 (44) 341
Wu, T.Y., 2, 3 (5) 40
Wysocyanski, W., 49 (21) 87

Yakovlev, B.S., 236, 237 (87) 256
Yamadagni, R., 275, 294, 295, 303, 304 (72)
  308; 303 (104) 309; 303 (105) 309; 303
  (106) 309
Yamamoto, M., 59 (49) 87
Yamobe, T., 13 (40) 41
Yoshida, H., 66 (92) 88; 66 (94) 88; 142, 145
  (10) 172; 142, 144, 145 (12) 172
Yoshida, Y., 179 (39) 209
Yound, R.A., 10 (29) 41
Young, G.R., 91 (1) 113

Zaklad, H., 213 (3) 254; 213 (4) 254
Zamaraev, K.I., 164, 167 (61) 173; 181, 183,
  185 (52) 209; 181, 184, 185 (53) 209;
  184 (62) 210
Zandstra, P.J., 132 (34) 138
Zasshi, K.K., 66 (91) 88
Zessoules, N., 227 (59) 255
Zhogolev, D.A., 267 (38) 308; 267, 274, 275
  (39) 308
Zimbrick, J., 68, 75 (126) 89; 68 (128) 89,
  85 (188) 90
Zolotarevski, V.I., 222 (45) 255
Zuazaga de Ortiz, C., 267 (35) 308
Zucker, U.F., 104, 105 (41) 114; 177 (31) 209
Zunger, A., 100, 110 (39) 114
Zuri, Z., 104, 105 (41) 114

SUBJECT INDEX